THE BIOORGANIC CHEMISTRY OF ENZYMATIC CATALYSIS

MYRON L. BENDER

Departments of Chemistry and Biochemistry
Northwestern University
Evanston, Illinois

RAYMOND J. BERGERON

Departments of Medicinal Chemistry and Medicine
University of Florida
Gainesville, Florida

MAKOTO KOMIYAMA

Department of Industrial Chemistry
Faculty of Engineering
University of Tokyo
Tokyo, Japan

A Wiley-Interscience Publication

JOHN WILEY & SONS

New York   Chichester   Brisbane   Toronto   Singapore

Library of Congress Cataloging in Publication Data:

Bender, Myron L., 1924–
  The bioorganic chemistry of enzymatic catalysis.

  "A Wiley-Interscience publication."
  Includes index.
  1. Enzymes. 2. Bioorganic chemistry. I. Bergeron,
Raymond J., 1945–   II. Komiyama, Makoto,
1947–   III. Title.
QP601.B374 1984   574.19'25   83-19857
ISBN 0-471-05991-9

Printed in the United States of America

10 9 8 7 6 5 4 3 2 1

To the National Institute of General Medical Sciences who have supported the work of MLB and RJB in this field for many years.

Preface

In a course in organic chemistry, a student is continually reminded of how well the subject material will serve him in a future course in biochemistry. However, on arriving in biochemistry, he or she very quickly becomes convinced that the only thing these two disciplines have in common is the asymmetric carbon atom. Most of what the student learned about organic reaction mechanisms seems totally useless and is soon forgotten. The situation becomes even more tragic when the young scholar is faced with advanced courses in biochemistry, for example, enzymology. At this point, the student isn't even sure of the value of his undergraduate biochemistry. Reaction mechanisms have returned to haunt him.

This textbook is designed to bridge the gap between organic chemistry and biochemistry and to point out to the student precisely how an understanding of organic reaction mechanisms is relevant to an understanding of how enzymes work. In so doing, it provides both advanced undergraduate students and beginning graduate students with the foundation for future studies.

Our sincere appreciation is extended to Greta Michael, Dr. Bender's secretary, Cynthia Jordan, Dr. Bergeron's secretary, and Carol Slingo for their diligent labor in typing and retyping the manuscript, to Philipp M. Chalabi and Dr. Motoo Yamaguchi for help in proofreading, to Dr. Hua-Lin Wu and Dr. Valerian D'Souza for help with figures and other assistance, and to our wives, Muriel S. Bender, Kathy Bergeron, and Mitsuko Komiyama, whose understanding was essential.

Of course many other people helped, and we thank all of them, but as usual, the final criticism must be borne by us.

MYRON L. BENDER
RAYMOND J. BERGERON
MAKOTO KOMIYAMA

Evanston, Illinois
Gainesville, Florida
Tokyo, Japan
February 1983

Contents

THE BIOORGANIC CHEMISTRY
OF ENZYMATIC CATALYSIS

1 | Introduction

Although enzymatic catalysis may be the ultimate in catalytic efficiency and specificity, it is not magical but rather based on straightforward principles of physical, organic, and inorganic chemistry. It is the purpose of this book to bridge the gap between these areas of chemistry and enzymology and to delineate those principles that are relevant to the student's understanding of this rapidly growing area.

Catalysis has been recognized almost since the study of chemistry began; it is presently undergoing a most intensive investigation, and promises not only to be better understood in times to come, but also to become more and more valuable in understanding other facets of chemistry.

Berzelius, over a century ago, alluded to a "catalytic force,"[1] but it was not until 1900 that the concept of catalysis by hydronium and hydroxide ions was placed on a firm experimental and theoretical basis.[2,3] During the twenties and thirties, with the advent of new ideas about acids and bases, the concept of general acid–base catalysis was developed.[4] Within the last decade, enzymes as chemical catalysts have received much attention.[5] At the same time, new ideas about catalysis by polyfunctional organic molecules and by metal ion complexes have arisen.

Although this book concentrates on the mechanisms of homogeneous catalysis in solution, a brief description of heterogeneous catalysis is also given. Questions as to how and why catalytic reactions proceed are asked here and a unified set of principles is presented that allows for an understanding of the mechanisms for all the reactions discussed, thus providing the student with the tools to better understand enzyme catalysis.

1

1.1. DEFINITIONS

In 1835 Berzelius first used the word "catalyst" to denote "substances [which] are able to awaken affinities which are asleep at [this] temperature by their mere presence and not by their own affinity."[1] He could not specify the nature of this force, but contented himself with saying that it was a special manifestation of the normal electrochemical properties of substances. Ostwald, in 1895,[2] defined catalysts as "substances which change the velocity of a given reaction without modification of the energy factors of the reaction." Later, he gave a definition more pointedly concerned with the kinetics of the reaction: "A catalyst is a substance which alters the velocity of the chemical reaction without appearing in the end product of the reaction."[3]

In each of these *qualitative* definitions, catalysis is intimately tied to velocity and stoichiometry, both of which are amenable to *quantitative* evaluation.[6,7] These two very important characteristics of a chemical transformation can indeed be expressed mathematically. A simple uncatalyzed reaction may be described by the equation:

$$A \longrightarrow \text{Products} \qquad (1.1)$$

The rate of the uncatalyzed reaction V_u, is

$$V_u = \frac{-d[A]}{dt} = k[A] \qquad (1.2)$$

where A is the reacting substance, [A] its concentration, t the time, and k the rate constant of reaction. The definitions of catalysis quoted in the preceding paragraph indicate that when the same transformation occurs under the influence of a catalyst C, then the catalyst appears in the equation as both reactant and product:

$$\text{Reactants} + C \longrightarrow \text{Products} + C \qquad (1.3)$$

Furthermore, since the catalyst increase the rate of the reaction, V_{cat} must incorporate the concentration of the catalyst,

$$V_{cat} = \frac{-d[A]}{dt} = k[A][C] \qquad (1.4)$$

where k_{cat} is the rate constant of the catalyzed reaction.† Since, by definition, the catalyst is not used up ([C] = a constant), its concentration may be combined with k_{cat} into a single constant characteristic of the catalytic process. On the basis

---

† The superficiality of these kinetics will be seen later in the detailed treatment of the kinetics of catalyzed reactions.

of these more quantitative expressions of the influence of catalysis on chemical change, a more quantitative definition of catalysis is possible: "A substance is said to be a catalyst for a reaction (in a homogeneous system) in which its concentration occurs in the velocity expression to a higher power than it does in a stoichiometric reaction."[4]

Alternatively, a catalyst may be said to increase the rate of approach of a system to any state that is kinetically and thermodynamically possible in its absence. The question then arises as to whether the free-energy change of the overall process from reactants to products can be changed at all by a catalyst. The answer is no. However, this definition of catalysis would be too rigorous. The use of a finite amount of catalyst will in general modify the free energy by a finite amount. A more realistic definition then requires the catalyst not to alter the standard free-energy change of reaction by more than a small fraction of its original value.[8]

A corollary of Eq. (1.3) is that a catalyst may be effective when present in small proportion relative to the reactants; this property follows from the fact that the catalyst is not used up in the reaction, and so is available to act again and again. From a mechanistic point of view, however, a more important property—also implicit in Eq. (1.3)—is that the catalyst functions by reacting chemically with one or more of the reactants.

If the catalyst indeed participates in the reaction and effects a change in its rate, then a modification of the reaction pathway must occur. Thus expressing catalysis in mechanistically meaningful and useful terms, the essential role of a catalyst that acts according to the scheme of Eq. (1.3) is to make available to the system an *alternate reaction pathway*. The requirements of this new pathway are that:

1. It circumvents the rate-limiting step of the original pathway.
2. It leads to products that differ from the original set only by the presence of the catalyst.
3. The catalyst can be easily removed from the products by exchange with reactants.

The first of these conditions (that the catalyzed reaction is faster than the uncatalyzed one) satisfies the definition of a catalyst as an accelerator. The second condition indicates that the effect of the catalyst is limited to the kinetics of the reaction: It does not affect the thermodynamics of the reaction. The third condition indicates that the catalytic substance is regenerated. Therefore, it is effective when present in small amount relative to the reactants (and is, at least in principle, recoverable after the reaction is complete).

The effectiveness of a catalyst is determined by the relative velocities of the original and alternate pathways. Each pathway usually has some particular step that controls the overall rate. The rate of this step is determined by its free energy of activation, which is the highest free-energy barrier along the route from reactants to products. The most important factor in catalytic efficiency is the amount by which the free energies of activation of the catalyzed and uncatalyzed reaction differ.

The frequent references to the function of a catalyst as a substance that lowers the activation energy of a reaction are somewhat misleading[9] (Fig. 1.1). The correct statement is that the catalyst changes the pathway of the reaction to a pathway having a lower free energy of activation (Fig. 1.2). Hinshelwood expressed this idea concisely, defining catalysis as a "process providing an *alternative and more speedy reaction route*."[10] A corollary of this definition is that catalysis consists of "normal" reactions involving the catalyst and the substrate (cf. Figs. 1.1 and 1.2).

Catalyzed reactions ordinarily proceed through the formation of intermediates that undergo further transformations yielding products and regenerating the catalyst. Thus a catalytic reaction may be described as a chain reaction in which the catalyst is used up in one step, but is regenerated in a succeeding one. Sometimes this chain process can be dissected into its individual component reactions. The chain character of catalysis is seen in many coenzyme reactions, in which the coenzyme is consumed but later regenerated.

A complication in this discussion concerns the definition of substances that accelerate chemical reactions, but are not regenerated. How, then, is one to treat these reactions? The usual answer to this question is to call such processes accelerated reactions and to call the chemical substances accelerators (or promoters). The attitude taken here is that these substances should be considered together with true catalysts when they lead to reactions of a common mechanistic pattern.

Some limitation, however, must be placed on the space devoted to substances that are not true catalysts, so that they do not divert us too far from a discussion of true catalysis. Many reactions are accelerated by acids or bases that are not regenerated because of a subsidiary prototropic equilibrium. For example, in "base-catalyzed" ester hydrolysis, the carboxylic acid product consumes the base in a reaction incidental to the hydrolysis. In other prototropic reactions, such as

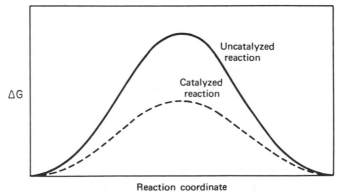

Reaction coordinate

Fig. 1.1. Hypothetical catalysis without changing the pathway of a reaction.

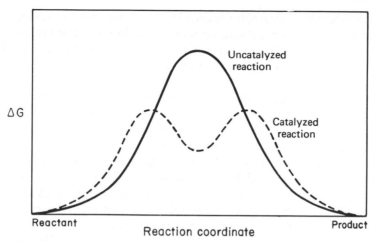

Fig. 1.2. Hypothetical catalysis with a change in the pathway of reaction.

the "base-catalyzed" bromination of acetone, the base is also converted to its conjugate acid. In the "aluminum chloride–catalyzed" acetylation of benzene, aluminum chloride forms a complex with the acetophenone product and so is not available for further catalytic action.

For some purposes, it is useful to regard any substance that changes the rate of a desired reaction as a catalyst, regardless of its fate. However, if the substance is not regenerated, it is not a catalyst but a reactant; it is then more appropriate to recognize two distinct, though similar, reactions than to speak of one reaction, catalyzed or uncatalyzed.

The reaction involving the new reactant, although the same as the old one in many respects, may differ profoundly in its standard free energy change. If the free energy changes from positive to negative, the effect of the added reactant changes a thermodynamically unfavorable process to a favorable one—a change of fundamental consequence. On the other hand, if a reaction is only slightly altered, then its standard free-energy change will likewise be little affected, and the accelerator or promoter will be included as part of a coherent mechanistic description.

Earlier we considered some substances that acclerate reactions but are not catalysts. It is worth mentioning some reactions accelerated by substances that actually are catalysts but are not always recognized as such. For example, the solvolysis of benzyl chloride in aqueous solution is markedly accelerated by mercuric ion. This acceleration is not caused by displacement of equilibrium by removal of the product chloride ion in the form of mercuric chloride complex, as sometimes described. Rather this reaction proceeds through a carbonium ion intermediate; mercuric ion (or complexes thereof) accelerates the formation of the benzyl cation in the rate-determining step by interaction with the halogen atom in the transition state. Thus stabilization of the transition state rather than the product leads to the acceleration by mercuric ion. Along the same lines, chemical folklore has repeatedly suggested that sulfuric acid speeds a reaction by removing water from the system, thereby displacing its equilibrium. In fact, however, sulfuric acid accelerates many reactions because of its great ability to donate protons in a medium of high dielectric constant. Thus catalysis by sulfuric acid is caused by stabilization of a transition state rather than of the product.

Some factors that lead to the acceleration of reaction will be excluded from this discussion. For example, light and heat are not substances and therefore cannot catalyze reactions; they have profound effects on reaction rates, but will not be considered here. The solvent poses a more difficult problem: Changes in the medium can also have profound effects on the rates of reaction. Whether the solvent participates in the transition state is often a moot question. However, any rigorous consideration of catalysis must include a discussion of solvent and salt effects on the grounds that these can represent important accelerations by substances that do not participate directly in the reaction.

On the basis of the foregoing survey of ideas concerning catalysis, it is possible and desirable at this point to state a definition of it upon which development of the subject can be rationally based. The following definition will serve as such a reference point for the remainder of this book: Catalysis of a chemical reaction is an acceleration brought about by a substance that is not consumed in the overall reaction. The catalyst usually functions by interacting with a starting material to yield an alternate species that can react by a pathway involving a lower free energy of activation to give the products and to regenerate the catalyst.

1.2. THE BASIS OF CATALYTIC ACTION

Although no general correlation exists between the tendency of reactions to occur (thermodynamics) and the rates at which they occur (kinetics), the free energy of a reaction is directly related to its free energy of activation often enough for exceptions to attract attention. Catalytic reactions constitute such exceptions, either because (1) the catalyzed reaction is fast compared to expectation from equilibrium, or (2) the uncatalyzed reaction is abnormally slow compared to expectation from equilibrium.[11] One such abnormally slow reaction is that of ceric ion with thallous ion

$$2Ce^{\oplus 4} + Tl^{\oplus 1} \longrightarrow Tl^{\oplus 3} + 2Ce^{\oplus 3} \qquad (1.5)$$

Since $Tl^{\oplus 2}$ does not exist, the direct reaction requires a three-body collision [Eq. (1.5)] and is therefore immeasurably slow. If manganous ion is added to this system, however, a series of two-body reactions can occur [Eq. (1.6)], each with a normal velocity; the effect of catalysis is to convert an abnormally slow reaction to one of normal speed. The alternate pathway afforded by the catalyst apparently derives its efficiency from the higher frequency factors of the individual steps compared to the lower frequency factor of the uncatalyzed reaction. This, of course, is reflected in the entropies of activation.

$$Ce^{\oplus 4} + Mn^{\oplus 2} \longrightarrow Ce^{\oplus 3} + Mn^{\oplus 3}$$

$$Mn^{\oplus 3} + Ce^{\oplus 4} \longrightarrow Ce^{\oplus 3} + Mn^{\oplus 4} \qquad (1.6)$$

$$Mn^{\oplus 4} + Tl^{\oplus 1} \longrightarrow Mn^{\oplus 2} + Tl^{\oplus 3}$$

Catalysis that is fast compared to expectation from equilibrium is exemplified by numerous reactions catalyzed by acids or bases. Acid–base catalysis is effective because it depends on proton transfers, which are fast compared to the making and breaking of most other chemical bonds. In fact, proton transfers are quite rapid compared to reactions of comparable free energy change. Consider the two reactions [Eqs. (1.7) and (1.8)]

$$CH_3OCH_3 + H_2O \longrightarrow CH_3O^{\ominus} + CH_3OH_2^{\oplus} \qquad \text{(slow)} \qquad (1.7)$$

$$CH_3OH + H_2O \longrightarrow CH_3O^{\ominus} + H_3O^{\oplus} \qquad \text{(fast)} \qquad (1.8)$$

There is only a 3 kcal/mole difference in the standard free energy of these two reactions, the equilibrium constants differing by only a hundredfold. The rates, however, differ by a tremendous amount: The former is too slow to measure, while the latter has a large rate constant that can be measured only with special equipment designed for fast reactions. The reaction of water with dimethyl ether is sterically hindered with respect to the reaction of water with methanol because of the repulsion of nonbonded atoms (the hydrogen atoms of the methyl groups).

Although many proton transfers to or from oxygen are fast, not all such processes occur at the same speed (Chapter 2). Consider the reaction involved in the autoprotolysis of water [Eq. (1.9)]

$$H_2O + H_2O \rightleftarrows H_3O^{\oplus} + OH^{\ominus} \qquad (1.9)$$

This reaction is endothermic as written. The heat of the reverse reaction is 13.8 kcal/mole. Therefore, the activation energy of the forward reaction must be at least this amount. Thus an unfavorable equilibrium constant necessarily makes this reaction slow, even though on other grounds it should proceed rapidly.

Easily polarizable nucleophiles react at faster rates than expected from thermodynamics. Nucleophiles such as bromide ion, thiosulfate ion, thiourea, and carbanions react with alkyl iodides at rates higher than their basicities predict. The explanation of this phenomenon has been given many times: The outer-shell electrons of these species are easily polarized. Their distribution can be distorted in the field of an electrophilic reagent; therefore, a new bond can be formed without bringing the rest of the system into close contact. Hence steric repulsion is reduced and reactivity enhanced. An easily polarizable nucleophile of high reactivity, together with an intermediate formed from its reaction which is more unstable than the reactants leads to the phenomenon of nucleophilic catalysis. Bromide ion is a highly polarizable nucleophile; moreover, it is an easily displaced leaving group. This duality of properties leads to the use of bromide ion as a catalyst in the hydrolysis of alkyl iodides [Eq. (1.10)]

$$Br^{\ominus} + RI \longrightarrow R\,Br + I^{\ominus}$$

$$R\,Br + H_2O \longrightarrow R\,OH + Br^{\ominus} + H^{\oplus}$$

(1.10)

Although its basicity is much lower, the bromide ion is a more active nucleophile than water because of polarizability. Furthermore, the alkyl bromide intermediate is more susceptible to hydrolysis than the alkyl iodide. Together these factors produce catalysis.

The reaction of metal ions with organic ligands is often fast, when they can be introduced into an organic molecule. If a metal ion can be easily introduced into an organic molecule through coordination to an oxygen or a nitrogen atom, this leads to an overall polarization of the system and catalysis.

These observations give an inkling of the basis of catalytic action. A more complete elucidation will, however, take a considerably more detailed discussion. It is, in fact, the subject of this book.

1.3. CATALYSIS AND THE EQUILIBRIUM CONSTANT

We have said before that a catalyst speeds up the attainment of equilibrium, but does not change the position of equilibrium. A catalyst cannot change the yield, since it cannot combine with the product, although it can combine with the reactant. These statements follow from thermodynamic arguments since the standard free energy of reaction and thus the equilibrium constant depend only on the initial and final stages and are independent of the path of reaction.

A proof of the independence of equilibrium from reaction mechanism comes from a consideration of the principle of microscopic reversibility.[12] Microscopic reversibility stems from statistical considerations that state that if a system is at equilibrium and there are a number of molecular transitions occurring between the various states within the system, each transition must separately be at equilibrium. Thus, for both an uncatalyzed reaction and the corresponding catalyzed reactions [Eqs. (1.11) and (1.12)], one can write the equilibrium constant 1.13. The equilibrium constant has not been changed in going from the uncatalyzed reaction to the catalyzed one.

$$A \underset{k_{-1}}{\overset{k_1}{\rightleftharpoons}} B$$

(1.11)

$$A + C \underset{k_{-2}}{\overset{k_2}{\rightleftharpoons}} X \underset{k_{-3}}{\overset{k_3}{\rightleftharpoons}} B + C$$

(1.12)

$$K_{eq} = \frac{a_B}{a_A}$$

(1.13)

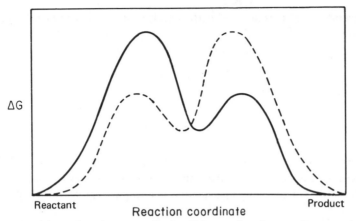

Fig. 1.3. A two-path isotopic exchange reaction in which two mirror image pathways each comprise 50% of the total reaction in each direction. Reprinted from R. L. Burwell, Jr. and R. G. Pearson, *J. Phys. Chem.*, **70**, 300 (1966). Copyright © 1966 by the American Chemical Society. Reprinted by permission of the copyright owner.

where *a* is activity. From Eq. (1.13), it follows that any change the catalyst brings about in the rate constant of the forward reaction is accompanied by a corresponding change in the rate constant of the reverse reaction.

Further consideration of microscopic reversibility indicates that if the mechanisms of the forward reaction is known, the mechanism of the reverse reaction must also be known. For example, if a reaction goes through an intermediate X in the forward direction, the reaction proceeds via the same intermediate X in the reverse direction. In other words, if the energy barriers are known in the forward direction, then they are known (in reverse) for the reverse direction.

The principle of microscopic reversibility, however, does not rule out multiple pathways.[13] Consider an isotopic exchange in which the reactant and product differ only by isotopic substitution. The system must proceed through a single symmetric pathway, but this symmetric pathway can be a combination of two asymmetric pathways. Microscopic reversibility will be satisfied for each system, as shown in Fig. 1.3.

1.4. CATALYSIS AND KINETICS

We have seen earlier that a catalyst accelerates a reaction by providing an alternate mechanism. Let us assume that in the catalytic pathway an intermediate between the catalyst and the substrate is formed and then reacts to yield the product and regenerate the catalyst. Using this description of the catalytic process, it is of interest to develop kinetic equations for it.

Let us assume the stoichiometry

$$R + R' \xrightarrow{C} P \tag{1.14}$$

(where R and R' are reactants, C is the catalyst, and P is the product), and the mechanism

$$R + C \underset{k_{-1}}{\overset{k_1}{\rightleftharpoons}} X$$

$$X + R' \overset{k_2}{\longrightarrow} P + C \tag{1.15}$$

(where X is the intermediate). For most systems represented by Eq. (1.15), useful kinetic expressions can be derived by employing the steady-state approximation; that is, the rate of change of [X] is negligible compared to a change in [R] or [P]

$$\frac{d[X]}{dt} \ll \frac{d[R]}{dt} \quad \text{or} \quad \frac{d[P]}{dt} \tag{1.16}$$

Therefore, we have

$$\frac{d[X]}{dt} = k_1[R][C] - k_{-1}[X] - k_2[X][R'] = 0 \tag{1.17}$$

Hence

$$[X] = \frac{k_1[R][C]}{k_{-1} + k_2[R']} \tag{1.18}$$

If k_{-1} is much larger than either k_1 or $k_2[R']$, X will exist in a low concentration and will be virtually at equilibrium with R and C; such an intermediate is called an Arrhenius intermediate, an equilibrium intermediate.

If k_{-1} and $k_2[R']$ are comparable and much larger than k_1, the concentration of X is low but not simply related to [R] and [C]; in this case, X is called a van't Hoff intermediate, a steady-state intermediate. Now, from Eq. (1.18),

$$\frac{d[P]}{dt} = k_2[X][R'] = \frac{k_1 k_2[R][R'][C]}{k_{-1} + k_2[R']} \tag{1.19}$$

Often it is more convenient to consider only the *beginning* of the reaction, when the approximations $[C] = [C]_0 - [X]$ and $[R] = [R]_0 - [X]$ are valid. Substituting these values into Eq. (1.17) and dropping terms in $[X]^2$ (because of their small magnitude compared to the other terms) gives

$$V_0 = \frac{d[P]}{dt_{t=0}} = \frac{k_1 k_2[R]_0[R']_0[C]_0}{k_{-1} + k_1([C]_0 + [R]_0) + k_2[R']_0} \tag{1.20}$$

In a catalyzed reaction, the catalyst is at a much lower concentration than the reactant; that is, $[C]_0 \ll [R]_0 \sim [R']_0$, and thus $k_1[C]_0 \ll k_1[R]_0$. Then

$$V_0 = \frac{k_1 k_2 [R]_0 [R']_0 [C]_0}{k_{-1} + k_1 [R]_0 + k_2 [R']_0} \qquad (1.21)$$

If $k_2[R']_0 \ll k_{-1} - k_1[R]_0$, a condition that may hold sometimes but need not always follow, the intermediate X is an Arrhenius intermediate and

$$V_0 = \frac{k_1 k_2 [R]_0 [R']_0 [C]_0}{k_{-1} + k_1 [R]_0} \qquad (1.22)$$

Defining $K = k_{-1}/k$,

$$V_0 = \frac{k_2 [R]_0 [R']_0 [C]_0}{K + [R]_0} \qquad (1.23)$$

Equation (1.23) has the form of the Michaelis–Menten equation in enzymatic catalysis,[14,15] in which the initial formation of an adsorptive complex between enzyme and substrate is usually assumed, or the Langmuir isotherm in heterogeneous catalysis.[16] As the Michaelis–Menten equation (1.23) indicates, the initial rate is directly proportional to $[C]_0$ and $[R']_0$ but is of variable order with respect to $[R]_0$. This can be most easily seen in Fig. 1.4 (a plot of Eq. (1.23) in the form of V_0 versus $[R]_0$). At very low $[R]_0$ ($\ll K$), the reaction is first order in $[R]_0$; at very high $[R]_0$ ($\gg K$), the reaction is zero order in $[R]_0$. This

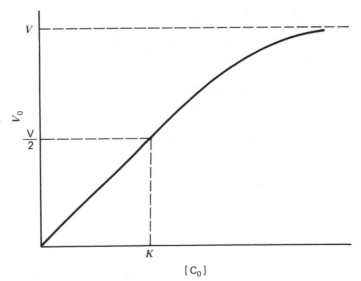

Fig. 1.4. A plot of Eq. (1.23).

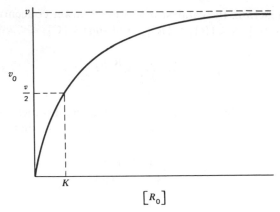

Fig. 1.5. A plot of Eq. (1.24).

phenomenon is usually described as a "saturation" of the *catalyst* by the substrate.

Alternatively, we could have assumed that $[C]_0 \gg [R]_0 \sim [R']_0$ and that $k_2[R']_0 \ll k_{-1}$. Then Eq. (1.24) results.

$$V_0 = \frac{k_2[R]_0[R']_0[C]_0}{K + [C]_0} \qquad (1.24)$$

Equation (1.24) also has the same form as Eq. (1.23). Thus, at very low $[C]_0$ $(\ll K)$, the reaction is first order in $[C]_0$; at very high $[C]_0$ $(\gg K)$ the reaction is zero order in $[C]_0$. This time a "saturation" of the *substrate* by the catalyst is seen (Fig. 1.5). Thus the phenomenon of "saturation" is not specific to the *catalyst*; it can occur with either *catalyst* or *reactant* [Fig. 1.4, which corresponds to Eq. (1.23) or Fig. 1.5, which corresponds to Eq. (1.24)] in any reaction forming an unstable intermediate in equilibrium with reactant.

This treatment implies that the catalytic process can proceed no faster than the first step. Alternatively, the catalyst must react with the reactant to form the intermediate more rapidly than the reactant normally collapses to products.

For a chain mechanism involving initiation, propagation, and termination reactions, the assumption of the stationary state can again be used to define the kinetics. For the system

$$\text{Catalyst} \xrightarrow{k_i} X \qquad \text{(initiation)}$$

$$X + R + R' \xrightarrow{k_p} P + X \qquad \text{(propagation)} \qquad (1.25)$$

$$X + X \xrightarrow{k_t} \text{Destruction of catalyst} \qquad \text{(termination)}$$

the assumption of the stationary state gives

$$\text{rate} = k_p k_i \frac{[C][R][R']}{k_t[I]} \tag{1.26}$$

Equation (1.26) shows that the rate of the overall reaction (due essentially to the propagation step) can be many times greater than the rate of initiation (first step). This conclusion is contrary to the conclusion derived earlier for a nonchain reaction. Thus a chain process should be the best kind of catalytic reaction from the point of view of acceleration. However, a propagation step is a phenomenon that occurs only under special mechanistic circumstances.

1.5. CLASSIFICATION OF CATALYSIS

With this brief survey of some of the basic questions involved in homogeneous catalysis, let us discuss catalysis by looking at some of its manifestations.

We shall begin with a consideration of what is perhaps the oldest and, superficially at least, the simplest of (homogeneous) catalyses: acid–base catalysis. We shall then advance through general acid–base catalysis, nucleophilic–electrophilic catalysis, coenzymes, catalysis by fields, and metal ion catalysis, bridging to enzymatic catalysis. The bridge includes multiple catalysis, intramolecular catalysis, and catalysis by complexation. Whenever appropriate, we will mention the enzyme systems relevant to our discussion and appropriate enzyme models.

REFERENCES

1. J. Berzelius, *Jahresberichte*, **15**, 237 (1835); *Ann. Chim. Phys.*, **61**, 146 (1836); quoted in J. E. Jorpes, *Jöns Jacob Berzelius, His Life and Work*, Almqvist and Wiksell, Stockholm, 1966, p. 112.

2. W. Ostwald, *Chemische Betrachungen, Die. Aula.*, no. 1 (1895); quoted in R. P. Bell, *Acid Base Catalysis*, Oxford University Press, Oxford, 1941, p. 2.

3. W. Ostwald, *Phys. Z.*, **3**, 313 (1902).

4. R. P. Bell, *Acid Base Catalysis*, Oxford University Press, Oxford, 1941, p. 3.

5. M. L. Bender, *Mechanisms of Homogeneous Catalysis from Protons to Proteins*, Wiley-Interscience, New York, 1971.

6. W. P. Jencks, *Catalysis in Chemistry and Enzymology*, McGraw-Hill, Book Co., New York, 1969.

7. H. Dugas and C. Penney, *Bioorganic Chemistry*, Springer Verlag, New York, 1981.

8. P. G. Ashmore, *Catalysis and Inhibition of Chemical Reactions*, Butterworth and Co., Ltd., Woburn, MA, 1963.

9. E. D. Hughes, F. Juliusberger, S. Masterman, B. Topley, and J. Weiss, *J. Chem. Soc.*, 1525 (1935).

10. C. N. Hinshelwood, *The Structure of Physical Chemistry*, Oxford University Press, Oxford, 1951, p. 369.

11. A. A. Frost and R. G. Pearson, *Kinetics and Mechanism*, 2nd ed., John Wiley & Sons, New York, 1963.

12. R. C. Tolman, *The Principles of Statistical Mechanics*, Oxford Unviersity Press, Oxford, 1938, pp. 163 and 165.

13. R. L. Burwell, Jr., and R. G. Pearson, *J. Phys. Chem.*, **70**, 300 (1966).

14. L. Michaelis and M. L. Menten, *Biochem. Z.*, **49**, 333 (1913).

15. K. J. Laidler and P. Bunting, *The Chemical Kinetics of Enzyme Action*, 2nd ed., Clarendon Press, Oxford, 1973.

16. I. Langmuir, *J. Am. Chem. Soc.*, **40**, 1361 (1918).

2 | Proton Transfer

Proton transfer is the essence of acid–base catalysis.[1,2] In recent years, the experimental description and theoretical interpretation of proton transfer (and therefore acid–base catalysis) has flowered with the advent of modern relaxation and spectroscopic methods for the measurement of the extremely fast reactions in which the proton participates.[3-8] Although a consideration of the elegant methods and theoretical framework used in the determination of the kinetics of these reactions, whose half-lives range from 10^{-5} to 10^{-11} sec is beyond the scope of this work, their results are essential to a proper consideration of acid–base catalysis.

2.1. PROTON TRANSFERS TO OR FROM HYDRONIUM AND HYDROXIDE IONS

Table 2.1 lists some representative rate constants for the reactions of acids with hydroxide ion (k_H) and of bases with hydroniun ion (k_{OH}). Many of these rate constants are of the order of magnitude calculated for diffusion control (the rate at which reactants can diffuse together in the solvent, about 10^{11} M^{-1} sec^{-1}). Bases including hydroxide ion, imidazole, fluoride ion, and water, representing a range of 16.7 pK units, react with hydronium ion with essentially identical rate constants. Likewise, water, imidazolium ion, hydrogen fluoride, and hydronium ion react with hydroxide ion with essentially identical rate constants. The rate of reaction of a hydronium ion with hydroxide ion (1.4×10^{11} M^{-1} sec^{-1}) is even faster than its reaction with a solvated electron (2.3×10^{10} M^{-1} sec^{-1}). The simplest explanation that the ion–ion reaction is faster than the ion–electron reaction is that proton transfers are not slowed by solvent reorganization,

Table 2.1. Rate Constants for the Reaction of Some Acids with Hydroxide Ion and of Their Conjugate Bases with Hydronium Ion[a]

Acid	Conjugate Base	pK_a	k_H $(M^{-1}\,sec^{-1})$	k_{OH} $(M^{-1}\,sec^{-1})$
H_2O	$HO^{\ominus}$	15.75	1.4×10^{11}	
H_2O (ice)	$HO^{\ominus}$	21.4	$8.6 \times 10^{12\,b}$	
D_2O	$DO^{\ominus}$	16.5	8.4×10^{10}	
$ImH^{\oplus}$	Imidazole	6.95	1.5×10^{10}	2.5×10^{10}
H_2S	$HS^{\ominus}$	7.24	7.5×10^{10}	
HF	$F^{\ominus}$	3.15	1.0×10^{11}	
(salicylate structure)	Salicylate dianion	11.05		$1.4 \times 10^{7\,c}$
(dianion structure, NO₂)	(Dianion)	11.90		4.75×10^{5}
Enol of acetylacetone	Enolate ion	8.24	3.1×10^{10}	1.9×10^{7}
Acetylacetone	Enolate ion	9.0	$1.2 \times 10^{7\,c}$	9×10^{4}
Acetone	$CH_3COCH_2^{\ominus}$	~20	$\sim 5 \times 10^{10}$	2.7×10^{-1}
H_2CO_3	$HCO_3^{\ominus}$	6.35	5.6×10^{4}	1×10^{4}
Tropylium ion ($+H_2O$)	Tropanol	4.75	6.6×10^{4}	

Source: M. L. Bender, *Mechanisms of Homogeneous Catalysis from Protons to Proteins*, Wiley-Interscience, New York, 1971, p. 20.
[a]Temperature = 298 K unless otherwise specified.
[b]Temperature = 263 K.
[c]Temperature = 285 K.

whereas electron transfers are impeded by this process. The mechanism of proton transfer is shown by Eq. (2.1), which is comprised of three basic steps: (1) The diffusion-controlled formation of a hydrogen bond between the acid H—Y and the base X:; (2) the transfer of a proton to make a new hydrogen-bonded complex; and (3) the diffusion-controlled dissociation of this hydrogen-bonded product, leading to an overall facile reaction

$$X: + H-Y \underset{2.1}{\rightleftarrows} X\text{---}H-Y \underset{2.2}{\rightleftarrows} X-H\text{---}Y \underset{2.3}{\rightleftarrows} X-H + :Y \qquad (2.1)$$

Equation (2.1) indicates that proton transfer is guided by a hydrogen bond, whereas electron transfer is not. In addition, the distance over which the proton is transferred is small compared with the electron jump distance, and the field of the surrounding solvent is relatively much less altered as a result of the smaller distance. As seen in Table 2.1, the reaction of a hydronium ion and a hydroxide ion in ice is faster than the corresponding reaction in water by a factor of approximately 60. This difference can be readily explained using Eq. (2.2), since ice contains a perfectly hydrogen-bonded lattice.

The ice crystal consists of a three-dimensional network of water molecules linked together by hydrogen bonds.[9] Ice conducts direct current; the charge carriers may be assumed to be hydronium and hydroxide ions formed by dissociation of the water molecule. Charge can be transported through the crystal by multiple proton jumps in hydrogen bonds in an electrical field.

Conduction of direct current requires that charge transport always be possible in the same direction. But after passage of a single charge, all chains of water molecules are in the state shown in the lower part of Eq. (2.2), unsuitable for further charge transport from left to right. To explain conduction of direct current, the chain must be renewed for proton passage: Rotation of the water molecules in the cyrstal will serve this purpose. Thus, by a combination of proton jumps in hydrogen bonds and reorientation of water molecules, a proton can migrate through the entire ice crystal.

$$
\begin{array}{l}
\quad\; \text{H} \qquad \text{H} \qquad \text{H} \qquad \text{H} \qquad \text{H} \qquad \text{H} \\
\quad\; | \qquad\;\; | \qquad\;\; | \qquad\;\; | \qquad\;\; | \qquad\;\; | \\
\text{---H} - \text{O} - \text{H} \text{---} \text{O} \text{---} \text{H} \text{---} \text{O} - \text{H} \text{---} \text{O} - \text{H} \text{---} \text{O} - \text{H} \text{---} \text{O} - \text{H} \text{---} \\
\qquad\quad \oplus
\end{array}
$$

$$
\begin{array}{l}
\qquad\qquad\;\; \text{H} \qquad \text{H} \qquad \text{H} \qquad \text{H} \qquad \text{H} \\
\qquad\qquad\;\; | \qquad\;\; | \qquad\;\; | \qquad\;\; | \qquad\;\; | \\
\text{---H} - \text{O} \text{---} \text{H} - \text{O} \text{---} \text{H} - \text{O} \text{---} \text{H} - \text{O} \text{---} \text{H} - \text{O} - \text{H} \\
\qquad\qquad\qquad\qquad\qquad\qquad\qquad\qquad\qquad \oplus
\end{array}
\qquad (2.2)
$$

The rate-determining step of the conduction in ice is the proton translation in a hydrogen bond. Here proton transfer follows a "nonclassical" mechanism, probably involving tunneling[10] (movement through the usual activation barrier because of the movement of a small ion through a thin barrier) effect. The mobility of a proton in ice (Table 2.1) is only one or two orders of magnitude less than than that of an electron in metals.

Although the structure of water does not present the perfect hydrogen-bonded structure that ice does for the chain proton jump mechanism of Eq. (2.2), the high mobility of the proton in water must be due to a similar process.

In discussing rate constants of the hydronium and hydroxide ion reactions of Table 2.1, the following factors need to be considered: (1) spatial (symmetry and steric) factors, (2) electrostatic interactions, (3) the hydrogen-bonded structure of the substrate, and (4) electronic redistribution during reaction.

The reaction of hydronium ion with fluoride ion is slightly faster than the reaction of hydronium ion with hydrosulfide ion. This small difference can be explained on a statistical basis since fluoride ion has four electron pairs that can accept a proton, whereas hydrosulfide ion has only three. Electrostatic interactions cause only small perturbations of the rate constant, presumably because of the high dielectric constant of water, the solvent for these proton transfers. Roughly speaking, the rate constant of proton transfer from the hydronium ion decreases by a factor of 2 for each positive charge introduced into systems of equal size. For example, the rate constants for the reaction of hydronium ion with metal ion complexes of varying charge are $HOCu(H_2O)_5^\oplus$, 10^{10} M^{-1} sec^{-1}; $HOCo(NH_3)_5^{2\oplus}$, 5×10^9 M^{-1} sec^{-1}; and $HNRPt(en)_2^{3\oplus}$, 1.9×10^9 M^{-1} sec^{-1}.

In contrast to these small effects, the presence of a hydrogen bond in the substrate leads to a large diminution in the rate constant of proton transfer to the solvent or another molecule. Two examples of this phenomenon may be seen in Table 2.1: the reactions of hydroxide ion with salicylate ion and with the *ortho*-hydroxyazobenzene derivative. The effect of the internal hydrogen bond on these reactions amounts to a 10^3–10^5 decrease in the rate constant compared to "normal" proton transfers. Mechanism (2.1) implies that proton transfer between these two substrates and hydroxide ion can occur only after the internal hydrogen bond is broken. The rate constant of proton transfer will then reflect the difference in stability constants of the internal and external hydrogen bonds, as compared to a rate constant of a diffusion-controlled reaction without this complication.

Another large effect on the rate constant of proton transfer reactions is seen in the reaction of an acid with hydroxide ion or the reaction of a conjugate base with hydronium ion. If these reactions are associated with a large redistribution of electronic charge through a resonance interaction, the reaction rate may be much lower than that of a diffusion-controlled reaction. Consider the reaction of the carbon acid acetylacetone of Table 2.1. Carbon acids, in contrast to "normal" oxygen and nitrogen acids, show rate constants much lower than those of diffusion-controlled reactions. The reaction of the enolate ion of acetylacetone with a hydronium ion, to give the enol, which does not involve appreciable redistribution of electronic charge, proceeds with a diffusion-controlled rate constant of 10^{10} M^{-1} sec^{-1}. On the other hand, the reaction of the enolate ion with hydronium ion to give the ketone, which involves considerable redistribution of electronic charge, has a rate constant of only 10^7 M^{-1} sec^{-1}. Furthermore, the reaction of hydroxide ion with the enol of acetylacetone, which contains an internally hydrogen-bonded proton, has a rate constant of 10^7 M^{-1} sec^{-1}, whereas the reaction of the ketone, which does not contain an internally hydrogen-bonded proton, but does involve with hydroxide ion considerable electronic redistribution, has a rate constant of only 4×10^4 M^{-1} sec^{-1}.

$$(2.3)$$

The low rate constants of proton transfer involving carbon acids are of special significance to organic chemistry. Two interrelated factors apparently account for these low rate constants: (1) A carbon acid does not form hydrogen bonds readily, and (2) considerable electronic redistribution is ordinarily associated with transfer of a proton from a carbon acid.

2.2. RELATIONSHIP BETWEEN RATE CONSTANTS AND EQUILIBRIA IN PROTON TRANSFER

Let us now consider the relationship between the ionization constants of acids and bases and their abilities to accept or donate protons. As a measure of this ability, we will use the rate constant for the donation of a proton from an acid to water or the rate constant for the abstraction of a proton from water by a base. Tables 2.2 and 2.3 give the rate constants of proton transfer between several bases and water, according to Eq. (2.4), and between several acids and water, according to Eq. (2.5).

$$B + H_2O \underset{k_r}{\overset{k_f}{\rightleftharpoons}} BH^{\oplus} + OH^{\ominus} \tag{2.4}$$

$$HA + H_2O \underset{k_r}{\overset{k_f}{\rightleftharpoons}} H_3O^{\oplus} + A^{\ominus} \tag{2.5}$$

The data in Table 2.2 show that a direct relationship exists between the pK_a of the conjugate acid of the base and the logarithm of the rate constant for proton abstraction from water by the base (k_f). Likewise, Table 2.3 and Fig. 2.1 show a direct relationship between the pK_a of the acid and the logarithm of the rate constant for proton donation to water by the acid (k_f'). These relationships are Brönsted relationships (see Chapter 5). The reason for both these relationships is that the reverse reactions are diffusion-controlled proton transfers having either a zero or negative free-energy change. The direct relationship between the logarithm of the rate constants of proton transfer and the pK_a's of the acids (k') or bases (k) may be expressed by Eqs. (2.6) and (2.7), respectively. In fact, these

Table 2.2. Rate Constants for the Reaction of Some Bases with Water

B	$k_f(M^{-1} sec^{-1})$	$k_r(M^{21} sec^{-1})$	pK_a of BH$^{\oplus}$	$\log k_f$
OH$^-$	5×10^9	5×10^9	15.74	9.69
NH$_3{}^b$	6×10^5	3.4×10^{10}	9.25	5.78
Imidazole	2.2×10^3	2.5×10^{10}	6.95	3.34
H$_2$O	2.5×10^{-5}	1.4×10^{11}	-1.74	-4.21

Source: References 5 and 11.
^{a}25°C unless otherwise noted.
^{b}20°C.

Table 2.3. Rate Constants for Reactions of Some Acids with Water[a]

HA	$k_f'(M^{-1}\,sec^{-1})$	$k_r'(M^{-1}\,sec^{-1})$	pK_a of HA	$\log k_f'$
$H_3O^{\oplus}$	1×10^{10}	1×10^{10}	-1.7	-10.0
H_2SO_4	$\sim 10^{9}$	$\sim 1 \times 10^{11}$	2	~ 9
HF	7.1×10^{7}	1×10^{11}	3.15	7.85
CH_3COOH	8.2×10^{5}	4.5×10^{10}	4.74	5.91
H_2S	4.3×10^{3}	7.4×10^{10}	7.24	3.63
$ImH^{\oplus}$	1.8×10^{3}	1.5×10^{10}	6.95	3.26
$(CH_3)_3NH^{\oplus}$	1×10^{1}	6.3×10^{10}	9.8	1.00
Glucose	$<5 \times 10^{-3}$	$>10^{10}$	12.3	<-2.7
H_2O	2.5×10^{-5}	1.4×10^{11}	15.74	-4.61
H_2SO_3	3.2×10^{6}	2×10^{3}	1.8	6.51
H_2CO_3	0.025	5.6×10^{4}	3.60	-1.20

Source: References 5 and 11.
[a] $25°C$.

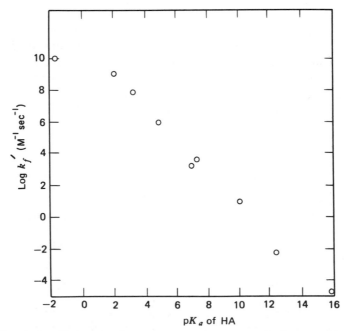

Fig. 2.1. The relationship between the rate constant of proton transfer from an acid to water (k_f') and the pK_a of the acid.

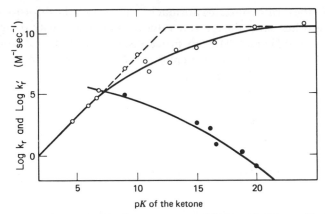

Fig. 2.2. Dependence of the recombination rate constants k_r' ($\bigcirc$) and k_r ($\bullet$) on pK in the protolysis of ketonic compounds. Experimental data are circles. The dashed line denotes the linear extrapolation. From M. Eigen, *Angew. Chem. Int. Ed.*, **3**, 12 (1964).

equations were used in calculating a few of the quantities shown in Tables 2.2 and 2.3.

$$\log k_{f'} = \log k_r' \times 10^{-pK_a} \tag{2.6}$$

$$\log k_f = \log k_r \times 10^{-(15.5 - pK_a)} \tag{2.7}$$

Table 2.3 includes two slow proton transfer reactions involving sulfurous acid and carbonic acid. The low values of these rate constants probably reflect the large electronic redistribution accompanying ionization.

The ionizations of carbon acids must be probed in more detail. The rates of prototropic reactions of a family of ketones are represented by Fig. 2.2, which shows a monotonic nonlinear variation of k_r and k_r' as a function of the pK of the ketone. There is, however, no simple relation between k_r' and the pK when the tabulation includes carbon acids other than ketones (Table 2.4). As mentioned

Table 2.4.   Rate Constants for Reactions of Some Carbon Acids

HA	pK^a	$\log k_f^b$	$\log k_r^b$
CH_3COCH_3	20	-9.0	10.7
$CH_2(CO_2C_2H_5)_2$	13.3	-4.6	8.7
$CH_2(CN)_2$	11.2	-1.8	9.4
$CH_3COCH_2COCH_3$	9.0	-1.8	7.2
$C_2H_5NO_2$	8.6	-7.4	1.2
$CH_2COCH_2NO_2$	5.1	-1.4	3.7
$CH_2(NO_2)_2$	3.6	-0.1	3.5

Source: References 5 and 11.
[a]k_f determined by bromination.
[b]Rate constants M^{-1} sec^{-1} at 18–25°C.

earlier, all of these rate constants are much smaller than the diffusion-controlled limit. On the other hand, the reaction rates of enolate ions with oxonium ion do approch this limit (Table 2.1); these processes, of course, do not strictly involve carbon acids, since they produce enols (proton transfer to oxygen) and not ketones (proton transfer to carbon).

2.3. GENERALIZED ACID–BASE SYSTEMS IN PROTON TRANSFER

Proton transfers in acid–base systems may be described by the equation

$$HX + Y^{\ominus} \underset{k_r}{\overset{k_f}{\rightleftharpoons}} X^{\ominus} + HY \tag{2.8}$$

For the reactions of hydroxide ion with HX or HY, or for the reactions of hydronium ion with $X^{\ominus}$ or $Y^{\ominus}$, one will expect diffusion-controlled rate constants, assuming that the species X and Y are "normal" (oxygen and nitrogen) acids and bases. For reactions proceeding according to Eq. (2.8), one would expect that k_r will likewise be diffusion-controlled when $pK_{HY} > pK_{HX}$; that is, when $Y^{\ominus}$ is a stronger base than $X^{\ominus}$. This statement may be alternatively expressed by saying that the reaction will be diffusion-controlled when the reaction as written is exergonic. Certainly, the faster proton transfer must occur in the direction of the exergonic reaction.

A proton transfer can be diffusion-controlled even in a symmetrical system when ΔpK between the two reactants is zero; that is, when the free energy of the reaction is zero. Tables 2.2 and 2.3, for example, show that the rate constants of the symmetrical reactions between hydronium ion and water or between hydroxide ion and water are essentially diffusion-controlled. Those prototropic reactions in which the free energy of the reaction is negative would certainly be expected to proceed with an even higher rate, which can only be diffusion-controlled.

When a whole series of prototropic reactions of noncarbon acids is exergonic, all the rates should be diffusion-controlled and therefore independent of the pK's of the bases involved. When this is true, $\log k_r$ must be a linear function of the difference between the pK's of the two reactants (a Brönsted relationship, as described in Chapter 4). This conclusion may be easily seen since the equilibrium constant for Eq. (2.9) can be expressed as either of two quantities:

$$K_{eq} = \frac{HX}{HY} = \frac{k_f}{k_r} \tag{2.9}$$

Equation (2.9) can be transformed into its logarithmic counterpart

$$\log k_f - \log k_r = pK_{HX} - pK_{HY} = \Delta pK \tag{2.10}$$

When $\log k_f$ is constant, as it must be in these exergonic reactions, Eq. (2.10) becomes

$$\log k_r = \text{constant} - \Delta p K_{HY} \qquad (2.11)$$

Thus a proton transfer reaction conforming to Eq. (2.11) can be represented by Fig. 2.3: When the pK of the acceptor is greater than the pK of the donor (when ΔpK is positive), the rate of the proton transfer is independent of ΔpK and diffusion-controlled; when ΔpK is approximately zero, a transition from independence of ΔpK to a linear dependence on ΔpK occurs, the forward reaction getting progressively slower and the reverse reaction becoming diffusion-controlled. Thus two limiting slopes of zero and ± 1 are reached for each reaction.

For a proton transfer reaction involving a separation of charge, such as

$$XH + Y \underset{k_r}{\overset{k_f}{\rightleftharpoons}} X^{\ominus} + HY^{\oplus} \qquad (2.12)$$

the relationship between $\log k$ and ΔpK is asymmetric, because the limiting value for the diffusion-controlled reaction will not be attained in the reaction direction involving separation of charge. A hypothetical example of the relationship between rate constants and equilibrium constants of reactions related to Eq. (2.12) is shown in Fig. 2.4,[11,12] which depicts a reaction diffusion-controlled in one direction but not in the other. Many other sets of reactions correspond closely to the hypothetical situation shown in Fig. 2.4. The curvature in the relationship between $\log k$ and ΔpK in both Figs. 2.3 and 2.4 can be expressed by a McLaurin series, as shown in Eq. (2.13).

$$\log k \simeq a\Delta pK + a'(\Delta pK)^2 + a''(\Delta pK)^3 + \cdots \qquad (2.13)$$

In proton transfer reactions involving normal oxygen and nitrogen acids and

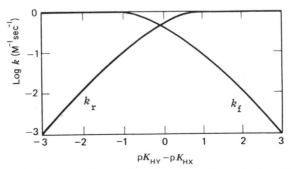

Fig. 2.3.   Idealized $\log k$ versus $p K_{HY} - p K_{HX}$ dependence for proton transfer for symmetrical charge type [Eq. (2.6)]. The rate constant for the diffusion controlled reaction is taken as the reference. From M. Eigen, *Angew. Chem. Int. Ed.*, **3**, 13 (1964).

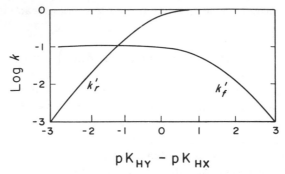

Fig. 2.4.   Idealized $\log k$ versus $\Delta pK = pK_{HY} - pK_{HX}$ dependence for proton transfer in "normal" acid–base systems: for charge neutralization [Eq. (2.10)]. The rate constant for the diffusion-controlled reaction is taken as the reference. From M. Eigen, *Angew. Chem. Int. Ed.*, **3**, 13 (1964).

bases, all the factors pertaining to proton transfers involving hydronium or hydroxide ions come into play. A simple relationship between $\log k$ and pK would be expected when the donor and acceptor atoms consist of classical hydrogen bond-forming atoms such as oxygen and nitrogen, when no internal hydrogen bond exists, and when electronic and spatial configurations are such that the acids and their conjugate bases are comparable.

Conversely, reactions of carbon acids and bases cannot be described by such a simple relationship. A typical example is the reaction of acetylacetone with various bases, as shown in Fig. 2.5. Qualitatively, proton transfers to or from carbon have the characteristics of reactions of normal acids and bases. However, there are quantitative differences; most importantly, the transition from a slope of zero to one occurs over a very much wider range of ΔpK than with normal acid and base reactions.

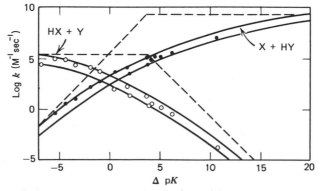

Fig. 2.5. $\log k$ versus ΔpK dependence for proton transfer between acetylacetone (HX) or its enolate (X) and a series of bases (Y) or acids (HY). Acetylacetone reacts in its keto form. From M. Eigen, *Angew. Chem. Int. Ed.*, **3**, 16 (1964).

2.4.   SOLVENT PARTICIPATION IN PROTON TRANSFER[13]

By means of proton magnetic resonance (pmr) it is possible in favorable cases to measure not only the rate of proton transfer between the solute and solvent, but also the number of solvent molecules (n) that participate in a given process.[14,15]

The pmr method was used to study the kinetics of proton exchange involving aqueous solutions of methylammonium ion. These studies have been extended to a number of amines and hydroxylic solvents, covering a wide range of dielectric constant.[13] In solvents of high dielectric constant, for example, water and methanol, the major fates of $BH^{\oplus}$ include: (1) transfer of a proton from $BH^{\oplus}$ to the solvent; (2) direct proton transfer from $BH^{\oplus}$ to B; and (3) proton transfer from $BH^{\oplus}$ to B through one solvent molecule, as shown in Eqs. (2.14), (2.15), and (2.16), respectively.

$$BH^{\oplus} + ROH \underset{k_{-1}}{\overset{k_1}{\rightleftharpoons}} B + ROH_2^{\oplus} \qquad (2.14)$$

$$BH^{\oplus} + B \xrightarrow{k_2} B + HB^{\oplus} \qquad (2.15)$$

$$BH^{\oplus} + \underset{R}{O}H + B \xrightarrow{k_3} B + \underset{R}{H}O + HB^{\oplus\dagger} \qquad (2.16)$$

In solvents of low dielectric constant, proton transfer processes are affected by ionic association. In t-butyl alcohol ($D = 12.47$ at 25°C), the reactive species is the solvated ion pair rather than the solvated free ion.

$$BH^{\oplus} X^{\ominus} + \underset{t\text{-Bu}}{O}H + B \xrightarrow{k_1} B + \underset{t\text{-Bu}}{H}O + X^{\ominus} HB^{\oplus} \qquad (2.17)$$

In this reaction, a proton hydrogen-bonded to the anion in the ion pair is transferred much less readily than a proton hydrogen-bonded to a solvent molecule. In glacial acetic acid ($D = 6.22$ at 25°C), methylamines are converted almost completely to their acetate salts, which exist in this solution largely in the form of ion pairs, $BH^{\oplus} OAc^{\ominus}$. Proton exchange is first order in this ion pair and probably involves a two-step mechanism with initial formation of undissociated amine.

Rate constants of proton exchange of methyl-substituted amines in high dielectric solvents such as water according to Eqs. (2.14)–(2.16) are shown in Table 2.5. The values of $k_{-1}(=k_1/K_a)$ for these substrates are of such large

---

[†] ^nROH was found to be one experimentally.

Table 2.5.   Rate Constants for Proton Transfer and Acid Dissociation Constants in Water at 25°C[a]

$BH^{\oplus}$	k_1 (sec^{-1})	$k_{-1} \times 10^{-10}$ (M^{-1} sec^{-1})	$k_2 \times 10^3$ (M^{-1} sec^{-1})	$k_3 \times 10^3$ (M^{-1} sec^{-1})	$K_a \times 10^{10}$ (M)
$NH_4^{\oplus}$	25	4.3	11.7	0.9	5.68
$CH_3NH_3^{\oplus}$	—	—	4.0	5.3	0.242
$(CH_3)_2NH^{\oplus}$	—	—	0.5	9.0	0.168
$(CH_3)_2NH^{\oplus}$	4.7	3.0	0.0	3.4	1.57
$HO_2CCH_2NH_2^{\oplus}CH_3$	110	3.2	—	—	30

Source: E. Grunwald, *Progr. Phys. Org. Chem.* **3**, 130 (1965).
[a]The rate constants in this table apply to Eqs. (2.12)–(2.14) and to dilute solutions.

magnitude as to suggest diffusion-controlled processes, whereas values of k_1 are much lower, as expected from the basicity of the amine and water.

The rate constant k_2 for direct proton transfer from $BH^{\oplus}$ to B decrease sharply with methyl substitution on nitrogen, as seen in Table 2.5. This could be due to increased steric hindrance or to increased energy of the desolvation step that must precede proton transfer. On the other hand, the rate constant k_3 for the reaction of $BH^{\oplus}$, water, and B is not highly sensitive to methyl substitution. One water molecule is involved in the proton transfer between trimethylamine and trimethylammonium ion. A single water molecule is likewise necessary for the proton exchange between dihydrogen phosphate and monohydrogen phosphate ions or between phenol and phenoxide ion in aqueous solution.

Proton exchange of carboxylic acids has also been investigated by pmr methods. Table 2.6 lists pseudo-first-order rate constants for proton exchange between a carboxylic acid and neopentyl alcohol in glacial acetic acid. The rate constants of these proton transfers increase systematically with the strength of the acid, the sequence being $CH_3CO_2H < NCCH_2CO_2H < Cl_3CCO_2H$. The exchange involves one molecule of acid and one of alcohol. The exchange reaction can then be written as

$$
\begin{array}{ccc}
\text{H---O} & & \text{H---O} \\
\diagup \quad \diagdown\!\!\diagdown & k_i & \diagup \quad \diagdown\!\!\diagdown \\
\text{R---O} \qquad \text{C---R'} & \rightleftharpoons & \text{R---O}^{\oplus} \quad {}^{\ominus}\text{C---R'} \\
\diagdown \quad \diagup & k_{-i} & \diagdown \quad \diagup\!\!\diagup \\
\text{H}^*\!\!-\!\!\text{O} & & \text{H}^*\!\!\text{---O}
\end{array}
\qquad (2.18)
$$

k_i of Eq. (2.18) is related to k_{exch} according to $k_i = 2k_{exch}$ using assumptions that (1) ionization (with rate constant k_i) is the rate-determining step for proton exchange; and (2) the carboxylate group in the ion pair is symmetrical; that is, the two oxygen atoms are equivalent. The rate constants k_i of Table 2.6 are quite high, but well below the range of diffusion-controlled rate constants that characterize the reverse reactions.

Table 2.6.   Analysis of Rate Constants for Proton Exchange

Carboxylic Acids with Neopentyl Alcohol in Glacial Acetic Acid, 25° C

RCO_2H	$k_{exch} (sec^{-1})$	k_i	$1/k_i$	$1/k_{-i}$
CH_3CO_2H	6.7×10^5	5×10^{-8}	13×10^5	3×10^{13}
$NCCH_2CO_2H$	8.8×10^6	1×10^{-5}	18×10^6	2×10^{12}
Cl_3CCO_2H	2.5×10^8	3×10^{-3}	5×10^8	2×10^{11}

Source: Reference 15.

2.5.   PROTON TRANSFERS IN NONAQUEOUS MEDIA[16]

Although in water, the rates of proton transfer from nitromethane, phenol, *p*-toluenesulfonic acid, or water to a base are vastly different, in dimethyl sulfoxide, proton transfers from these acids to triphenylmethyl or fluorenyl anions proceed at identical rates. Furthermore, the proton exchange between dimethyl sulfoxide and its conjugate base occurs with a rate constant of 7 M^{-1} sec^{-1}, approximately a millionfold faster than the proton exchange of the carbon base fluorenyllithium with fluorene in ether solution (10^{-5} M^{-1} sec^{-1}). These observations emphasize the importance of changes in hybridization on the rates of proton transfer. When rehybridization (e.g., of a carbon acid) occurs in a solvent where anionic stabilization by the solvent is important, solvent reorganization will contribute a major part of the activation energy of the reaction. When rehybridization is unimportant, such as in dimethyl sulfoxide, which has more electron donating ability than ethers, solvent reorganization should not be necessary for ionization and proton transfer rate constants should be large and less dependent on structure.

In solvents of low dielectric constant, such as benzene, proton transfer will lead not to separated ions but rather to ion pairs, triplets, and quadruplets.[17]

The rate constants of proton transfer in nonaqueous solution may possibly be related to the equilibrium constants of these reactions. In the solvent cyclohexylamine, the acidities of a number of hydrocarbons were determined by competition with methoxide ion. For a series of reactions represented by

$$R\text{---}H + OMe^{\ominus} \underset{k''_r}{\overset{k''_f}{\rightleftharpoons}} R^{\ominus} + HOMe \tag{2.19}$$

a linear relationship was found between log k''_f and log K, with a slope of 0.4. These results mean that k''_r is a diffusion-controlled process and is the same for all members of the series. The results further mean that the forward reaction is not diffusion-controlled.

It appears that proton transfers in hydroxylic solvents always involve hydrogen bonding to the solvent or to another molecule. They must involve at least one and possibly a very large number of solvent molecules. The rate of proton transfers, as those of other reactions, are susceptible to structural variation in the substrate, and to considerable solvent effects. In chapter 4, we will consider the participation of proton transfers in acid and base catalysis.

2.6.   PROTON TRANSFERS IN ENZYMATIC REACTIONS

Proton transfers from enzymes to substrates are essential parts of many enzymatic reactions. Proton transfers between prototropic agents, that is, general acids and bases, described in Chapters 5 and 6, are of major importance. In addition, proton transfers also occur from many coenzymes (Chapter 8). Although enzymes are water soluble, enzymatic active sites are apolar (hydrophobic). Therefore, Section 2.5 is most pertinent to enzymatic proton transfers. All principles enunciated in this chapter will apply to many enzymatic reactions.[18]

REFERENCES

1.   R. P. Bell, *Acid-Base Catalysis*, Clarendon Press, Oxford, 1941.

2.   R. P. Bell, *The Proton in Chemistry*, 2nd ed., Cornell University Press, Ithaca, NY, 1973.

3.   M. Eigen and L. DeMaeyer, in *Techniques of Organic Chemistry*, A. Weissberger, Ed., Vol. 8, Part 2, 2nd ed., Interscience Publishing Co., New York, 1963, p. 895.

4.   H. Strehlow, in *Techniques of Organic Chemistry*, A. Weissberger, Ed., Vol. 8, Part 2, 2nd ed., Interscience Publishing Co., New York, 1963, pp. 799 and 865.

5.   R. M. Noyes and A. Weller, in *Techniques of Organic Chemistry*, A. Weissberger, Ed., Vol. 8, Part 2, 2nd ed., Interscience Publishing Co., New York, 1963, p. 845.

6.   G. Porter, in *Techniques of Organic Chemistry*, A. Weissberger, Ed., Vol. 8, Part 2, 2nd ed., Interscience Publishing Co., New York, 1963, p. 1055.

7.   E. S. Caldin, *Fast Reactions in Solution*, John Wiley and Sons, New York, 1964.

8.   L. K. Patterson, *Chem. Br.*, **4**, 24 (1968).

9.   M. Eigen and L. DeMaeyer, *Proc. R. Soc.*, A247, 505 (1958).

10.   E. S. Lewis, in *Proton Transfer Reactions*, E. S. Caldin and V. Gold, Eds., Chapman and Hall, London, 1975, Chapter 10.

11.   M. Eigen, *Angew Chem. Int. Ed.*, **3**, 1 (1964).

12.   For measurement of very high reaction rates by Raman line broadening, see M. M. Kreevoy and C. A. Mead, *J. Am. Chem. Soc.* **84**, 4596 (1962).

13.   E. Grunwald and D. Eustace, in *Proton Transfer Reactions*, E. S. Caldin and V. Gold, Eds., Chapman and Hall, London, 1975, Chapter 4.

14.   R. M. Noyes, *Disc. Faraday Soc.*, **39**, 130 (1965).

15.   E. Grunwald, *Prog. Phys. Org. Chem.*, **3**, 344 (1965).

16.   B. H. Robinson, in *Proton Transfer Reactions*, E. S. Caldin and V. Gold, Eds., Chapman and Hall, London, 1975, Chapter 5.

17.   E. M. Arnett, in *Proton-Transfer Reactions*, E. S. Caldin and V. Gold, Eds., Chapman and Hall, London, 1975, Chapter 3.

18.   F. J. Kezdy and M. L. Bender, in *Proton-Transfer Reactions*, E. S. Caldin and V. Gold, Eds., Chapman and Hall, London, 1975, Chapter 12.

3 | Catalysis by Fields

3.1. INTRODUCTION

In this chapter, we discuss catalysis of chemical reactions by substances whose interaction cannot be specified as discretely as those given earlier. For this reason, we refer to this phenomenon as catalysis by fields. The substance of this chapter is, in fact, catalysis by salts and solvents. Both of these are external to the substrate of the reaction and both lead to important rate accelerations. Neither salts nor solvents appear simply in the rate expression, as do other catalysts of the general acid–base (Chapter 4) or nucleophilic–electrophilic (Chapter 7) variety. However, they do affect the standard free energy of the ground state and/or transition state, and therefore they can profoundly affect the rate constant of a reaction. (We will not consider effects on equilibria.) In general, the pathway of a reaction is not changed by salts and solvents, in contrast to many of the catalyses described earlier. But from both a practical and theoretical point of view, these effects must be considered in describing the acceleration of chemical reactions.

Sometimes these accelerations are called medium effects, while at other times they are called catalytic effects. The gradation between medium and catalytic effects is a subtle one, but the magnitude of the effects, whatever they are called, is far from subtle. If the rate enhancements are significant, we shall refer to them as catalyses.

The fundamental basis of the catalyses discussed here can be seen in a comparison of the rates of chemical reactions in solution compared to those in the gas phase, using the transition state theory of reaction rates.[1] Consider the reaction

$$A + B \rightleftharpoons M^{\ddagger} \longrightarrow \text{Product} \tag{3.1}$$

where A and B are reactants and $M^{\ddagger}$ is the activated complex; the specific rate constant for this reaction in the gas phase is $k_g = RT/Nh) K_g^{\ddagger}$, where $K_g^{\ddagger}$ is the "equilibrium constant" between ground and transition states. R is the gas constant, N is Avogadro's number, and h is Planck's constant. In solution, the analogous equation must be corrected to account for derivations from ideal behavior. The thermodynamic "equilibrium constant" between the ground and transition states should be defined as a ratio of activities

$$K_s^{\ddagger} = \frac{a^{\ddagger}}{a_A b_B} = \frac{C^{\ddagger}}{C_A C_B} \frac{\gamma_{\ddagger}}{\gamma_A \gamma_B} \tag{3.2}$$

Consequently, if the rate of reaction is proportional to the concentration of the activated complex, the rate constant is dependent on the ratio of activity coefficients, as shown in Eqs. (3.3) and (3.4).

$$\text{rate} = \left(\frac{RT}{Nh}\right) C^{\ddagger} = \left(\frac{RT}{Nh}\right) K_s^{\ddagger} C_A C_B \frac{\gamma_A \gamma_B}{\gamma_{\ddagger}} \tag{3.3}$$

$$k_{obs} = \left(\frac{RT}{Nh}\right) K_s^{\ddagger} \frac{\gamma_A \gamma_B}{\gamma_{\ddagger}} \tag{3.4}$$

The activity coefficients can be referred to any convenient standard state, the usual one being that of infinite dilution of the solutes. The rate constant in the standard state will then be equal to $(RT/Nh)K_s^{\ddagger}$ and can be defined as k_0, yielding

$$k_{obs} = k_0 \frac{\gamma_A \gamma_B}{\gamma^{\ddagger}} \tag{3.5}$$

3.2. SALT EFFECTS

3.2.1. The Effect of Ionic Strength on Rate Constants

The most important application of Eq. (3.5) occurs when one or more of the reactants are ions. According to the Debye–Hückel theory, the relation between the activity coefficient of an ion and the ionic strength in dilute solutions (less than 0.01 M) is given by

$$-\ln f_i = \frac{Z_i^2 \alpha \sqrt{\mu}}{1 + \beta r_i \sqrt{\mu}} \tag{3.6}$$

where μ is the ionic strength, r_i is the distance of closest approach of an ion to the ith ion, α and β are constants for a given solvent and temperature, and Z is the

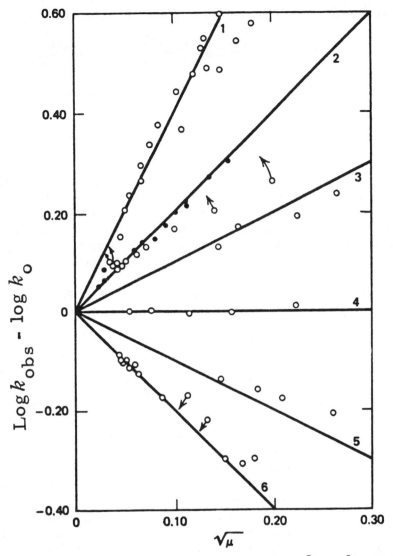

Fig. 3.1. Primary salt effects in ionic reactions. (1) $2[Co(NH_3)_5Br]^{\oplus 2} + Hg^{\oplus 2} + 2H_2O \longrightarrow$ $2[Co(NH_3)_5H_2O]^{\oplus 3} + HgBr_2$ (2) (○) $CH_2BrCOO^{\ominus} + S_2O_3^{\ominus 2} \longrightarrow CH_2S_2O_3COO^{\ominus 2} + Br^{\ominus}$ (sodium salt; no foreign salt added); (●) $S_2O_8^{\ominus 2} + 2I^{\ominus} \longrightarrow I_2 + 2SO_4^{\ominus 2}$ (bimolecular). (3) $[NO_2{=}N{-}COOC_2H_5]^{\ominus} + OH^{\ominus} \longrightarrow N_2O + CO_3^{\ominus 2} + C_2H_5OH$. (4) Inversion of cane sugar, catalyzed by $OH^{\ominus}$ ions. (5) $H_2O_2 + 2H^{\oplus} + 2Br^{\ominus} \longrightarrow 2H_2O + Br_2$ (first order with respect to $H^{\oplus}$ and $Br^{\ominus}$). (6) $[Co(NH_3)_5Br]^{\oplus 2} + OH^{\ominus} \longrightarrow [Co(NH_3)_5OH]^{\oplus 2} + Br^{\ominus}$. From R. P. Bell, *Acid Base Catalysis*, Clarendon Press, Oxford, 1941, p. 33.

charge. Using Eq. (3.6), Eq. (3.5) can be written as

$$\ln k_{obs} = \ln k_0 - \frac{Z_A^2 \alpha \sqrt{\mu}}{1 + \beta r_A \sqrt{\mu}} - \frac{Z_B^2 \alpha \sqrt{\mu}}{1 + \beta r_B \sqrt{\mu}} + \frac{(Z_A + Z_B)^2 \alpha \sqrt{\mu}}{1 + \beta r_i^{\ddagger} \sqrt{\mu}} \qquad (3.7)$$

or, if we assume a mean value of r for the distance of closest approach,

$$\ln k_{obs} = \ln k_0 + \frac{2 Z_A Z_B \alpha \sqrt{\mu}}{1 + \beta r \sqrt{\mu}} \simeq \ln k_0 + 2 Z_A Z_B \alpha \sqrt{\mu} \qquad (3.8)$$

Equation (3.8) predicts a linear relationship between $\log k_{obs}$ and the square root of the ionic strength, with a slope proportional to the product of charges $Z_A \cdot Z_B$. Figure 3.1 shows excellent quantitative agreement of a number of ionic reactions with this equation.[2] This figure indicates that reactions between ions of the same sign are accelerated by an increase of ionic strength, whereas reactions between ions of opposite sign are retarded. The former may be viewed as catalyses; the latter, as inhibitions.

Equation (3.8) has been much abused by attempting to apply it to concentrated solutions where it is not valid. Even in dilute solutions, complex formation between ionic compounds of opposite sign can invalidate the relationship.[3] Certainly at higher concentrations specific effects of added ions will be found.

If one of the reactants is a neutral molecule, so that $Z_A \cdot Z_B = 0$, Eq. (3.8) predicts that ionic strength will have no effect. This appears to be true for very dilute solutions. But at higher ionic concentrations, the rate constants may change because the activity coefficients do not conform to the Debye–Huckel theory and because the activity coefficients of neutral molecules are affected by higher ionic strength. For example, the acid-catalyzed hydrolysis of γ-butyro-lactone is affected by the presence of salts. The logarithm of the rate constant is a linear function of the first power of the ionic strength for a number of salts; sodium sulfate and sodium chloride increase the rate constant, whereas sodium iodide and sodium perchlorate decrease it.

Likewise Eq. (3.8) predicts that ionic strength will not affect reactions of two neutral molecules, except when the activities of the neutral molecules are affected by higher ionic strength. But when neutral molecules react to form oppositely charged ions, as in the hydrolysis of an alkyl halide, the transition state may be considered a strong dipole, and salt effects will again occur.

3.2.2. Electrolyte Catalysis

In the solvolyses of secondary and tertiary halides, reactions that proceed through carbonium ion intermediates, the dipolar nature of the transition state has been elucidated. A theoretical treatment suggests dependence of the logarithm of the rate constant on the ionic strength, with a slope dependent on the square of the dipolar charge and on the distance separating the assumed point dipoles. An experimental study in 90% acetone–water solvent bears out these predictions semiquantitatively, but the rate enhancements are quite small, 0.1 M salts leading to only 30–100% rate increases. However, in aprotic solvents, salt effects on reactions proceeding through carbonium ion intermediates lead to rate enhancements greater than a millionfold. The rate of ionization of p-methoxy-neophyl p-toluenesulfonate in diethyl ether is accelerated by a factor of 10^5 by 0.1 M lithium perchlorate, while the same reaction in acetic acid is accelerated by a factor of only 2.5-fold by this salt, as seen in Fig. 3.2.[4] The pattern of salt effects in the range shown in the figure is different for the two solvents: It is linear for acetic acid [Eq. (3.9)], whereas it is complex for diethyl ether [Eq. (3.10)]

$$k_{obs} = k_0\{1 + b[\text{LiClO}_4]\} \tag{3.9}$$

$$k_{obs} = k_0\{1 + b[\text{LiClO}_4] + c[\text{LiClO}_4]^n\} \tag{3.10}$$

Although the rate constant of ionization in acetic acid in the absence of lithium perchlorate exceeds that in ether by a factor of 2×10^4, ether becomes a better

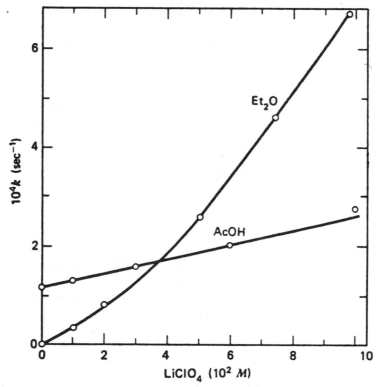

Fig. 3.2. The effect of lithium perchlorate on the ionization of p-methoxyneophyl p-toluene-sulfonate. From S. Winstein, S. Smith, and D. Darwish, *J. Am. Chem. Soc.*, **81**, 5511 (1959). Copyright © 1959 by the American Chemical Society. Reprinted by permission of the copyright owner.

ionizing medium than acetic acid at concentrations of lithium perchlorate above 0.0236 M.

Even larger salt effects are seen in the ionization of a spirodienyl p-nitrobenzoate in ether solution, where 0.05 M lithium perchlorate accelerates the rate by 10^{87}. The effect of lithium perchlorate on the rate constant again fits an equation of the form of Eq. (3.10). With both this substrate and the p-toluenesulfonate, different salts show different accelerations, cations showing the order, $Li^{\oplus} > Na^{\oplus} > Bu_4N^{\oplus}$, identical with the order of increasing ionic radius. Considerable anion specificity is also evident. The magnitude of the salt effects and the observed cation order suggest that the chief role of the salt in this catalysis is to provide specific electrophilic assistance to ionization. Assistance to ionization by an ion pair, $M^{\oplus} Y^{\ominus}$, may be

$$RX + M^{\oplus}Y^{\ominus} \rightleftharpoons R^{\oplus}X^{\ominus}M^{\oplus}Y^{\ominus} \longrightarrow Products \qquad (3.11)$$

Whereas uncatalyzed ionization leads to an ion pair in a medium of low dielectric, electrolyte-catalyzed ionization leads to an ion quadruplet. Estimated interaction energies between two ion pairs are consistent with very large assistance to ionization provided by an ion pair in this medium.

The catalytic effect of uncharged bases alone in the base-catalyzed mutarotation of tetramethylglucose and tetraacetylglucose in nitromethane solution is very low. This effect is greatly enhanced by the addition of a wide variety of salts. The mutarotation of tetraacetylglucose in pyridine solution is accelerated tenfold by 0.02 M lithium perchlorate. Other rate accelerations vary widely with both the anionic and cationic component of the salt and are smaller than the lithium perchlorate acceleration. The mechanism of catalysis by an uncharged base requires the intermediate formation of an ion pair

$$\text{(3.12)}$$

since the reaction proceeds through the aldehydic form of glucose. The electrolyte catalysis can then be explained by the stabilization of the transition state leading to this charge-separated intermediate by means of anion-quadruplet interaction of the kind proposed in Eq. (3.13).

$$\text{(3.13)}$$

Although the electrolyte catalysts described earlier show specificity with respect to both the cation and the anion of the salt, they must be described in terms of catalysis by the salt as a whole rather than by either of the component parts. In addition to these electrolyte catalysts, some salts act catalytically as anions while others act catalytically as cations. One can therefore define cationic and anionic catalysis. Many, although not all, of the metal ion catalyses described in Chapter 9 may be viewed in this context. For example, alkaline earth cations catalyze the hydrolysis of many phosphate esters and anhydrides. They do so by stabilization of the transition state of the hydrolytic reactions through an electrostatic interaction made favorable by the chelation possible with these metal ions. Likewise, lithium counterions may be viewed as a cationic catalysts in the carbenoid reactions of $LiCCl_3$ in aprotic solvents.

In electrophilic substitution reactions, "one-" and "two"-anion catalyses have been found as shown in the isotopic exchange reaction of a simple alkylmercuric halide with mercuric halide, labeled with radiomercury. A plot of the second-order rate constant for the exchange reaction against the added salt concentration is biphasic with a linear first-order dependence on the salt concentration up to the equivalence point of the salt and the substituting agent, followed by a change in the slope to a new linear (second-order) dependence on the salt concentration. These two slopes imply two processes, one depending on a higher order of the salt concentration than the other, as shown in Eqs. (3.14) and (3.15) and Fig. 3.3.

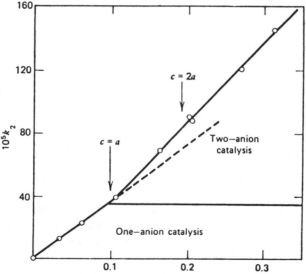

Fig. 3.3. One-anion and two-anion catalysis from H. B. Charman, E. D. Hughes, Sir C. Ingold, and H. C. Volger, *J. Chem. Soc.*, **1961**, 1142.

$$CH_3HgBr \rightleftharpoons CH_3Hg\overset{Br}{\underset{Br^{\ominus}}{\diagdown}} \overset{*}{\underset{}{HgBr_2}}$$

$$\left[CH_3\overset{HgBr_2}{\underset{\overset{*}{HgBr_2}}{\diagup}} \right]^{\ominus} \underset{HgBr_2}{\rightleftharpoons} CH_3\overset{*}{HgBr_2}{}^{\ominus} \underset{Br^{\ominus}}{\rightleftharpoons} CH_3\overset{*}{HgBr} \qquad (3.14)$$

$$CH_3HgBr \overset{Br^{\ominus}}{\rightleftharpoons} CH_3HgBr_2{}^{\ominus} \overset{Br^{\ominus}, \overset{*}{HgBr_2}}{\rightleftharpoons}$$

$$\left[CH_3\overset{HgBr_2}{\underset{\overset{*}{HgBr_2}}{\diagdown}}Br \right]^{\ominus} \underset{HgBr_2, Br^{\ominus}}{\rightleftharpoons} CH_3\overset{*}{HgBr_2}{}^{\ominus} \underset{Br^{\ominus}}{\rightleftharpoons} CH_3\overset{*}{HgBr} \qquad (3.15)$$

3.3. SOLVENT EFFECTS

3.3.1. The Effect of the Dielectric Constant on Kinetics

The influence of the dielectric constant of the medium on the rate constants of reactions between two ions can be considered in terms of Eq. (3.5). The electrostatic free energy for bringing two ions from an infinite separation to the equilibrium distance in the activated complex ($r^{\ddagger}$) is

$$\Delta F^{\ddagger}_{el} = \frac{Z_A Z_B e^2}{D r^{\ddagger}} \qquad (3.16)$$

where D is the dielectric constant of the medium and e is the charge on the electron. From Eq. (3.16), it is possible to write an expression for the dependence of the rate constant on the dielectric constant.[5]

$$\ln k = \ln k_0' - \frac{N Z_A Z_B e^2}{D R T r^{\ddagger}} \qquad (3.17)$$

where k_0' is a specific rate constant in the medium of infinite dielectric constant, Avogadro's number N, gas constant, R, and temperature T. This equation predicts a linear dependence of $\log k$ on $1/D$, with a negative slope if the charges on the ions are of the same sign and a positive slope if the charges are of opposite sign. Figure 3.4 shows data on the alkaline fading of bromphenol blue in ethanol–water mixtures and on the reaction between hydronium ion and the doubly negative azodicarbonate ion in mixtures of water and dioxane. The first reaction is between univalent negative and divalent negative ions, while the second is between a univalent positive ion and a divalent negative ion. In agreement with predictions of theory, the slope of $\log k$ versus $1/D$ is negative for the former reaction and positive for the latter. The slopes lead to values of $r^{\ddagger}$ of 2.81 Å for the former and 3.42 Å for the latter reaction, values of the expected

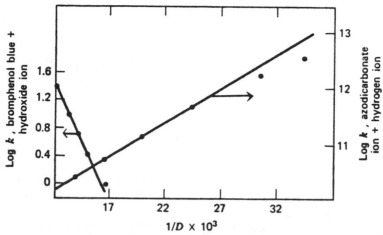

Fig. 3.4.   Influence of dielectric constant on rate of reaction between two ions of same sign (left axis) and of opposite sign (right axis). From E. S. Amis and V. K. LaMer, *J. Am. Chem. Soc.*, **61**, 905 (1939). C. V. King and J. J. Josephs, *J. Am. Chem. Soc.*, **66**, 767 (1944). Copyright © 1939 and 1944 by the American Chemical Society. Reprinted by permission of the copyright owner.

order of magnitude. In the reaction of hydronium ion and the azodicarboxylate ion in dioxane–water mixtures, the enthalpy of activation remains sensibly constant while the entropy of activation increases with increasing dioxane content. A similar but exaggerated phenomenon is seen in the hydrolysis of alkyl hydrogen sulfates in dioxane–water mixtures. The rate constant of this reaction in 98% dioxane–water is 10^7 times larger than in water solely because of a difference in entropy of activation of about 30 units. The logarithm of the rate constant is again linearly dependent on $1/D$; the slope of this line leads to a value of $r = 16 Å$, an extraordinarily large value whose meaning is not clear. The latter two reactions, which involve oppositely charged ions, have transition states that are less highly solvated than the ground states. This phenomenon should lead to an increase in entropy. When the solvent is changed to one of lower dielectric constant, this effect is markedly enhanced.

A corollary of this treatment is that when neutral molcules form a transition state in which separation of charge occurs, an equal but opposite solvent effect should occur. The most familiar example of this phenomenon is in reactions following the S_N1 mechanism in which an ionization leads to a transition state resembling a carbonium ion–anion pair. The solvolytic rates of several secondary and tertiary halides increase by approximately a thousandfold over the solvent range from 60% ethanol–water to water. For these and other reactions, a useful qualitative theory was expounded long ago: "An increase in the ion-solvating power of the medium will accelerate the creation and concentration of charges and inhibit their destruction and diffusion".[6] Quantitatively, the following equation correlates the rates of solvolysis of compounds reacting via the S_N1 mechanism

$$\log \frac{k}{k_0} = mY \tag{3.18}$$

where k and k_0 are the solvolysis rates of a compound in a given solvent and in 80% ethanol. Y is a measure of the "ionizing" power of the medium and is defined by the relation

$$Y = \log \frac{k_{BuCl}}{k_{0\,BuCl}} \tag{3.19}$$

where k_{BuCl} and $k_{0\,BuCl}$ are the rate constants for the solvolysis of t-butyl chloride in the given solvent and in the reference solvent (80% ethanol). In Eq. (3.18), m is a constant characteristic of the compound solvolyzed.[7]

Both theoretically and experimentally, the polarity of the solvent per se has little effect on other reactions such as the reactions of ions with neutral (polar) molecules. The theoretical prediction that the rate constant of reaction is higher in a medium of lower dielectric constant is sometimes obeyed (alkyl halides + hydroxide ion), but other times disobeyed (esters + hydroxide ion—see Fig. 3.5).

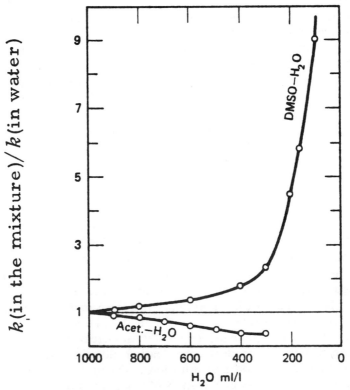

Fig. 3.5.   Effect of dimethyl sulfoxide and acetone on the saponification of esters. From E. Tommila and M.-L. Murto, *Acta Chem. Scand.*, **17**, 1947 (1963).

3.3.2. Solvent Effects on Reactions of Ionic Species

The solvent can exert enormous effects on the rate constants of reactions of ions with organic molecules, whether they are nucleophiles or bases. For example, the change in solvent from water to acetone multiplies the second-order rate constant of the chloride ion–methyl iodide reaction by about 10^7. In addition, the rate of racemization of optically active 2-methyl-3-phenylpropionitrile by methoxide is 10^9 times faster in dimethyl sulfoxide than in methanol.[8] The dielectric constant of the medium can cause some small part of these effects, but the major cause must certainly be some specific solvent effect. As implied earlier, the most profound differences are found when comparing protic and aprotic solvents. A solvent change of this kind can produce two effects, one concerned with perturbation of the equilibrium between ion pairs and free ions, the other concerned with specific solvation of ions, usually by the protic solvent. The importance of ion association in controlling apparent nucleophilicity is evident when reactivities of lithium or tetra-n-butylammonium halides are compared. In the reaction of n-butyl p-bromobenzenesulfonate with 0.04 M halide salts in acetone solution, the relative reactivity pattern is

$$\text{LiI, 6.2} \quad > \quad \text{LiBr, 5.7} \quad > \quad \text{LiCl, 1.0}$$
$$\vee \qquad\qquad\qquad \wedge \qquad\qquad\qquad \wedge$$
$$(n\text{-Bu})_4\text{NI, 3.7} < (n\text{-Bu})_4\text{NBr, 18} < (n\text{-Bu})_4\text{NCl, 68}$$

Thus lithium or tetra-n-butylammonium halides show exactly opposite orders of S_N2 reactivity. Tetra-n-butylammonium halides are known to be more dissociated than lithium halides in acetone solution. Hence these data suggest that the nucleophilicity order of anions in acetone is actually $Cl^\ominus > Br^\ominus > I^\ominus$, that observed with tetra-n-butylammonium salts, and the reactivity pattern of lithium halides is governed largely by the extreme variation in the dissociation constants of the salts in acetone solution. In fact, when the dissociation constants are taken into account, assuming the reactivity of the ion pair to be negligibly small, good agreement between the reactivities of the lithium and tetra-n-butylammonium salts is found.

When association of ions to ion pairs (and higher complexes) is taken into account, it is then possible to evaluate the reactivity of dissociated ions in different solvents. These reactivities appear to be intimately associated with the specific solvations of the ions. In protic solvents, anions are solvated by ion–dipole interactions, superimposed on a strong hydrogen-bonding component, greatest for small anions on electrostatic grounds. This solvation by protic solvents decreases strongly in the series $OH^\ominus$, $F^\ominus \gg Cl^\ominus > Br^\ominus > N_3^\ominus > I^\ominus > SCN^\ominus > $ picrate$^\ominus$. In dipolar aprotic solvents, the solvation of anions occurs to a much smaller extent, due only to ion–dipole interactions. There is no significant contribution to solvation by hydrogen bonding. Solvation of anions by dipolar aprotic solvents is thus relatively insensitive to the structure of the anion.

Two important effects are noted in comparing the rate constants of S_N2 reactions of halides in aprotic and protic solvents. First, the order of nucleophilicity changes from $Cl^{\ominus} > Br^{\ominus} > I^{\ominus}$ in dimethylformamide and acetone solvents to $I^{\ominus} > Br^{\ominus} > Cl^{\ominus}$ in aqueous solution. Second, the absolute rate constants of chloride ion reactions are of the order of 10^6–10^7 larger and those of iodide ion reactions are of the order of 10^4 larger in aprotic solvents than in water. These effects are related to one another because the solvent has a profound effect on the nucleophilicity of different halide ions.

It is thus concluded that both solvation of anions and ion-pairing have a retarding effect on bimolecular substitution reactions.[9] Interestingly, both retardations are most effective with small anions with their high charge density. It is a matter of choice whether one considers aprotic solvents to have an accelerating effect on bimolecular substitution reactions or whether one considers protic solvents to have a retarding effect. The choice is really which solvent is the standard. Since this discussion is oriented toward aqueous systems, the changes occurring in aprotic solvents, such as would occur in the active site of an enzyme, are regarded as accelerations.

The rate of any reaction involving a metal cation–carbanion ion pair in an aprotic solvent is increased by the addition of a solvent that will preferentially solvate the cation and thus lead to a higher concentration of dissociated anions. The alkylation of enolate ions is approximately one hundredfold faster in diglyme than in diethyl ether. The reaction of sodium diethyl n-butylmalonate with alkyl halides is 1420 times faster in dimethyl sulfoxide than in benzene; the addition of even 5% dimethylformamide to a benzene solution results in a twentyfold rate enhancement of this reaction. The effect of cryptands, increasing the nucleophilicity of anions by preferential complexation of the cation, such as with the crown ethers, is related to this category (see Chapter 12).

Anion solvation through hydrogen bonding to protic solvents is most effectively reduced by the addition of solvents such as dimethyl sulfoxide, which have a reasonably high dielectric constant but no proton with which to form H bonds. For example increasing the dimethyl sulfoxide concentration of an aqueous solvent increases the rate constant of the saponification of esters, markedly at higher dimethyl sulfoxide concentrations. However, adding comparable concentrations of acetone to the aqueous solvent depresses the rate (Fig. 3.5).

A very large kinetic effect of a solvent change occurs in the methoxide ion-catalyzed racemization of 2-methyl-3-phenylpropionitrile in methanol–dimethyl sulfoxide solutions, as seen in Fig. 3.6.[8] From pure methanol to 98.5% dimethyl sulfoxide–methanol, the rate of racemization increases by a factor of 5×10^7. Extrapolation of the curve of Fig. 3.6 to 100% dimethyl sulfoxide gives a rate increase over that in methanol of about nine powers of ten. These two solvents have dielectric constants that do not differ widely from one another (34 for methanol and 49 for dimethyl sulfoxide), and the bulk of evidence supports

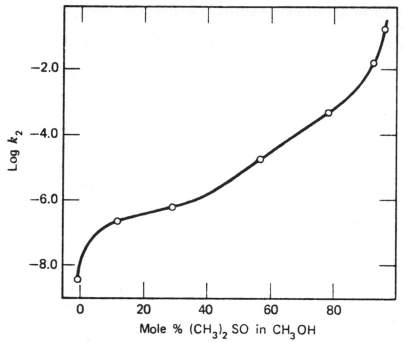

Fig. 3.6. Plot of mol% (CH₃)₂SO in CH₃OH against logarithm of k_2 (M^{-1} sec^{-1}) for CH₃OM-catalyzed racemization of 2-methyl-3-phenylpropionitrile at 25.0°C. From D. J. Cram, B. Rickborn, C. A. Kingsbury, and P. Haberfield, *J. Am. Chem. Soc.* **83**, 3678 (1961). Copyright © 1961 by the American Chemical Society. Reprinted by permission of the copyright owner.

the hypothesis that at low base concentration the metal alkoxides are dissociated. The big difference in activity of methoxide anion in the two solvents is attributed to the presence of solvent–anion hydrogen bonds in methanol (CH₃OH · · · OCH₃⁻) that are absent in dimethyl sulfoxide. Although methoxide ions are in a sense "buffered" by methanol solvent, they are less solvated in dimethyl sulfoxide. In methanol, the driving force for dissociation of a metal methoxide derives from solvation of the methoxide anion by the hydrogen and of the metal cation by the oxygen of the hydroxyl group. In dimethyl sulfoxide, positive sulfur solvates the methoxide anion and the negative oxygen atom solvates the positive metallic cation. Since the charge on sulfur is shielded by three shells of electrons and is also sterically hindered, the main impetus for dissociation probably derives from solvation of the metal cation that leaves the methoxide anion relatively poorly solvated and highly reactive. This is schematically shown in Eq. (3.20).

$$CH_3O^{\ominus}\text{---}HOCH_3 + 2(CH_3)_2SO \rightleftarrows$$

$$CH_3O^{\ominus}\text{---}\underset{\underset{O^{\ominus}}{|}}{S}^{\oplus}(CH_3)_2 + (CH_3)_2S^{\oplus}\text{---}O^{\ominus}\text{---}HOCH_3 \qquad (3.20)$$

3.3.3.   Other Special Solvent Effects on Reactions of Neutral Species

Implicit in the preceding discussion was the assumption that solvent effects on reactions involving ions are primarily associated with solvent effects on the ion in the ground/or transition state, but not on neutral molecules in these states. This assumption in general must be incorrect. But since the effects on ions so far outweigh other effects, ordinarily the latter may be neglected.

A closer analysis of reactions in dimethyl sulfoxide solutions reveals the following points:

1.   Rate increases occur even at low dimethyl sulfoxide concentration, with no sharp change ordinarily noted as it passes from the minor to the major solvent species.

2.   The rate increases per increment of dimethyl sulfoxide is due mainly to a change in $\Delta H^{\ddagger}$.

3.   Addition of dimethyl sulfoxide to a system increases the rate irrespective of the charge character of the reaction, with either a negatively-charged or a neutral nucleophile, and with either a neutral or positively charged substrate.

These data do not support the simple picture of Eq. (3.18) for the action of dimethyl sulfoxide. The observation of large rate increases in solvent mixtures where there is sufficient methyl alcohol present to form hydrogen bonds with both dimethyl sulfoxide and the nucleophile shows that some other factor must be influencing the reaction rates. Moving from solvent mixtures in which dimethyl sulfoxide is the minor component to mixtures where it predominates results in only a modest increase; if the anion were poorly solvated at high dimethyl sulfoxide concentrations, a much larger increase might have been expected. Also expected would be a difference in the sensitivity of oxygen and sulfur anions to dimethyl sulfoxide concentration, since the former smaller anion has a considerably higher solvation energy. Changes in solvation should be reflected in $\Delta S^{\ddagger}$, but were observed in $\Delta H^{\ddagger}$. Changes in the state of solvation of the nucleophile apparently do not become important until the concentration of the hydroxylic species in the solvent is very low and the concentration of dimethyl sulfoxide is very high. Also, desolvation of halide anions may occur more readily than desolvation of alkoxides in going to highly concentrated dimethyl sulfoxide solutions.

At low dimethyl sulfoxide concentrations, the solvent must lower the energy of the transition state. One attractive interpretation suggests enhanced dipolar character of the substrate upon interaction with a random dimethyl sulfoxide molecule. Attack of a nucleophile upon such a dipolar species may well be facile compared to an unpolarized substrate molecule.

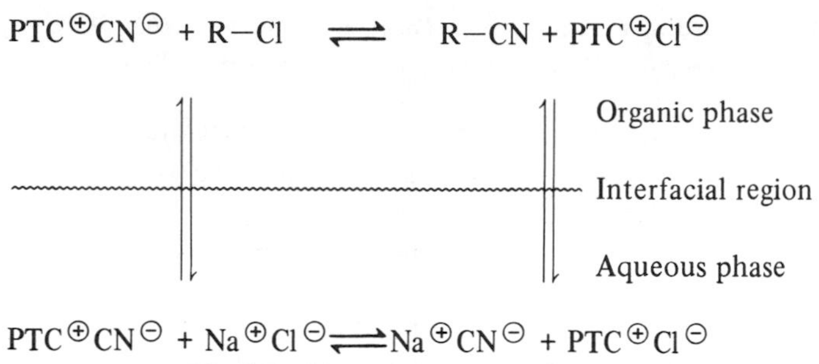

$$PTC^{\oplus}CN^{\ominus} + R{-}Cl \ \rightleftharpoons \ R{-}CN + PTC^{\oplus}Cl^{\ominus}$$

	Organic phase
	Interfacial region
	Aqueous phase

$$PTC^{\oplus}CN^{\ominus} + Na^{\oplus}Cl^{\ominus} \rightleftharpoons Na^{\oplus}CN^{\ominus} + PTC^{\oplus}Cl^{\ominus}$$

Fig. 3.7. Phase transfer catalysis.

3.3.4. Phase Transfer and Triphase Catalysis

A problem frequently encountered by organic chemists is the reaction of anions that are insoluble in traditional organic solvents with organic soluble lipophilic electrophiles. This difficulty has now been overcome with the advent of phase transfer catalysts.[10-15] The mechanism by which these catalysts function is best indicated by the example in Fig. 3.7.

The phase transfer catalyst (PTC) is able to shuttle charge reactants between the organic and aqueous phase, thus bringing them close enough to react.

These catalysts are usually polyethers or onium salts. Examples of the former systems are the glymes, crowns, and cryptands, and of the latter systems, ammonium and phosphonium salts.

The salts function by exchanging their anions with the reactant anion resulting in salts that are lipophilic enough to enter the organic phase. The neutral ethers operate by chelating the metal ion also making it more lipophilic. The complexed metal ion and its weakly bound anion are then soluble in the organic phase.

The technique of triphase catalysis has recently been introduced.[10] This method is characterized by three separate phases, each containing a unique reagent or catalyst. The technique has been shown to be potentially useful in a wide variety of reactions.

REFERENCES

1. A. A. Frost and R. G. Pearson, *Kinetics and Mechanism*, 2nd ed., John Wiley & Sons, New York, 1961, p. 127.
2. R. P. Bell, *Acid Base Catalysis*, Clarendon Press, Oxford, 1941, p. 33.
3. R. P. Bell, *The Proton in Chemistry*, 2nd ed., Cornell University Press, Ithaca, NY, 1973.
4. S. Winstein, S. Smith and D. Darwish, *J. Am. Chem. Soc.*, **81**, 5511 (1959).
5. G. Scatchard, *Chem. Rev.*, **10**, 229 (1932).
6. C. K. Ingold, *Structure and Mechanism in Organic Chemistry*, Cornell University Press, Ithaca, NY, 1953, p. 349.

7. A. H. Fainberg and S. Winstein, *J. Am. Chem. Soc.*, **79,** 1597, 1602 (1957).

8. D. J. Cram, B. Rickborn, C. A. Kingsbury, and P. Haberfield, *J. Am. Chem. Soc.*, **83,** 3678 (1961).

9. A. J. Parker, *Quart. Rev.*, **16,** 163 (1962); *Chem. Rev.*, 69, 1 (1969).

10. D. W. Armstrong and M. Godat, *J. Am. Chem. Soc.*, **101,** 2490 (1979).

11. S. L. Regen, *J. Am. Chem. Soc.*, **97,** 5956 (1975).

12. E. V. Dehmlow and S. S. Dehmlow, *Phase Transfer Catalysis*, Verlag Chemie, New York, 1980.

13. S. L. Regen, *J. Org. Chem.*, **42,** 875 (1977).

14. S. F. Jacobson et al., *J. Am. Chem. Soc.*, **101,** 6938 (1979).

15. W. R. Gilkerson and M. D. Jackson, *J. Am. Chem. Soc.*, **104,** 1218 (1982).

4 | Hydronium and Hydroxide Ions in Catalysis

Acid–base catalysis was originally identified as catalysis by strong acids or bases in aqueous solution.[1] It is now recognized that the catalytic species present in these solutions are hydronium or hydroxide ions. Acid–base catalysis has now been extended to include Brönsted and/or Lewis acids or bases as well. In the present chapter, the original concept of catalysis by hydronium and hydroxide ions, that is, specific acid–base catalysis, usually referred to now as specific hydronium and hydroxide ion catalysis is treated. Other lyonium and lyoxide ion catalysis will be briefly described also.

In water, $pH(= -\log [H_3O^{\oplus}])$ can affect the rate of a chemical reaction in many ways, as shown in Fig. 4.1. The logarithm of the rate constant can be independent of pH (curve a, slope 0). It can also be directly dependent on the logarithm of the hydronium ion concentration (curve b, slope -1) or on the logarithm of the hydroxide ion concentration (curve c, slope $+1$). It can vary with one or more combinations of these: Curve i is a typical combination curve showing dependence on the logarithm of the hydronium ion concentration, independence of pH, and dependence on the logarithm of the hydroxide ion concentration with ascending pH. The log of the rate constant corresponding to curve i is

$$\log k_{obs} = \log k_0 + \log k_H[H_3O^{\oplus}] + \log k_{OH}[OH^{\ominus}] \qquad (4.1)$$

In Eq. (4.1), k_0 is a rate constant of a reaction insensitive to pH, usually called a

45

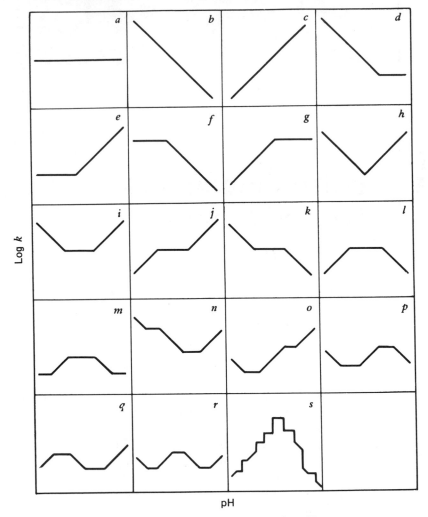

Fig. 4.1.   Some hypothetical pH–log k profiles.

spontaneous reaction or, more accurately, a water reaction, while k_H and k_{OH} are rate constants for hydronium ion- and hydroxide ion–catalyzed reactions. A reaction can also be dependent on one or more forms of the prototropic group(s) of one of the reactants. Rates can be dependent on the prototropic groups not only of the reactants, but also of the intermediates. The slopes of log k–pH profiles involving dependence on the ionization of a prototropic group can also have values of $+1, -1$, and 0, leading to considerable ambiguity in the interpretation of such profiles, which will be considered later. One ambiguity, however, that must always be met directly is shown by curve s; although these curves are meaningless since they are due to experimental error, they have been published repeatedly.

4.1.   ACIDITY AND BASICITY FUNCTIONS

Prior to the discussion of hydronium and hydroxide ion catalysis, a brief review of several fundamental concepts regarding acid–base equilibrium in solution seems appropriate. In dilute aqueous solutions of strong acids, the proton-donating power of the medium is defined as the concentration of hydronium ion. However, in concentrated acid solution, where the solvent is no longer essentially water, the proton-donating power of the solvent cannot be so easily expressed. The problem is simply that the apparent acidity of the medium varies because of the change in activity coefficients.

Consider the protonation of an indicator:

$$HIn^{\oplus} \rightleftharpoons H^{\oplus} + In \tag{4.2}$$

The equilibrium constant for this reaction is expressed by

$$K_{InH} = \frac{a_{H^{\oplus}} a_{In}}{a_{HIn}} = a_{H^{\oplus}} \frac{C_{In}}{C_{HIn^{\oplus}}} \frac{\gamma_{In}}{\gamma_{HIn^{\oplus}}} \tag{4.3}$$

where a is activity, C is concentration, and γ is the activity coefficient. Alternatively, the pK_a of the acid HIn can be written as

$$pK_a = -\log a_{H^{\oplus}} + \log \frac{C_{HIn^{\oplus}}}{C_{In}} + \log \frac{\gamma_{HIn^{\oplus}}}{\gamma_{In}} \tag{4.4}$$

In dilute solution, $\lim_{c \to 0}(\gamma_{HIn^{\oplus}}/\gamma_{In}) = 1$ and thus $\lim_{c \to 0}\log(\gamma_{HIn^{\oplus}}/\gamma_{In})$ is zero; thus Eq. (4.4) becomes

$$pK_a = pH + \log \frac{C_{HIn^{\oplus}}}{C_{In}} \tag{4.5}$$

However, in strong acid, the activity coefficients must be reckoned with.

Let us see how we first measure the proton-donating ability of very concentrated acid solutions and then, using this information, describe a scale for quantitating the strength of these very concentrated acid solutions. The measurement of any such acid strength is predicted on our ability to evaluate the extent to which an indicator is protonated, specifically the ratio $C_{HIn^{\oplus}}/C_{In}$. This means in comparing various acid solutions, we must not use an indicator or indicators that are completely protonated simply because the ratio $C_{HIn^{\oplus}}/C_{In}$ is immeasurable. If, however, we can determine the pK_a of a series of indicators of varying acidity and relate them to each other, we can then use them to relate the proton-donating power of strongly acidic media to a thermodynamic standard state in water.

Consider, for example, the dye aniline, which has a $pK_a = 2.80$ in dilute aqueous acid where the γ values approach 1. If we now mix both this indicator

(In) and a second, more weakly basic indicator, (In′) (p-nitroaniline) in solution and add sufficient HCl to protonate In′ but not so much as to completely protonate In, we may relate the indicators by

$$pK_a - pK_a' = \log\left(\frac{C_{HIn^{\oplus}}C_{In'}}{C_{In}\,C_{HIn'^{\oplus}}}\right) + \log\left(\frac{\gamma_{HIn^{\oplus}}\gamma_{In'}}{\gamma_{In}\,\gamma_{HIn'^{\oplus}}}\right) \qquad (4.6)$$

This equation derives from a simple subtraction of the pK_a equation for each of the indicators, making no assumptions about activity coefficients. In Eq. (4.6), the concentration terms can be measured, and the pK_a value of the more basic indicator In is known; however, the activity coefficient terms remain a question. Fortunately, these terms are nearly zero. Although the ratio of $\gamma_{In}/\gamma_{HIn^{\oplus}}$ does, in fact, vary with the medium, it varies in a similar way for $\gamma_{In'}/\gamma_{HIn'^{\oplus}}$ in the same medium. This means the term $(\gamma_{In}/\gamma_{HIn^{\oplus}})(\gamma_{HIn'^{\oplus}}/\gamma_{In'})$ is nearly 1 and its log is therefore approximately zero. The concentration term has been shown to remain constant for the same two indicators in a variety of different acid solutions. Thus the value of pK_a' is easily determined and is, in fact, 1.11 for the indicators discussed. We could, of course, use an even weaker third basic indicator and determine its pK_a relative to the second basic indicator In′. We can now construct an acidity scale using these indicators. Going back to Eq. (4.3), we can rewrite it in the following way

$$a_{H^{\oplus}} = K_{HIn^{\oplus}}\frac{C_{HIn^{\oplus}}}{C_{In}}\frac{\gamma_{HIn^{\oplus}}}{\gamma_{In}} \qquad (4.7)$$

The quantity $a_{H^{\oplus}}$ cannot be calculated because the value $\gamma_{HIn^{\oplus}}/\gamma_{In}$ is not accessible. Furthermore, it should be pointed out that this term is nearly independent of the indicator used but rather characteristic of the medium. However, the quantity $a_{H^{\oplus}}(\gamma_{In}/\gamma_{HIn^{\oplus}})$, hereafter referred to as h_0, a measure of the medium's proton-donating ability, can be determined using the indicator system we developed earlier to measure the acidity of concentrated acids.[2]

$$h_0 = \frac{a_{H^{\oplus}}\gamma_{In}}{\gamma_{HIn^{\oplus}}} = K_{HIn^{\oplus}}\frac{C_{HIn^{\oplus}}}{C_{In}} \qquad (4.8)$$

The acidity function h_0 is more conveniently expressed as its logarithmic function

$$H_0 = -\log h_0$$

Therefore, in proton-donating power measured by H_0, solutions of strong acids of the same titratable acidity differ from one another markedly, as seen in Fig. 4.2.[11]

Variants of the Hammett acidity function have been devised to treat compounds the acidity of which cannot be described in terms of H_0. One obvious variant is an acidity function for a series of bases or indicators of a charge type

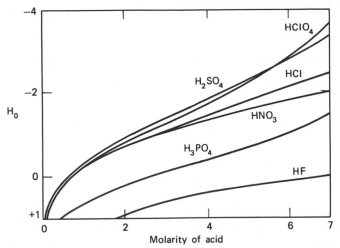

Fig. 4.2. H_0 and related acidity functions. From M. A. Paul and F. A. Long, *Chem. Rev.*, **57**, 13 (1957).

other than the neutral aniline bases. For prototropic equilibria involving carbonium ions, acid dependency is so markedly different from that of a Hammett base that it has been necessary to devise a new acidity function called H_R (Ref. 3) or J_0 (Ref. 4). The operational definition of H_R is similar to that of H_0.

$$H_R = pK_{R^\oplus} + \log \frac{[ROH]}{[R^\oplus]} \qquad (4.9)$$

or

$$h_R = \frac{a_{H^\oplus} \gamma_{ROH}}{a_{H_2O} \gamma_{R^\oplus}}$$

The relationship of h_R to the hydronium ion activity differs from the definition of h_0 principally by the inclusion of a term involving the activity of water.

Corresponding to the acidity functions in strong acid solutions,[5] basicity functions have been devised for strongly basic solutions, including those of sodium hydroxide, potassium hydroxide, hydrazine, ethylenediamine, and ethanolamine in water. Studies of a series of weakly acidic organic indicators in these basic solutions led to the basicity function $H_\ominus$. The indicator equilibrium observed is

$$HIn + OH^\ominus \rightleftharpoons In^\ominus + H_2O \qquad (4.10)$$

In analogous fashion to the definition of the Hammett acidity function, the

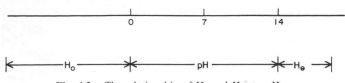

Fig. 4.3. The relationship of H_0 and $H_\ominus$ to pH.

basicity function $H_\ominus$ is defined by

$$H_\ominus = -\log \frac{a_{H_2O}\gamma_{In\ominus}}{a_{OH\ominus}\gamma_{HIn}} = -\log a_{H\ominus} \frac{\gamma_{In\ominus}}{\gamma_{HIn}} \tag{4.11}$$

In dilute solution, the acidity function $H_\ominus$ becomes identical with pH, so that the pH scale can be extended on the alkaline side of neutrality as it is extended on the acid side (see Figure 4.3). Values of $H_\ominus$ as high as 19 can be achieved in solutions of a tetralkylammonium hydroxide in water, sulfolane–water, pyridine–water, and dimethyl sulfoxide–water solutions. Of the media tested, the dimethyl sulfoxide–water mixture gives the highest $H_\ominus$ at given stoichiometric water and hydroxide ion concentrations.

Two basicity scales are needed to reflect the ability of a base either to remove a proton from an organic compound or to add to the organic compound to give a tetrahedron addition intermediate. To this end, a $J_\ominus$ basicity function has been defined [Eq. (4.12)] to supplement $H_\ominus$. Indicators that reflect the $H_\ominus$ and $J_\ominus$ scales are 2,4-dinitroaniline

$$J_\ominus = -\log \frac{\gamma_{InOH}}{\gamma_{In}\,a_{OH\ominus}} \tag{4.12}$$

and N, N-dimethyl-2,4,6-trinitroaniline, respectively.[6]

4.2.   THE PROTON AS CATALYST

The proton is a convenient and powerful agent for the distortion of the electronic configuration of a substrate in order to facilitate reaction. The mechanism by which this process occurs has many variants. For example, a covalent bond may be more easily broken after protonation of one of the bonded atoms:

$$ROH_2^\oplus \longrightarrow R^\oplus + H_2O$$

is easier than

$$ROH \longrightarrow R^\oplus + OH^\ominus \tag{4.13}$$

Another possibility is that addition of a nucleophile to an unsaturated linkage may occur more readily after protonation of one of the atoms (the more basic) of the double bond:

$$\overset{\overset{\displaystyle OH^{\oplus}}{\|}}{RCOR} + N^{\ominus} \longrightarrow R\overset{\overset{\displaystyle OH}{|}}{\underset{\underset{\displaystyle N}{|}}{C}}OR$$

is easier than

$$\overset{\overset{\displaystyle O}{\|}}{RCOR} + N^{\ominus} \longrightarrow R\overset{\overset{\displaystyle O^{\ominus}}{|}}{\underset{\underset{\displaystyle N}{|}}{C}}OR \qquad (4.14)$$

Finally, the abstraction of a proton from a molecule can be facilitated by the introduction of a proton at a different position in the molecule:

$$\overset{\overset{\displaystyle OH^{\oplus}}{\|}}{RC}-CH_3 + B \longrightarrow R\overset{\overset{\displaystyle OH}{|}}{C}=CH_2 + BH^{\oplus}$$

is easier than

$$\overset{\overset{\displaystyle O}{\|}}{RC}-CH_3 + B \longrightarrow R\overset{\overset{\displaystyle O^{\ominus}}{|}}{C}=CH_2 + BH^{\oplus} \qquad (4.15)$$

Instead of distorting the electronic configuration, a proton can stabilize a leaving group and thus facilitate reaction. This description of protonic catalysis is similar to that in Eq. (4.15) and can be described by

$$\overset{\overset{\displaystyle O}{\|}}{RC}X + H^{\oplus} \longrightarrow \overset{\overset{\displaystyle O}{\|}}{RC}{}^{\oplus} + HX \qquad (4.16)$$

Distinguishing between these two descriptions of protonic catalysis may not be meaningful. The difference pertains to whether a preequilibrium protonation of the substrate occurs, for certainly the transition states of Eqs. (4.15) and (4.16) are identical. Finally, a proton is an electrophile that can add to a π-electron system to produce a reactive entity capable of further processes, as in Eqs. (4.17) and (4.18),

$$\text{olefin} + H^{\oplus} \longrightarrow \text{Carbonium ion} \longrightarrow \text{Product} \qquad (4.17)$$

$$
\begin{array}{c}
\text{Y-C}_6\text{H}_5 + H^\oplus \longrightarrow \left[\text{ring}^\oplus\right] \longrightarrow \text{Product} \qquad (4.18)
\end{array}
$$

where Y can be any electron-donating substituent.

Equation (4.17) schematically represents an enormous number of reactions, including the polymerization and hydration of olefins and alkylation of aromatics. A particularly important subsequent reaction step of the reactive carbonium ion intermediate is hydride transfer. This occurs when the stability of the carbonium ion initially generated can be stabilized. For example, a hydride ion can move from the primary carbonium ion of an alkane to a secondary or tertiary carbonium ion, generating a more stable carbonium ion intermediate.

Thus the rapid combination of a proton with almost any organic molecule leads to a plethora of mechanistic possibilities for electronic transition and facilitation of reaction.

Probably the simplest analysis of protonic catalysis is an electrostatic one. The introduction of a proton into an organic molecule is little more than the introduction of a single positive charge. When a positive charge is introduced into an organic molecule, normal electrostatic effects must ensue. Reaction of the protonated species with a negatively-charged nucleophile will be much faster because of coulombic interaction and slower if the nucleophile is positively-charged. If the nucleophile is a dipole, an ion–dipole interaction will occur. Thus any of the qualitative observations of Eqs. (4.13)–(4.18) may be quantitatively explained on the basis of electrostatic theory. In addition to simple electrostatic perturbation, the introduction of a positive charge into a π-electron system can cause a considerable electron delocalization that can also alter reactivity profoundly.

Hydronium ion catalysis can be generally represented by an equilibrium between the substrate and hydronium ion, followed by a rate-determining transformation of the protonated substrate. The two variants of this scheme are differentiated by the fact that one represented by Eq. (4.19) does not involve a nucleophile (N) in the rate-determining step, whereas the other, Eq. (4.20), does.

$$
S + H^\oplus \underset{}{\overset{K}{\rightleftharpoons}} SH^\oplus \xrightarrow{k \text{ (slow)}} \text{Products} \qquad (4.19)
$$

$$
S + H^\oplus \underset{}{\overset{K}{\rightleftharpoons}} SH^\oplus \xrightarrow[N]{k \text{ (slow)}} \text{Products} \qquad (4.20)
$$

Equations (4.19) and (4.20) identify the intermediates in these reactions as protonated substrates. All organic compounds can be considered bases, although variation in base strength of different compounds is quite profound. Essentially every organic compound can be protonated to some extent; even the extremely weak base methane can be converted to $CH_5^{\oplus}$ in the mass spectrometer, and evidence for this species in very strong acids has been reported. Protonation of organic compounds has been investigated in great depth; some results concerning the basicities of important classes of organic compounds are summarized in Table 4.1.

Since protonation often can occur in more than one position in an organic molecule, the protonated species cannot always be specified uniquely. However, it is possible to gain some information about the position of the proton. For example, protonation of a carboxylic acid derivative can lead to either **4.1** or **4.2**

$$R-C\overset{\oplus OH}{\underset{X}{\diagdown}}\qquad\qquad R-C\overset{O}{\underset{XH^{\oplus}}{\diagdown}}$$

$$\text{4.1}\qquad\qquad\qquad\text{4.2}$$

where X is any group with an unshared electron pair. Usually the most abundant species is **4.1**. The effect of substituents on the pK_a's of *para*-substituted acetophenones is the same as that on the pK_a's of *para*-substituted benzoic acids. Since the protonation of an acetophenone leads unambiguously to **4.1** (X = CH_3), the protonation of benzoic acid must lead to **4.1** (X = OH) also.

For amides in acidic media, nmr spectroscopy provides strong evidence that the principal protonated species is **4.1** (X = NR_2).

Although a species such as **4.1** is present in the highest concentration, the less abundant species is generally the reactive one.

4.3. THE EFFECT OF ACIDITY ON HYDRONIUM ION–CATALYZED REACTIONS

The effect of the acidity of the medium on the rate of an acid-catalyzed reaction must be discussed in terms of its effect on the equilibrium protonation of organic compounds considered earlier. On the basis of the general equations (4.19) and (4.20), two systems were proposed to account for the effects of highly acidic media on the rates of hydronium ion–catalyzed reactions.

For the first of these, let us consider a reaction analogous to Eq. (4.19)

$$S + H^{\oplus} \underset{}{\overset{K}{\rightleftharpoons}} SH^{\oplus} \xrightarrow{k_2 \text{ (slow)}} \text{products} \qquad (4.21)$$

Applying the Brönsted equation for the effects of change in the solvent to the rate-determining step of Eq. (4.21) (which is equivalent to assuming the transition state theory) yields

$$\text{rate} = k_2[SH^{\oplus}]\frac{\gamma_{SH^{\oplus}}}{\gamma_{\ddagger}} \qquad (4.22)$$

Table 4.1. Basicities in Aqueous Acid of Representative Bases of Important Classes of Organic Compounds

Functional Group	Compound	$pK_a{}^a$	% H_2SO_4 to Half-Ionize
Aldehyde	Alkyl aldehydes	$(\sim -8)^b$	(~ 88)
	Benzaldehyde	-7.1	81
Amide	Acetamide	~ 0.0	6.5
	Benzamide	~ -2.0	34
Amine	Methylamine	10.6	
	Dimethylamine	10.6	
	Trimethylamine	9.8	
	Aniline	4.6	
Amine oxide	Trimethylamine oxide	4.7	
Aromatic hydrocarbon	Hexamethylbenzene	c	90.5
Carboxylic acid	Acetic acid	-6.1	74
	Benzoic acid	-7.2	82
Carboxylic ester	Ethyl acetate	$(\sim -6.5)^b$	(~ 77)
	Ethyl benzoate	-7.4	83
Ether	Diethyl ether	-3.6	52
	Tetrahydrofuran	-2.1	36
	Anisole	-6.5	77
Hydroxyl	Methanol	-2	34
	Phenol	-6.7	78
Ketone	Acetone	-7.2	82
	Acetophenone	-6.2	74
Mercaptan	Methyl mercaptan	-6.8	78
Nitro	Nitromethane	-11.9	>100
	Nitrobenzene	-11.13	>100
Olefin	1,1-Diphenylethylene	c	71
Phosphine	n-Butylphosphine	0.0	6.5
	Dimethylphosphine	3.9	—
	Trimethyphosphine	8.7	—
Phosphine oxide	Trimethylphosphine oxide	0	6.5
Sulfide	Dimethyl sulfide	-5.4	68
Sulfoxide	Dimethyl sulfoxide	0	6.5

Source: E. M. Arnett, *Prog. Phys. Org. Chem.*, **1**, 324 (1963).

[a] The pK_a refers to a thermodynamic equilibrium constant only in the case of weak bases that exactly obey the activity coefficient postulate. Hence most of these values refer only to the H_0 at half-ionization.

[b] No measured values are available for these compounds. Values shown are estimated by analogy from suitable compounds of known basicity.

[c] Protonation of these compounds probably follow H_R instead of H_0. Therefore, no pK_a on the pH–H_0 scale is given.

where $\gamma_{\ddagger}$ is the activity coefficient for the transition state. Assuming a fast equilibrium, in the formation of $SH^{\oplus}$, Eq. (4.22) may be transformed with the help of (4.23)

$$K = \frac{[H^{\oplus}][S]\gamma_{H^{\oplus}}\gamma_S}{[SH^{\oplus}]\gamma_{SH^{\oplus}}} \quad \text{or} \quad [SH^{\oplus}] = \frac{1}{K}[H^{\oplus}][S]\frac{\gamma_{H^{\oplus}}\gamma_S}{\gamma_{SH^{\oplus}}} \tag{4.23}$$

to give (4.24)

$$\text{rate} = \frac{k_2}{K}[S]a_{H^{\oplus}}\frac{\gamma_S}{\gamma_{\ddagger}} \tag{4.24}$$

If we assume that the activity coefficient ratio $\gamma_S/\gamma_{\ddagger}$ equals the activity coefficient ratio found in the Hammett acidity function $\gamma_{In}/\gamma_{HIn^{\oplus}}$ then Eq. (4.24) may be transformed to

$$\text{rate} = \frac{k_2}{K}[S]h_0 \tag{4.25}$$

Since the observed rate law shows a first-order dependence on substrate, Eq. (4.25) leads to

$$k_{obs} = \frac{k_2}{K}h_0 \tag{4.26}$$

Equation (4.26) may be expressed in the more usual log form

$$\log k_{obs} = -H_0 + \text{constant} \tag{4.27}$$

which indicates that a reaction following Eq. (4.21) should show a linear dependence of $\log k_{obs}$ on H_0 with a slope of -1.0.

A reaction following Eq. (4.28), whose rate expression is shown in Eq. (4.29), leads to a completely different dependence of the rate on acidity

$$S + H^{\oplus} \underset{K}{\rightleftharpoons} SH^{\oplus} \xrightarrow[k_2(\text{slow})]{H_2O} \text{Products} \tag{4.28}$$

Applying the Brönsted equation to the rate-determining step of Eq. (4.28) yields

$$\text{rate} = k_2[SH^{\oplus}][H_2O]\frac{\gamma_{SH^{\oplus}}\gamma_{H_2O}}{\gamma_{\ddagger}} \tag{4.29}$$

Assuming a fast equilibrium between S and $SH^{\oplus}$, we may write

$$\text{rate} = \frac{k_2}{K}[S][H_3O^{\oplus}]\frac{\gamma_S\gamma_{H_3O^{\oplus}}}{\gamma_{\ddagger}} \tag{4.30}$$

The species S and $H_3O^{\oplus}$ are tantamount to the (highest) transition state of the reaction. Since both the reactant and transition state have identical charges, it is reasonable to assume that the ratio of the activity coefficients in Eq. (4.23) is equal to unity. If this condition holds, the rate of reaction will be directly proportional to the hydronium ion concentration, as shown in Eq. (4.31).

$$\text{rate} = \frac{k_2}{K}[\text{S}][H_3O^{\oplus}] \qquad (4.31)$$

and the log of the observed rate constant will be linearly related to the log of the concentration of hydronium ion with a slope of $+1.0$:

$$\log k_{\text{obs}} = \log[H_3O^{\oplus}] + \text{constant} \qquad (4.32)$$

A large number of acid-catalyzed reactions follow Eq. (4.27), whereas a number of other acid-catalyzed reactions follow Eq. (4.32). (See Tables 4.2 and 4.3, respectively.) The mechanisms of reactions (4.27) and (4.32) have been designated A-1 and A-2, respectively, to indicate their apparent order in species other than hydronium ion.[10]

Table 4.2.   Some Acid-Catalyzed Reactions that Follow Mechanism A-1

Hydrolysis of t-butyl acetate
Hydrolysis of methyl mesitoate
Hydrolysis of alkoxymethyl esters
Lactonization of γ-hydroxybutyric acid
Hydrolysis of β-propiolactone
Esterification of alcohols with sulfuric acid
Hydrolysis of sucrose
Depolymerization of paraldehydes
Hydrolysis of tertiary ethers
Hydrolysis of epoxides
Hydrolysis of acetic anhydride
Hydrolysis of acetals, ketals, and glucosides
Dehydration of tertiary alcohols
Hydrolysis of carboxylic anhydrides
Pinacol rearrangement
Hydrolysis of benzyl fluoride
Beckman rearrangement of acetophenone oximes
Decarbonylation of formic acid
Decarboxylation of mesitoic acid
Hydrolysis of diisopropyl phosphorofluoridate
Hydrolysis of ethyl vinyl ether
Intramolecular migration during aromatic hydroxylation
Hydrolysis of neopentyl phosphate
Hydrolysis of dihydroxy and/or methoxy benzenes

Source: Reference 10.

Table 4.3.   Some Acid-Catalyzed Reactions Following Mechanism A-2

Hydrolysis of dimethyl ether
Hydrolysis of ethylenimine
Hydrolysis of carboxamides
Hydrolysis of esters
cis–trans isomerization of benzalacetophenone
Hydrolysis of pyrophosphoric acid
Hydrolysis of trimethylacetic anhydride
Hydrolysis of aryl phosphate
Hydrolysis of aryl phosphinates
Hydration of carbonyl groups
Hydration of alkenes

Source: Reference 10.

Although the use of Eqs. (4.27) and (4.32) as criteria of mechanism of acid-catalyzed reactions became widespread, difficulties arose with this treatment. Unfortunately, few reactions fit either category perfectly; even when a reasonable fit was observed, a slope of precisely 1.0 was rare.

Two obvious reasons may account for the difficulties with Eqs. (4.27) and (4.32):

1. The activity coefficients of only a limited number of substrates conform to the requirements set forth earlier; that is, many substrates are not Hammett bases.

2. The dependence of the rate of reactions on the activity of water may not be correctly described by either Eq. (4.27) or (4.32).

An empirical method to distinguish between the mechanisms of Eqs. (4.27) and (4.32) (and to obtain more information about the mechanism) has been proposed.[8] This method takes advantage of the slope (ω) of the plot of (log $k + H_0$) versus log a_{H_2O}, which was found to be approximately linear. Plotting (log $k + H_0$) versus log a_{H_2O} is equivalent to considering the extent to which a plot of log k versus $-H_0$ deviates from the ideal slope of 1.0, as a function of log a_{H_2O}. It is an empirical fact that the deviation is often linear in log a_{H_2O}. If the slope of log k versus H_0 is greater than one, ω is negative; if it is less than one, ω is positive. The ω plots magnify the deviations from the H_0 line. Presumably, the magnitude of ω correlates with the mechanism of the reaction. If ω is negative, no water participates in the reaction; if $\omega = 0$, water participates as a proton transfer agent in the reaction; and if ω is positive, water participates as a nucleophile.

A rigorous treatment of the effect of acidity on the rate of an acid-catalyzed reaction was developed by considering the effect of acidity both on equilibrium protonation and on the rate of reaction of the substrate.[9] Let us consider a generalized acid-catalyzed reaction involving hydrated species, such as

$$S(H_2O)_s + H(H_2O)_n^{\oplus} \xrightleftharpoons{K_{SH}^{\oplus}} SH(H_2O)_p^{\oplus} + (s + n - p)H_2O$$

$$SH(H_2O)_p^{\oplus} + rH_2O \left[\rightleftharpoons S^{\ddagger}(H_2O)_{\ddagger}^{\oplus} \right] \xrightarrow{k} \text{Products} \tag{4.33}$$

Assuming transition state theory and a fast preequilibrium protonation of the substrate, we can write the rate equation for an acid-catalyzed reaction rate:

$$\frac{k}{K_{SH}^{\oplus}} [S(H_2O)_s] \frac{\gamma_{S(H_2O)_s}}{\gamma_{\ddagger(H_2O)_{\ddagger}}} \times a_{H(H_2O)_n^{\oplus}} \times a_H^{r-(s+n-p)} = k_{obs}[S]_{st} \tag{4.34}$$

Equation (4.35) is a generalized equation for an indicator equilibrium:

$$In(H_2O)_b + H(H_2O)_n^{\oplus} \rightleftharpoons InH(H_2O)_c^{\oplus} + (b + n - c)H_2O \tag{4.35}$$

leads to Eq. (4.36), a general expression for an acidity function in aqueous solution

$$h = \frac{a_{H(H_2O)_n^{\oplus}}}{a_{H_2O}^{(b+n-c)}} \times \frac{\gamma_{In(H_2O)_b}}{\gamma_{InH(H_2O)_c^{\oplus}}} \tag{4.36}$$

Substitution for the hydronium ion activity in Eq. (4.34) in terms of the relation in Eq. (4.36) leads to

$$\text{rate} = \frac{k}{K_{SH}^{\oplus}} [S_{aq}] \frac{\gamma_{S_{aq}} \gamma_{InH_{aq}^{\oplus}}}{\gamma_{\ddagger aq} \gamma_{In_{aq}}} \times h \times a_{H_2O}^{r+(b-c)-(s-p)} \tag{4.37}$$

where the subscripts aq, refer to the fully hydrated species. If the substrate and the indicator are of the same family of organic compounds, γ_s and $\gamma_{In_{aq}}$ refer to the same species and thus will cancel. Likewise, the $\gamma_{\ddagger aq}$ and $\gamma_{inH_{aq}^{\oplus}}$ contributions, which are determined chiefly by electrostatic and volume expansion terms, will cancel. Furthermore, $(b-c) - (s-p) = 0$, since these two quantities refer to the same process. Finally $h = h_{amide}$ (where h_{amide} refer to the Hammett acidity function for the amide family specifically). Thus Eq. (4.38) holds for the hydrolysis of amides in strongly acidic solution.

$$\text{rate} = \frac{k}{K_{SH}^{\oplus}} [S_{aq}] h_{amide} (a_{H_2O})^r \tag{4.38}$$

For weakly basic amides or for a solution in which the amide is not appreciably protonated, $[S_{aq}] = [S_{aq}]_{st}$ (where st = stoichiometric). Thus the simple equation (4.39) is derived, which describes the dependence of the rate on the acidity

$$k_{obs} = \frac{k}{K_{SH}^{\oplus}} h_{amide} (a_{H_2O})^r \tag{4.39}$$

or

$$\log k_{obs} + H_A = r \log a_{H_2O} + constant \qquad (4.40)$$

Equation (4.39) has the same form as Eq. (4.26) except that an acidity function strictly appropriate to the substrate is involved. Also, r can be directly related to the number of water molecules needed to convert the protonated substrate to its transition state. For the acid-catalyzed hydrolyses of benzamide and p-nitrobenzamide, plots of Eq. (4.40) are linear, with r equal to 2.6 and 2.7, respectively, indicating that at least three water molecules may be involved in the transition state of the acid-catalyzed hydrolysis of benzamides. One of these water molecules is the nucleophile. The other two water molecules may be concerned with proton transfer or some process that cannot be specified other than as solvation.

Although the preceding analyses of proton-catalyzed reactions work well for the systems considered they are, unfortunately, not of general applicability

4.4. MECHANISMS OF SOME HYDRONIUM ION-CATALYZED REACTIONS

Many organic reactions exhibit hydronium ion catalysis. One mechanistic classification parallels the distinction made by Eqs. (4.32) and (4.27) between the participation or nonparticipation of water in the rate-determining step of the reaction. As mentioned earlier, Eqs. (4.27) and (4.32) have been designated A-1 and A-2 reactions.

As has already been pointed out, another classification utilizes not only the participation or nonparticipation of water in the rate-determining step of the reactions, but also the function of the water when it does participate either as a nucleophile or as a proton transfer agent. A third classification utilizes the number of hydronium ions participating in forming the unstable intermediate. As might be expected, these classifications overlap.

Some acid-catalyzed reactions that conform to Eq. 4.27 (and thus are A-1 reactions) are listed in Table 4.2, as mentioned above. Essentially, all the reactions of Table 4.2 show a linear dependence of $\log k$ on $-H_0$. Few of these linear relationships have the theoretical slope of 1.00. The reactions of Table 4.2 also show ω values of -2.5 to 0, indicating that water is not involved in the rate-determining step.

The hydrolysis of acetals is probably the most thoroughly studied reaction of Table 4.2. The most notable characteristics[11,12] of this reaction are:

1. The reaction rate at very low acidities is proportional to the hydronium ion concentration.
2. The reaction rate at high acidities is proportional to h_0.
3. ω is close to zero.
4. The entropy of activation is near zero.

5. The reaction rate is independent of steric effects.

6. The rate of disappearance of the methoxy protons of a methyl ketal and the appearance of the hydroxylic protons of methanol from a methyl ketal are identical.

These observations are in accord with the mechanism:

$$R_1-\underset{\underset{OR}{|}}{\overset{\overset{R_2}{|}}{C}}-OR \overset{H^{\oplus}}{\rightleftharpoons} R_1-\underset{\underset{OR}{|}}{\overset{\overset{R_2}{|}}{C}}-OR \underset{ROH}{\overset{H^{\oplus} \atop slow}{\rightleftharpoons}} R_1-\underset{\underset{OR}{|}}{\overset{\overset{R_2}{|}}{C}}{}^{\oplus} \underset{-H^{\oplus}}{\overset{+H_2O}{\rightleftharpoons}}$$

$$R_1-\underset{\underset{OR}{|}}{\overset{\overset{R_2}{|}}{C}}-OH \underset{-H^{\oplus}}{\overset{H^{\oplus}}{\rightleftharpoons}} R_1-\overset{\overset{R_2}{|}}{C}=O + ROH \quad (4.41)$$

Several acid-catalyzed reactions depend on the acidity raised to a power greater than one. A classical example of multiprotonation is the nitration of nitrobenzene:

$$C_6H_5NO_2 + HNO_3 \overset{H_2SO_4}{\longrightarrow} O_2NC_6H_4NO_2 + H_2O \quad (4.42)$$

The rate of this reaction in 80–90% sulfuric acid follows H_R rather than any of the other measures of acidity discussed earlier.[13] This is explicable in terms of the preequilibrium formation of nitronium ion according to Eq. (4.43), which involves two protons. The reaction (4.43) is exactly analogous to the equilibrium defining the H_R function [Eq. (4.9)]. The dependence of the rate on H_R means a great sensitivity of the rate to the stoichiometric concentration of acid, the greatest sensitivity found in an organic reaction.

$$HONO_2 + 2H_2SO_4 \rightleftharpoons NO_2^{\oplus} + H_3O^{\oplus} + 2HSO_4^{\ominus} \quad (4.43)$$

$$ROH + 2H_2SO_4 \rightleftharpoons R^{\oplus} + H_3O^{\oplus} + 2HSO_4^{\ominus} \quad (4.44)$$

Many acid-catalyzed reactions conform to Eq. (4.28) [or (4.32)] and are spoken of as mechanism A-2; some examples of these reactions are shown in Table 4.3. These reactions usually show the following characteristics:

1. The rate of reaction is proportional to the hydronium ion concentration up to fairly high acidities.

2. The rate of reaction of the protonated substrate may be dependent on the activity of water.

3. The value of ω is positive.

4. The entropy of activation is negative and usually large.

5. The volume of activation is positive.

6. The rate of reaction is increased by added nucleophiles or bases.

In addition to these general criteria for A-2 reactions, other mechanistic features can occur.

Not included in Table 4.3 are those reactions that formally fit the A-2 mechanism of Eq. (4.32) but where the chemistry, the ω value, or other criteria indicate that water acts as a proton transfer agent. These reactions are better discussed as general acid-catalyzed reactions and will be treated in Chapter 5.

To illustrate the A-2 mechanism of Table 4.3, we will use ester hydrolysis. The acid-catalyzed hydrolysis of esters, like other acid-catalyzed reactions, proceeds via a rather complex pathway. In this instance, the scheme of Eq. (4.28) can be amplified to (4.45) to specify the five intermediates of the reaction (see Fig. 4.4):

$$
\underset{4.5}{\overset{\overset{\displaystyle O}{\parallel}}{RCOR}} \overset{H^{\oplus}}{\rightleftharpoons} \underset{4.5}{\overset{\overset{\displaystyle OH^{\oplus}}{\parallel}}{R\!-\!C\!-\!OR}} \underset{slow}{\overset{H_2O}{\rightleftharpoons}} \underset{\underset{4.6}{\overset{\displaystyle \,}{OH_2{}^{\oplus}}}}{\overset{\overset{\displaystyle OH}{|}}{R\!-\!C\!-\!OR}} \overset{-H^{\oplus}}{\rightleftharpoons} \underset{\underset{4.7}{\overset{\displaystyle \,}{OH}}}{\overset{\overset{\displaystyle OH}{|}}{R\!-\!C\!-\!OR}} \overset{+H^{\oplus}}{\rightleftharpoons}
$$

$$
\underset{\underset{4.3}{\overset{\displaystyle OH}{|}}}{\overset{\overset{\displaystyle OH}{|}}{R\!-\!C\!-\!O\!-\!R}}\,{}_{H^{\oplus}} \underset{ROH}{\overset{Slow}{\rightleftharpoons}} \underset{4.4}{\overset{\overset{\displaystyle OH^{\oplus}}{\parallel}}{RCOH}} \overset{}{\underset{H^{\oplus}}{\rightleftharpoons}} \overset{\overset{\displaystyle O}{\parallel}}{RCOH} \quad (4.45)
$$

The mechanism given for this reaction is based on the following:

1. First-order dependence of the rate on the hydronium ion concentration.

2. Spectrophotometric observation of protonated esters and acids.

3. Acyl–oxygen bond fission.

4. Sensitivity of the rate to structural change in either the alkyl or acyl portion of the molecules.

5. Simultaneous hydrolysis and carbonyl oxygen exchange.[22]

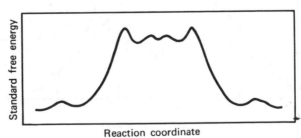

Reaction coordinate

Fig. 4.4. Standard free energy versus reaction coordinate for the acid-catalyzed hydrolysis of an ester. From M. L. Bender, *Chem. Rev.*, **60**, 68 (1960).

Intermediates **4.4** and **4.5** may be postulated from observations 1 and 2 since one proton is known to be involved from the kinetics, and one can actually "see" the protonated ester and acid spectrophotometrically. Fission of an ester can take place by acyl–oxygen fission ($RC \overset{O}{\overset{\|}{}} \!\!\mid\! O\ R'$) or alkyl–oxygen fission ($RCO \overset{O}{\overset{\|}{}} \!\mid\! R'$); isotopic oxygen labeling shows acyl–oxygen fission. This finding plus the finding of isotopic exchange from the carbonyl oxygen position concurrent with hydrolysis (observation 5) leads to the postulation of intermediates **4.3, 4.6,** and **4.7.** The exchange reaction requires addition intermediates rather than a straight displacement reaction at the carbonyl carbon atom, for the latter would not lead to an exchange whereas the former would (this postulate is consistent with observation 3 but is demanded by observation 5). Intermediate **4.7** is demanded by the symmetry required by the exchange reaction and by the fact that identical exchange occurs in both acidic and basic catalysis.[22]

One of the reactions of Table 4.3, the hydrolysis of benzamide, merits further consideration. Although in dilute acid solution, the rate of hydrolysis is proportional to the hydronium ion concentration, a rate maximum occurs, as shown in Fig. 4.5, at acidities somewhat larger than the pK_a of the amide. The hypothesis of a protonated amide intermediate effectively accounts for all the experimental facts.

$$R-\overset{O}{\overset{\|}{C}}-NH_2 + H^{\oplus} \rightleftharpoons R-\overset{OH^{\oplus}}{\overset{\|}{C}}-NH_2 \xrightarrow{H_2O} \text{Hydrolysis products} \quad (4.46)$$

Below the acid concentration necessary for maximal rate, the effect of increasing

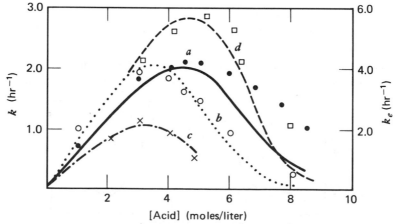

Fig 4.5.   The effect of acidity on hydrolysis of benzamide. From J. T. Edward and S. C. R. Meacock, *J. Chem. Soc.*, **1957**, 2000.

acid strength is chiefly to increase the concentration of the protonated intermediate; above the concentration for maximal rate, the effect is chiefly to decrease the concentration or activity of water. This hypothesis can account quantitatively for the changes in the rate of hydrolysis with concentration of the acid.[14] The experimental rate constant k_{obs} can be related to the second-order rate constant k_2 by the expression

$$k_{obs} = \frac{k_2 K[(H_3O^{\oplus})]}{K + h_0} \tag{4.47}$$

where K is the equilibrium constant for the protonation of the amide and h_0 is the Hammett acidity function.[†] For very weak bases or at low acidity, $K \gg h_0$ and Eq. (4.47) simplifies to

$$k_{obs} = k_2[(H_3O^{\oplus})] \tag{4.48}$$

Equation (4.47) predicts a linear dependence of the experimental first-order constant on the concentration of hydronium ion; such a relationship is found in dilute acid solution. However, when $K \ll h_0$, Eq. (4.47) simplifies to

$$k_{obs} = \frac{k_2 K[(H_3O^{\oplus})]}{h_0} \tag{4.49}$$

Since h_0 increases with acid concentration much more rapidly than $[H_3O^{\oplus}]$ when concentrations of mineral acid exceed about 2 M, k_{obs} decreases in high concentrations of acid. When $K \ll h_0$, protonation of the amide will be substantially complete; the decrease in rate of hydrolysis can then be regarded as due to the decreasing availability of water since $[H_2O] = K_a[H_3O^{\oplus}]/h_0$, where K_a is the equilibrium constant for the reaction $H_3O^{\oplus} \overset{K_a}{\rightleftharpoons} H_2O + H^{\oplus}$.

   Since the acid-catalyzed hydrolysis of nitriles does not show a rate maximum in concentrated acid solution as the hydrolysis of amides does, it is possible to convert a nitrile to an isolable amide in strong acid solution. Conversely, in dilute acid solutions, the hydrolysis of a nitrile will proceed directly to carboxylic acid, since under these conditions the rate of hydrolysis of the amide is faster than that of the nitrile.

   In concentrated sulfuric acid solutions, the rate of hydrolysis of ethyl acetate decreases rapidly and then rises again with increase in acidity. The decrease in rate may be explained in terms of low activity of water in these solutions, in analogous fashion to the earlier discussion of the hydrolysis of amides. The rise can then only be explained by a change in mechanism from A-2 to A-1 in extremely concentrated sulfuric acid. The dependence of the rate of hydrolysis of the protonated ester on the activity of water is consistent with this interpretation.

[†] More rigorously, h_{amide} should be used here in light of the earlier discussion.

Such changes in mechanism are met in many other acid-catalyzed reactions. For example, the hydrolyses of both methyl mesitoate and methyl benzoate follow Eq. (4.32) (A-2) in dilute acid solution. However, in moderately strong acid solutions, the hydrolysis of the former changes to Eq. (4.27) (A-1), as reflected in dependence on H_0, lack of oxygen exchange, a Hammett rho constant of -3.7, and a greater rate of hydrolysis than unhindered methyl benzoate.

The hydrolysis of methyl and ethyl benzoates in 99.9% sulfuric acid illustrates another mechanistic change in acid-catalyzed hydrolysis of esters. The effect of structure on the rate constant for these reactions is shown in Fig. 4.6. The Hammett rho constant for methyl benzoates is -3.7, the reaction probably following acyl–oxygen fission via an acylium ion [Eq. (4.50)]. Ethyl benzoates appear to follow the same pattern from a Hammett sigma of -0.3 to 0.7, but above 0.7, the slope of the line changes abruptly to a reaction facilitated by electron-withdrawing substituents (see Fig. 4.6). Since this break occurs with an ethyl ester but not with a methyl ester, it is reasonable to assume that on the left-hand side of the break, ethyl benzoate reacts via an acylium ion intermediate, and on the right-hand side of the break with alkyl–oxygen fission.

Despite the definition of specific hydroxide ion catalysis, primary and secondary amines catalyze the reaction, while tertiary amines do not. This behavior cannot be attributed to general basic catalysis, for all bases should catalyze the reaction if that were the case. Therefore, the effect of primary and secondary amines must represent some form of specific damine catalysis. As will

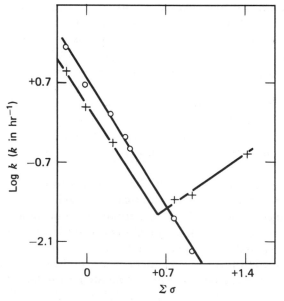

Fig. 4.6.   The hydrolysis of methyl and ethyl benzoates in 99.9% sulfuric acid: (O) methyl benzoate; (+) ethyl benozate. From D. N. Kershaw and J. A. Leisten, *Proc. Chem. Soc.*, **1960**, 84.

$$
\begin{array}{c}
\underset{\substack{\|\\ \text{O}}}{\text{R}'-\text{C}-\text{O}-\text{R}} + \text{H}^{\oplus} \rightleftharpoons \underset{\substack{\|\\ \text{H}-\text{O}^{\oplus}}}{\text{R}'-\text{C}-\text{OR}}
\end{array}
$$

$$
\underset{\substack{|\\ \text{H}}}{\text{R}'-\text{C}-\text{O}^{\oplus}-\text{R}} \qquad (4.50)
$$

$$
\underset{\substack{\|\\ \text{O}}}{\text{R}'-\text{C}^{\oplus}} + \text{R}-\text{O}-\text{H} \qquad \underset{\substack{\|\\ \text{O}}}{\text{R}'-\text{C}-\text{OH}} + \text{R}^{\oplus}
$$

be discussed in Chapter 7, this special situation is explicable in terms of imine (Schiff base) formation between diacetone alcohol and the amines.

4.5.   THE HYDROXIDE ION AS CATALYST

Although the hydroxide ion does not have the unique characteristics of small size per unit charge that the proton has, it serves a powerful and important function in catalysis. The usual function of the hydroxide ion is the reversible removal of a proton from the substrate, ordinarily an organic molecule. Alternatively, hydroxide ion can add into the substrate, forming an unstable addition intermediate that decomposes to form products [Eq. (4.59)]. In contrast to hydronium ion catalysis, acceleration by hydroxide ion is often not true catalysis since the hydroxide ion is usually not regenerated.

Carbanions are formed in many hydroxide ion reactions. These carbanions can either proceed directly to products or lead to other reactive intermediates, such as carbenes (4.51) or benzynes (4.52) prior to product formation.

$$
\underset{\substack{|\\ \text{H}}}{\cdot \text{R}_2\text{CX}} \xrightarrow{\text{OH}^{\ominus}} \text{R}_2\overset{\ominus}{\text{CX}} \xrightarrow{-\text{X}^{\ominus}} \underset{\text{Carbene}}{\text{R}_2\text{C}:} \longrightarrow \text{Products} \qquad (4.51)
$$

$$
\xrightarrow{\text{OH}^{\ominus}} \qquad \xrightarrow{-\text{X}^{\ominus}} \qquad \underset{\text{benzyne}}{} \longrightarrow \text{Products} \qquad (4.52)
$$

As in acid catalysis, two types of base catalysis are seen: (1) specific hydroxide (lyoxide) ion catalysis and (2) general base catalysis. Only in the former is the hydroxide (lyoxide) ion independent of the source of the ion.

Some kinetic schemes that satisfy specific hydroxide ion catalysis are given below. Preequilibrium removal of a proton from a substrate by a base may occur, followed by reaction of the substrate anion or subsequent intermediate to give products, with or without any other nucleophilic reactant.

$$SH + B \xrightleftharpoons{K} S^{\ominus} + BH^{\oplus} \xrightarrow[k]{slow} Products \qquad (4.53)$$

$$SH + B \xrightleftharpoons{K} S^{\ominus} + BH^{\oplus} \xrightarrow[k]{N \ (slow)} Products \qquad (4.54)$$

The kinetic equation for reaction (4.53) is given by

$$rate = k[S^{\ominus}] \qquad (4.55)$$

Assuming a fast preequilibrium, (4.55) can be transformed to

$$rate = \frac{k\,K[SH][B]}{[BH^{\oplus}]} \qquad (4.56)$$

The ratio $[B]/[BH^{\oplus}]$ is, of course, related to the hydroxide ion concentration through

$$B + H_2O \xrightleftharpoons{K_S} BH^{\oplus} + OH^{\ominus} \qquad (4.57)$$

Thus the rate expression for Eq. (4.53) can be written as

$$rate = \frac{kK}{K_S}\,[SH][OH^{\ominus}] \qquad (4.58)$$

The rate equation corresponding to mechanism (4.54) is analogous to Eq. (4.58) with an added term in the concentration of N.

In addition to the mechanisms of specific hydroxide ion catalysis involving the removal of a proton from the substrate, a mechanism involving the addition of hydroxide ion to the substrate is possible (see Chapter 7).

$$SH + OH^{\ominus} \xrightleftharpoons[k_{-1}]{k_1} SH{-}OH^{\ominus} \xrightarrow{k_2} Products \qquad (4.59)$$

The rate equation corresponding to Eq. (4.59) is

$$rate = \frac{k_1 k_2}{k_{-1} + k_2}\,[SH][OH^{\ominus}] \qquad (4.60)$$

For both rate equations, (4.58) and (4.60), the observed rates are proportional to the hydroxide ion concentration of the solvent; other bases present do not influence the experimentally determined rates. This is exactly the definition of specific catalysis.

Compared to the many correlations between rate and acidity for acid-catalyzed reactions, there are few correlations between rate and basicity for base-catalyzed reactions. One correlation describes the rate of racemization of (+)-2-methyl-3-phenylpropionitrile in dimethyl sulfoxide–methanol solutions containing methoxide ion.[15] Increases by factors as great as 10^9 in the rate of this racemization were found on substituting dimethyl sulfoxide for methanol as solvent (Fig. 3.6). The basicity function $H_\ominus$ was determined over the entire range of methanol–dimethyl sulfoxide compositions. An excellent linear correlation was found between $\log k_{\text{racemization}}$ and $H_\ominus$ with a slope of 0.87, over a range of 10^6-fold change in rate constant.

The mechanism of this reaction may be written as

$$\text{AH} + \text{B}^\ominus \underset{k_{-a}}{\overset{k_a}{\rightleftharpoons}} \underset{\substack{\text{Optically}\\\text{active}}}{\text{A}^\ominus\text{---HB}} \overset{k_b}{\longrightarrow} \underset{\text{Racemic}}{\text{A}^\ominus\text{---HB}} \overset{k_{-a}}{\longrightarrow} \text{A—H} + \text{B}^\ominus \qquad (4.61)$$

The rate constant for mechanism (4.61) is given by

$$k_{\text{obs}} = k_a \frac{k_b}{k_{-a} + k_b} \qquad (4.62)$$

If $k_{-a} \gg k_b$, an asymmetrically solvated carbanion is formed in a preequilibrium step, followed by rate-determining racemization. Then Eq. (4.62) reduces to

$$k_{\text{obs}} = \frac{k_a}{k_{-a}} k_b \qquad (4.63)$$

If, however, $k_b \gg k_{-a}$, a carbanion is produced by slow proton transfer, leading directly to racemic product, and Eq. (4.62) reduces instead to

$$k_{\text{obs}} = k_a \qquad (4.64)$$

The preceding correlation between $\log k_{\text{obs}}$ and $H_\ominus$ (and other independent kinetic evidence), favors transition state **4.8**, since it resembles in structure the nitrogen anions used as indicators in the establishment of the $H_\ominus$ scale. Conversely, the transition state of the reaction corresponding to Eq. (4.63) involves an anion whose charge is distributed on both carbon and oxygen atoms [see structure **4.9**].

$$[\text{C}^\ominus\text{---HOCH}_3] \qquad\qquad [\text{C}^{\delta\ominus}\text{---H}^\oplus\text{---}^{\delta\ominus}\text{OCH}_3]$$
$$\textbf{4.8} \qquad\qquad\qquad\qquad \textbf{4.9}$$

$$[-\text{N}^\ominus\text{---HOCH}_3]$$
Hydrogen-bonded anionic
form of nitrogen indicator

Specific hydroxide (lyoxide) catalysis is seen in many classical organic reactions, including the retrograde aldol reaction of diacetone alcohol, the Claisen, Michael, Perkin, and aldol condensations, and the bromination of β-disulfones and β-dinitriles. All of these reactions follow (specific hydroxide) ion catalysis according to Eqs. (4.53) or (4.54). The alkaline hydrolysis of an ester follows (specific) hydroxide ion catalysis according to Eq. (4.59).

Aldol condensation illustrates some of the features of (specific hydroxide) ion catalysis. The mechanism of this reaction may be written as

$$B + CH_3CHO \underset{k_{-1}}{\overset{k_1}{\rightleftharpoons}} {}^{\ominus}CH_2{-}CHO + BH^{\oplus}$$

$$ {}^{\ominus}CH_2CHO + CH_3CHO \xrightarrow{k_2} \text{Products} \qquad (4.65)$$

At low acetaldehyde concentrations, the second step is rate-determining. Evidence for this mechanism includes the following:

1. The rate is proportional to the second power of the acetaldehyde concentration and to the first power of the hydroxide ion concentration.

2. When the reaction is carried out in D_2O, deuterium atoms are incorporated into the reactant (hence ultimately into the products) at a rate greater than that of the overall reaction.

3. Variation in structure indicates that the rate of reaction is a function of both the ease of enolate ion formation and the ease of addition to various carbonyl compounds.

4. At very high (10 M) acetaldehyde concentrations, however, the first step is rate-determining and thus no deuterium is incorporated into the reactant in D_2O solution, since the reverse reaction of the first step rarely occurs.[6]

The retrograde aldol reaction of diacetone alcohol shows another complication affecting specific hydroxide ion-catalyzed reactions. The mechanism of this reaction is given by

$$ (4.66)$$

In these processes, the function of hydroxide ion is the same, that is, the initial removal of a proton from the substrate, thereby introducing a negative charge into the molecule. The effect of this negative charge on subsequent reactivity may be considered from the electrostatic and resonance points of view given earlier for the introduction of a positive charge into a substrate by hydronium ion.

4.6. LYOXIDE ION CATALYSIS IN SOLVENTS OTHER THAN WATER

Lyoxide ion has much greater nucleophilicity in aprotic solvents than in aqueous solution. Although most salts are poorly soluble in aprotic solvents, they can be solubilized by use of cryptands,[16] which can complex with metal ions of salts and also dissociate (or separate) the cation–anion pair leading to anion activation.[17] For example, the pK_a to the t-amylate anion in benzene (with the potassium cation complexed by cryptand **4.10**), is approximately 37 or greater. Besides, highly hindered esters, such as alkyl 2,4,6-trimethylbenzoates can be saponified by potassium hydroxide in toluene in the presence of cryptand **4.11**, although they are very resistant to hydrolysis otherwise.

4.10   4.11

Amide ion in liquid ammonia is a powerful basic catalyst whose action parallels that of hydroxide ion in water. Amide ion is both a stronger base and a more reactive nucleophile than hydroxide ion; these properties make it valuable in many of the reactions discussed earlier where a stronger catalytic species is needed. This system has been of considerable utility in the polymerization of styrene and especially of acrylonitrile and of methyl methacrylate, whose substituents stabilize the anionic charge of the intermediate formed by addition of amide ion to the monomer of the substrate and analogous intermediates formed by the addition of this species to other monomers.

In dimethyl sulfoxide or hexamethylphosphoramide, t-butoxide-catalyzed additions of aromatic heterocycles to conjugated hydrocarbons (homogeneous alkylation) have been achieved. In addition, the base-catalyzed isomerization of alkynes in ethanol containing some potassium hydroxide is known. The proton-abstracting ability of these systems is midway between the hydroxide ion–water and the sodium amide–ammonia systems. Another proton-abstracting species is the anion of dimethyl sulfoxide.

Although the use of strong bases as catalysts for organic reaction has been a powerful synthetic tool, the mechanisms of these reactions, especially in nonaqueous solution, are only now being investigated in a definitive manner.[18-21] Organometallic compounds or alkali metals, for example, have strongly basic properties. They are able to abstract allylic or benzylic protons from hydrocarbons. The carbanions thus formed are capable of undergoing profound changes such as isomerization, reaction with other olefins, and cyclizations. The future holds promise for further profound rate accelerations in this area of organic chemistry.

REFERENCES

1. W. Ostwald, *Phys. Z.*, **3**, 313 (1902).

2. L. P. Hammett and A. J. Deyrup, *J. Am. Chem. Soc.*, **54**, 2721 (1932); L. P. Hammett, *Physical Organic Chemistry*, McGraw-Hill Book Co., New York, 1940, p. 263.

3. N. C. Deno, P. T. Grove, and G. Sims, *J. Am. Chem. Soc.*, **81**, 5790 (1951); N. C. Deno, J. Jaruzelski, and A. S. Schriesheim, *ibid.*, **77**, 3044 (1955).

4. V. Gold and B. W. V. Hawes, *J. Chem. Soc.*, **1951**, 2102; V. Gold, *ibid.*, **1955**, 1263.

5. A. Streitwieser, Jr., W. C. Langworthy, and J. I. Brauman, *J. Am. Chem. Soc.*, **85**, 1761 (1963).

6. M. R. Crampton and V. Gold, *Proc. Chem. Soc. London*, 298 (1964).

7. L. Zucker and L. P. Hammett, *J. Am. Chem. Soc.*, **61**, 2791 (1939).

8. J. F. Bunnett, *J. Am. Chem. Soc.*, **83**, 4956, 4968, 4973, 4978 (1961).

9. K. Yates and J. B. Stevens, *Can. J. Chem.*, **43**, 529 (1965).

10. C. K. Ingold, *Structure and Mechanism in Organic Chemistry*, Cornell University Press, Ithaca, NY, 1953, Chap. XIV.

11. F. A. Long and M. A. Paul, *Chem. Rev.*, **57**, 935 (1957).

12. A. M. Wenthe and E. H. Cordes, *J. Am. Chem. Soc.*, **87**, 3173 (1965).

13. F. H. Westheimer and M. S. Kharasch, *J. Am. Chem. Soc.*, **68**, 1871 (1946).

14. J. T. Edward and S. C. R. Meacock, *J. Chem. Soc.*, **1957**, 2000.

15. R. Stewart, J. P. O'Donnell, D. J. Cram, and B. Rickborn, *Tetrahedron*, **18**, 917 (1962).

16. J. M. Lehn, in *Frontiers of Chemistry*, K. J. Laidler, Ed., Pergamon Press, Oxford, 1982, pp. 265-274.

17. J. M. Lehn, *Acc. Chem. Res.*, **11**, 49-57 (1978).

18. H. Pines and L. Schaap, *Adv. Catal.*, **12**, 117 (1960).

19. E. Grovenstein and G. Wentworth, *J. Am. Chem. Soc.*, **89**, 1852 (1967).

20. H. Pines and W. M. Wackher, *Base-Catalyzed Reactions of Hydrocarbons and Related Compounds*, Academic Press, New York, 1977.

21. H. Pines, *The Chemistry of Catalytic Hydrocarbon Conversions*, Academic Press, New York, 1981.

22. M. L. Bender, *Chem. Rev.*, **60**, 53 (1960).

5 | General Acid–Base Catalysis: Organic Reactions

Up to this point, we have confined our discussion to catalysis of reactions by ions characteristic of the solvent; since we usually refer to aqueous solutions, these catalytic entities are hydronium and hydroxide ions. The simplest examples of such catalyses are those effected by strong acids or bases, such as hydrochloric acid or potassium hydroxide, where the only acidic or basic species present in these solutions are the solvent and the ions related to it.

Catalysis by weak acids or bases can occur in the same way; the actual catalysts are, again, the solvent-related ions resulting from the partial dissociation of the acid or base. The only important consequence of the dissociation of these acids and bases is that they do not provide as high a concentration of hydronium or hydroxide ions as do the strong electrolytes. However, the aqueous system now contains acidic or basic species (the unionized acid or base) not related to the solvent. These species may also participate in either general acid or base catalysis.

In numerous reactions catalyzed by weak acids or bases, the rates can be shown to depend not only on the concentration of the solvent ions, but also upon the concentrations of all other proton donors or acceptors present. Classical examples of such reactions include the decomposition of nitramide, the iodination of acetone, and the mutarotation of glucose. Catalysis of this kind is defined as general acid or base catalysis; it is experimentally distinguished from specific catalysis by the presence in the rate expression of terms involving the concentration of every proton donor or acceptor present. The general form of such rate expressions is shown in Eqs. (5.1) and (5.2):

$$\text{rate} = \Sigma k_i^b [\text{B}_i][\text{substrate}] \qquad \text{for general base catalysis} \qquad (5.1)$$

$$\text{rate} = \Sigma k_j^a [\text{HA}_j][\text{substrate}] \quad \text{for general acid catalysis} \qquad (5.2)$$

For example, the rate law for a general acid–base catalyzed reaction in an acetate buffer solution is

$$\text{rate} = k_{\text{obs}}[\text{S}]$$

$$= \{k_0 + k_{\text{H}}[\text{H}_3\text{O}^\oplus] + k_{\text{OH}}[\text{OH}^\ominus] + k_{\text{HOAc}}[\text{HOAc}]$$
$$+ k_{\text{OAc}}[\text{OAc}^\ominus]\}[\text{S}] \qquad (5.3)$$

and the rate thus depends on the buffer *concentration* rather than the buffer *ratio*. The term k_0 involves the solvent, water, presumably acting either as an acid or base catalyst. A complicated system such as Eq. (5.3) can be analyzed experimentally in the following way. For reactions carried out in solutions sufficiently acidic that $[\text{OH}^\ominus]$ and $[\text{OAc}^\ominus]$ are negligible, a plot of the observed rate constant versus the hydronium ion concentration leads to a straight line of slope k_{H} and intercept k_0. For reactions carried out in solutions sufficiently alkaline that $[\text{H}^\oplus]$ and $[\text{HOAc}]$ are negligible, a plot of k_{obs} versus hydroxide ion concentration leads to a straight line of slope k_{OH} and intercept k_0. For reactions carried out in acetate buffer solutions, the remaining two rate constants, k_{HOAc} and k_{OAc} can be determined. A plot of the observed rate constants versus the ratio of acetic acid over the acetate ion concentration for such a reaction is shown in Fig. 5.1 at two different values of r, defined as the ratio $[\text{HOAc}]/[\text{OAc}^\ominus]$. The slopes of the two lines in Fig. 5.1 are $r_1(k_{\text{HOAc}} + k_{\text{OAc}})$ and $r_2(k_{\text{HOAc}} + k_{\text{OAc}})$.

In simple systems where only the acidic species HA has catalytic activity and the basic species $\text{A}^\ominus$ has none (or vice versa), the involvement of a general acid or general base catalyst in a reaction may be deduced from the pH dependence of the reaction (Fig. 5.1). The conservation equation of a prototropic species may be expressed as $[\text{HA}]_T = [\text{HA}] + [\text{A}^\ominus]$. Combining this equation with that for the ionization of the acid leads to

$$[\text{HA}] = \frac{[\text{HA}]_T}{1 + \dfrac{K_A}{[\text{H}_3\text{O}^\oplus]}} \qquad (5.4)$$

and

$$[\text{A}^\ominus] = \frac{[\text{HA}]_T}{1 + \dfrac{[\text{H}_3\text{O}^\oplus]}{K_A}} \qquad (5.5)$$

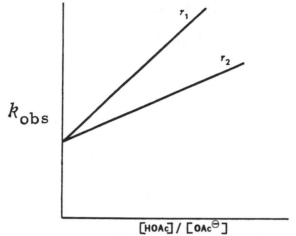

Fig. 5.1.   The rate constant of a reaction catalyzed by both acetic acid and acetate ion.

where K_A is the dissociation constant of HA and T = total. Consequently, a reaction rate dependent on HA (general acid catalysis), when graphed as k_{obs} versus pH, will have the sigmoid shape of a titration curve for an acid; likewise a rate dependent on $A^\ominus$ (general base catalysis), when graphed as k_{obs} versus pH, has the sigmoid shape of a titration curve for a base. In logarithmic terms, these curves have the form of Fig. 4.1, f and g.

5.1.   THE BRÖNSTED CATALYSIS LAW

A fundamental question in catalysis is the relationship between the electronic structure of a catalyst and its catalytic activity. In general acid or base catalysis, a pertinent measure of electronic structure is the pK_a of the catalyst. It is therefore of interest to consider the relationship between the pK_a of a catalyst and its catalytic rate constant. Very early in the development of general acid–base catalysis theory, a relationship of this kind was proposed by Brönsted and Pedersen.[1] These workers considered the decomposition of nitramide, shown below.

$$\begin{array}{c} H \\ \diagdown \\ \diagup \\ H \end{array} N{-}N \begin{array}{c} \diagup O \\ \diagdown O \end{array} \xrightarrow{\ k\ } N_2O + H_2O$$

They were able to show an excellent linear relation between the log of the observed rate constant and the pK_a of some fourteen bases. However, it is important to point out that nitramide is a hard acid and that all of the bases considered were hard bases, that is, oxyanions. Therefore, in the transition state it is unlikely that the proton will be any closer to the base than the acid for the entire series. The Brönsted catalysis law is expressed either by

$$k_A = G_A K_a^\alpha \tag{5.6a}$$

where A and α refer to an acid and G is a constant, or

$$k_B = G_B \left(\frac{1}{K_a} \right)^\beta \tag{5.6b}$$

where B and β refer to a base, K_a is the conventional acid dissociation constant of the acid A or of the acid conjugate to the base B.

Equations (5.6a) and (5.6b) are frequently expressed in linear logarithmic form:

$$\log k_A = -\alpha p K_a + \text{constant} \tag{5.7a}$$

$$\log k_B = \beta p K_a + \text{constant} \tag{5.7b}$$

A plot of $\log k_A$ versus $p K_a$ should be linear with slope $-\alpha$, while a plot of $\log k_B$ versus $p K_a$ should be linear with slope $+ \beta$.

For general acids or bases containing more than one ionizable group, it must be realized that the ionization constants are usually influenced by each other. When such compounds are to be treated by the Brönsted relationship, statistical corrections must be applied; details of such corrections were discussed by Brönsted and Pedersen.[1]

Figure 5.2 shows a plot of Eq. (5.7b) for some original data on the general

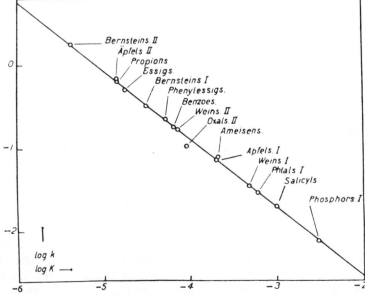

Fig. 5.2.   The first Brönsted relation. Catalysis in the decomposition of nitramide by 14 anions (reproduced from Ref. 2).

Table 5.1. Some Linear Bronsted Relations

Reaction	β or α	Maximal Rate Constant ($M^{-1}\,sec^{-1}$)	Range of pK_a	Strongest Acid (Base)	Weakest Base (Acid)
$CH_3CH(OH)_2 \overset{HA}{\rightleftharpoons} CH_3CHO + H_2O$	0.54	11	10	Chloroacetic acid	Hydroquinone
α-Glucose $\overset{B}{\rightleftharpoons}$ β-glucose	0.40	90	17	$OH^{\ominus}$	H_2O
α-Glucose $\overset{HA}{\rightleftharpoons}$ β-glucose	0.30	6.7×10^{-3}	17	$H_3O^{\oplus}$	H_2O
$H_2C(OH)_2 \overset{B}{\rightarrow} H_2C{=}O + H_2O$	0.40	1600	17	$OH^{\ominus}$	H_2O
$H_2C(OH)_2 \overset{HA}{\rightarrow} H_2C{=}O + H_2O$	0.10	2.7	17	$H_3O^{\oplus}$	H_2O
$Cl_2CHCO_2Et + OH^{\ominus} \overset{B}{\rightarrow} Cl_2CHCO_2^{\ominus} + EtOH$	0.47	13×10^{-4}	9	Imidazole	H_2O
4-Cl-butanol $\overset{B}{\rightarrow}$ tetrahydrofuran	0.25	2×10^{-2}	17	$OH^{\ominus}$	H_2O

Source: Reference 2.

base catalysis of the decomposition of nitramide. A fair linear correlation is observed over three pK units and approximately two powers of 10 in the rate constants. The slope β of the line is 0.66. Many other Brönsted plots are known, some with considerably more data than Fig. 5.2. Table 5.1 shows some examples of reactions that show linear Brönsted relations together with their α (or β) values.

Linear Brönsted relationships are found in catalytic reactions involving (1) oxygen or nitrogen acids and bases either as substrates or catalysts and (2) a rate constant not approaching diffusion-control. However, linear Brönsted relations do not hold in all reactions. For example, nonlinear Brönsted relations occur among reactions involving (1) carbon acids or bases either as substrates or catalysts and (2) rate constants near diffusion-control (see below).

In addition to nonlinearity of the Brönsted relation, other deviations may exist. In the general acid-catalyzed dehydration of acetaldehyde hydrate, several carbon acid catalysts show a negative deviation (of one or two logarithmic units) from the line defined by carboxylic acids and phenols while several oxime catalysts show a positive deviation of the same magnitude. Both the negative and the positive deviations probably reflect a reorganization, either within the molecule or in the solvation shell, that affects the kinetics of proton transfer, but not the equilibrium acidity. These deviations point out the arbitrary nature of the defined Brönsted line. Other deviations from the Brönsted relation include reactions catalyzed by primary, secondary, and tertiary amines that require different Brönsted lines either because of solvation differences or because of steric strain, the presence of steric hindrance in the catalyst and/or the substrate, special solvation effects, and changes in mechanism. The implication of all this is that the application of Brönsted plots [Eq. (5.7)] and their interpretation must be used with great caution.

5.2. THE MEANING OF THE BRÖNSTED CATALYSIS LAW (ITS RELATION TO PROTON TRANSFER)

Let us consider how the Brönsted relationship was initially regarded and then point out the pitfalls as they are currently recognized. For the sake of simplicity, let us assume a specific mechanism for the base-catalyzed dehydration of acetaldehyde hydrate and then show how the existence of the Brönsted relation implies the existence of a rate–equilibrium relationship for one of the individual steps of the reaction. The assumed mechanism is given by

$$CH_3-CH(OH)_2 + B \overset{K}{\rightleftharpoons} CH_3-\overset{\overset{\displaystyle O^{\ominus}}{|}}{C}H(OH) + BH^{\oplus} \underset{k_{-c}}{\overset{k_{c}}{\rightleftharpoons}}$$

$$CH_3-\overset{\overset{\displaystyle O}{||}}{C}H + H_2O + B \qquad (5.8)$$

The overall catalytic rate corresponding to this mechanism is given by

$$\text{Overall rate} = k_c[BH^{\oplus}][CH_3\overset{\overset{O^{\ominus}}{|}}{C}H(O^{\ominus})(OH)] = \nu \qquad (5.9)$$

Equation (5.9) may be transformed to (5.10) using K_b, the ionization constant of $BH^{\oplus}$, and K_a, the ionization constant of acetaldehyde hydrate

$$\nu = k_c \frac{K_a}{K_b}[B][CH_3CH(OH)_2] \qquad (5.10)$$

Since the observed rate of the reaction is $k_{obs}[B][CH_3CH(OH)_2]$, k_{obs} may be given by

$$k_{obs} = \frac{k_c K_a}{K_b} \qquad (5.11)$$

Now we wish to relate the equilibrium constant K_c, corresponding to step k_c, to the ionization constant of the catalyst, K_b. To do this, we will utilize the equilibrium definition

$$K_c = \frac{[CH_3CHO][B]}{[CH_3CH(O^{\ominus})(OH)][BH^{\oplus}]} \qquad (5.12)$$

which can be transformed to Eq. (5.13) and then to Eq. (5.14)

$$K_c = \frac{[CH_3CHO]}{[CH_3CH(O^{\ominus})(OH)][H^{\oplus}]} \times \frac{[B][H^{\oplus}]}{[BH^{\oplus}]} \qquad (5.13)$$

$$K_c = K_d \times K_b \qquad (5.14)$$

where K_d is the equilibrium constant of the reaction $H^{\oplus} + CH_3CHO^{\ominus}(OH) \rightleftharpoons CH_3CHO + H_2O$.

We can now demonstrate that the relationship between k_{obs} and K_b implies the existence of a *relationship between k_c and K_c, the rate and equilibrium constants of the same rate-determining proton transfer reaction*. The experimental Bronsted relationship for general base catalysis can be written as

$$\log k_{obs} = -\beta \log K_b + c \qquad (5.15)$$

The logarithmic form of Eq. (5.11) is

$$\log k_{obs} = \log k_c + \log K_a - \log K_b \qquad (5.16)$$

Combining Eqs. (5.15) and (5.16), we obtain

$$\log k_c = (1 - \beta) \log K_b - \log K_a + c \qquad (5.17)$$

Substituting Eq. (5.14), we obtain

$$\log k_c = (1 - \beta)[\log K_c - \log K_d] - \log K_a + c \qquad (5.18)$$

Since the last three terms of Eq. (5.18) are constants, the equation may be written as

$$\log k_c = \beta' \log K_c + \text{constant} \qquad (5.19)$$

Equation (5.19) shows that k_c and K_c, the *rate* and *equilibrium* constants of the same step, are related to one another logarithmically. This conclusion is independent of the assumed mechanism, with the proviso that the rate and equilibrium constants refer to a rate-determining proton transfer step. In fact, the illustrative mechanism used earlier may be mechanistically incorrect (see below). It was used in order to illustrate a mechanism where one cannot intuitively see any direct relationship between k_c and K_c. However, only in simpler systems is the relationship between rate and equilibrium constants straightforward.

The Brönsted equation has been interpreted in terms of molecular potential energy diagrams.[3] If one assumes that a proton transfer is a three-center reaction, that the only important repulsion terms are those between the two heavy atoms involved in the reaction, and that the proton moves between the two heavy atom centers that remain stationary at a fixed distance, then the potential energy–reaction coordinate diagram of Fig. 5.3 can be drawn. In this diagram, the resonance interaction lowering the transition state and the repulsion energies have been omitted for the sake of clarity. Curve I represents the reactants SH + B, where SH is a substrate and B a basic catalyst, while curve II represents the reaction products $S^\ominus + BH^\oplus$. The activation energy is then E°, and the free-energy change in the reaction ϵ°. If the basic catalyst is modified by the introduction of a substituent producing a weaker base, the change can be represented by the replacement of curve II by II', which has the same shape and position along the reaction coordinate, but is displaced vertically in the direction of higher energy. The changes δE° and $\delta \epsilon^\circ$ are related to one another by

$$\delta E^\circ = \frac{S_1}{S_1 + S_2} \delta \epsilon^\circ = \beta \, \delta \epsilon^\circ \qquad (5.20)$$

where S_1 and S_2 are the slopes of the two curves at the point of intersection. Equation (5.20) is related to the Bronsted catalysis law since the potential

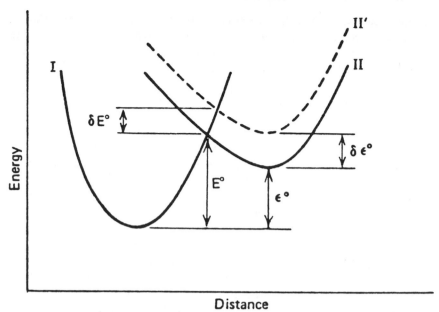

Fig. 5.3. The molecular basis of the Brönsted relation in terms of a potential energy–reaction coordinate diagram. Curve I represents SH + B; curve II represents $S^{\ominus}$ + $BH^{\oplus}$; curve II' represents $S^{\ominus}$ + $BH^{\oplus}$, a slightly weaker base. From R. P. Bell, *The Proton in Chemistry*, Cornell University Press, Ithaca, NY, 1959, p. 169.

energy differences δE° and $\delta \epsilon^{\circ}$ can be related to the observed equilibrium and velocity constants.

Figure 5.3 assumes that a change in catalyst will not change the shape of the curve. This assumption is valid only if the electronic redistribution in the substrate and/or catalyst upon proton transfer is constant. Since much evidence indicates that this is not always so, especially with carbon acids and bases, Fig. 5.3 is of limited validity.

Another limitation of Fig. 5.3 is the assumption of a three-center proton transfer. In organic chemical reactions, many proton transfers are intimately connected with other changes in the substrates. Numerous examples include proton transfers in β-elimination reactions, the removal of a proton alpha to a carbonyl group, and additions to carbonyl groups. In such systems, the maximal free energy of the system as a whole is not necessarily identical with the maximal free energy concerned with the proton transfer. We shall discuss this point in some detail later.

5.3. THE MEANING OF α AND β

It was shown above that the Brönsted catalysis law can be stated in terms of the relation between the *free energy of reaction* and the *free energy of activation* of the catalytically important step.

Initially, Bell suggested that α (β) is always between 0 and 1 and that the values represented a measure of the selectivity of reaction. There are many facts consistent with this view: Brönsted α's of 0.42–0.89 are observed for many carbon acids (Bell[3]). In addition, Brönsted α's of 0–1.0 are seen for oxygen acids (Eigen[4]). In defining the meaning of the terms α and β, the following assumptions were made: (1) The free-energy quantities can be described as linear combinations of the free energies associated with variations such as substituent changes. (2) A transition state has considerable resemblance to the reactant (R) and product (P) (both in composition and structure); therefore, any change in its free energy can be represented as a linear combination of the corresponding changes in the free energies of the reactant and product. These two assumptions lead to

$$\delta G^{\ddagger} = a\ \delta \overline{G}_P^{\circ} + b\ \delta \overline{G}_R^{\circ} \tag{5.21}$$

where δ is an operator representing a substituent change and G is the Gibbs free energy. Going beyond the extrathermodynamic assumptions made earlier, a further assumption was made. Since the reaction coordinate of a transition state lies between that of the reactant and product, one assumes that other properties of the transition state will be intermediate as well. Although the standard free energy of the transition state is at a maximum, it is likely that any changes in its value due to substituent changes will be intermediate between corresponding changes for reactant and product. This assumption allows us to convert Eq. (5.21) to

$$\delta G^{\ddagger} = \alpha\ \delta \overline{G}_P^{\circ} + (1 - \alpha)\ \delta \overline{G}_R^{\circ} \tag{5.22}$$

where $0 < \alpha < 1$. Equation (5.22) may be readily transformed to

$$\delta \Delta G^{\ddagger} = \alpha\ \delta \Delta \overline{G}^{\circ} \tag{5.23}$$

where $\Delta G^{\ddagger} = \overline{G}^{\ddagger} - \overline{G}_R^{\circ}$ and $\Delta \overline{G}^{\circ} = \overline{G}_P^{\circ} - \overline{G}_R^{\circ}$.

These considerations indicate that $\alpha(\beta)$ is a measure of the resemblance of the transition state to the reactant or product. That is, as α increases, the transition state more closely resembles the product than the reactant.

However, although this classical view was held for many years, this view has been hotly contested more recently. There are now many facts inconsistent with the preceding ideas. For example, Brönsted α's which are greater than 1.0 or less than 0 are now known (all of these involve nitroalkanes).[5,6] Some Brönsted α's are constant over a 10^8-fold range in rate constant, whereas the earlier treatment implies that they should vary due to the change in resemblance of the transition state to the reactant or product with change of catalyst. Likewise, the nucleophilic β for pyridines is constant over a range in rate constant of 10^8-fold, and the (anionic) nucleophilic β for 9-substituted fluorenyl anions is constant over a 10^{14} range in rate constants.

After reading the earlier discussion, it should be obvious that the arguments are laden with assumptions. The problem can be summarized in the following way. The Brönsted relation is a differential one as expressed in its basic form

$$\delta \Delta G^{\ddagger} = \beta \delta \Delta \overline{G}^{\circ} \qquad (5.24)$$

and the integrated forms expressed as (5.7) are valid only over a limited range. This means β is only a "local" constant and Brönsted plots can be expected to exhibit curvature over an extended range.[7,8] Furthermore, the assumption implicit in all the preceding derivations that $\Delta G^{\circ \ddagger}$ remains constant within a series cannot be justified. It is likely to fall apart with changes in the substrate as variation near the seat of the reaction will certainly affect the rate of reaction.

The errors inherent in the above assumptions are probably best reflected in Fig. 5.4 (Ref. 9), which represents the decomposition of diazodihexylmethane catalyzed by 19 general acids with acid strengths covering a range of 10 log units. The Brönsted plot of this reaction is very strongly curved and the catalytic constant becomes independent of acid strength with a value of only about 11 M^{-1} sec^{-1} compared with 10^9–10^{10} M^{-1} sec^{-1} for a diffusion-controlled reaction.

Finally, the assumption that the measured values of $\Delta G^{\ddagger}$ and ΔG° refer to

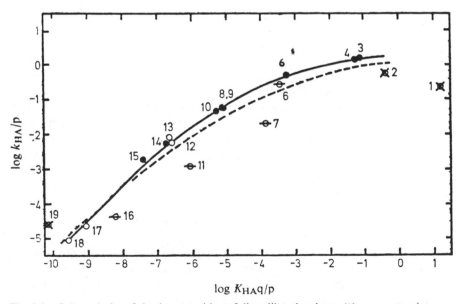

Fig. 5.4. Brönsted plot of the decomposition of diazodihexylmethane. (○) represent points generated by phenols; (⊖) points generated by nitrophenols; (●) points generated by carboxylic acids; (⊗) points generated by other acids. The dashed line is that using the data points for neutral acids. The solid line is generated using only the points representing carboxylic acids and phenols other than nitrophenols (Ref. 9).

Diffusion	$XH + Y$ $\rightleftharpoons$	$XH \parallel Y$	(encounter complex)
Reorganization	$XH \parallel Y$ $\rightleftharpoons$	$XH \cdot Y$	(reaction complex)
Proton transfer	$XH \cdot Y$ $\xrightarrow{\Delta G^{\ddagger}}$	$X \cdot HY$	(reaction complex)
Reorganization	$X \cdot HY$ $\rightleftharpoons$	$X \parallel HY$	(encounter complex)
Diffusion	$X \parallel HX$ $\rightleftharpoons$	$X + HY$	(products)

Fig. 5.5.   The steps involved in a proton transfer (Ref. 10).

the process in which the proton is actually being transferred is probably faulty. It is believed that solvent reorganization can be critical. Current thinking about what was once considered to be a simple reaction, proton transfer, is shown in Fig. 5.5.

Notwithstanding the difficulty in the interpretation of the magnitude of α or β, the relationship between the Brönsted $\alpha(\beta)$ and the pK_a of the catalyst is an empirically useful one. For example, it is possible to distinguish between transition states **5.1** and **5.2** in the glycine (glyNH$_3^{\oplus}$)-base (B) pair-catalyzed enolization of acetone by consideration of the Brönsted β of the process.

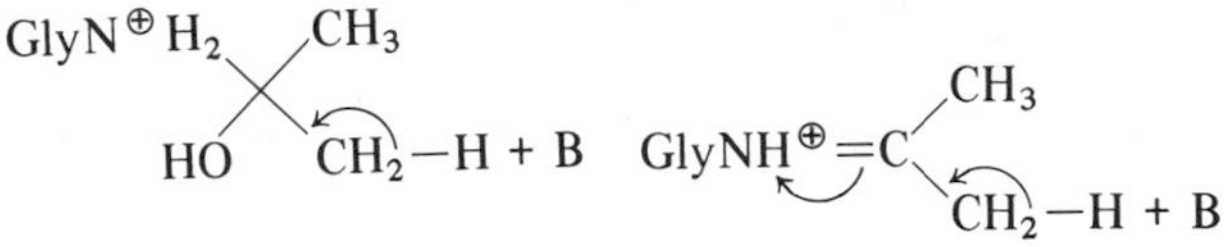

$$\text{GlyN}^{\oplus}\text{H}_2 \diagdown \diagup \text{CH}_3 \qquad\qquad \text{CH}_3 \tag{5.25}$$

$$\text{HO} \qquad \text{CH}_2{-}\text{H} + \text{B} \quad \text{GlyNH}^{\oplus}{=}\text{C} \diagdown \diagup \text{CH}_2{-}\text{H} + \text{B}$$

$$\textbf{5.1} \qquad\qquad\qquad\qquad \textbf{5.2}$$

The experimentally determined β equals 0.4. The β predicted for transition state **5.1** should be very close to one, since the proton removed in transition state **5.1** should have a pK_a much greater than 20, indicating that the transition state shown resembles the product. On the other hand, transition state **5.2** contains a proton that is quite acidic because of its protonated nitrogen atom that can act as an electron sink, and the transition state should resemble the reactant, with a β less than one. Consideration of the Brönsted β therefore clearly eliminates **5.1** and strongly supports **5.2** as the transition state for the reaction.[11]

5.4.   ISOTOPE EFFECTS IN GENERAL CATALYSIS

The kinetic isotope effect, which is defined as the magnitude in the effect of isotopic substitution on the reaction rate, gives important information about the reaction mechanism. Here, isotopes are regarded as substances different only in mass, but the same in all other properties.

The primary kinetic isotope effect stems from the difference of the zero-point level in the vibrational energy of the bond ($\frac{1}{2}h\nu$) due to isotopic substitution, where ν, the zero point energy of vibration, is proportional to $1/(\text{mass})^{1/2}$ and h is Planck's constant. Secondary isotope effects are caused by isotopic

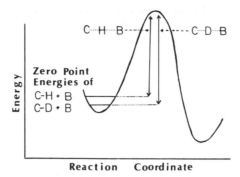

Fig. 5.6. The origin of the primary deuterium isotope effect.

substitution of atoms that do not react. These effects can have their origin in frequency changes in bonds to nonreacting atoms, inductive effects, hypercon- jugation, steric effects, and solvent effects. However, we will focus on the application of primary isotope effects to the elucidation of reaction mechanisms.

Isotopes often used in primary isotope effect studies include $^2H(D)$, $^3H(T)$, ^{13}C, ^{15}N, ^{18}O, and ^{19}F. However, the most widely investigated isotope effect is, of course, the deuterium isotope effect. Consider the manifestation of the primary isotope effect in the case of deuterium. When a hydrogen atom is transferred to a base B, the hydrogen's stretching vibration is lost. The stretching vibration changes to a translation as the hydrogen moves toward the base at the high point in the energy profile of the reaction (see Fig. 5.6). The difference in the energy required to attain the C---H---B transition state is less than that required to attain the C---D---B transition state. Both reach the same energy maximum, but the zero-point energy for the C–D bond is lower and thus the ΔG for the breakage of the C–D bond is higher. Although this analysis does not consider bending vibrations, new stretching vibration in the transition state, or vibra- tional couplings, nonetheless, it deals with all that is normally important in reactions that demonstrate large isotope effects.

The position of the proton in the transition state may be determined by the use of deuterium isotope effects.[12] The picture of proton transfer given earlier, together with the concept that maximal isotope effects will occur when the proton is half transferred,[4] leads to the prediction that when $\alpha(\beta)$ in the Brönsted relation is zero or one, only a small deuterium kinetic isotope effect should be seen, while a maximal kinetic isotope effect should be observed when $\alpha(\beta) = 0.5$. Although such an experiment has not been attempted, an analogous experiment has been carried out by determining the magnitude of the kinetic deuterium isotope effect as a function of ΔpK_a (the difference be- tween the pK_a of the substrate and that of the catalyst). This experiment, in- volving the general base-catalyzed proton abstraction from a series of organic substrates (Fig. 5.7), shows a maximal kinetic isotope effect when ΔpK_a ap- proximates zero. This is in accord with the hypothesis that a maximal kinetic isotope effect should occur when the proton is half-transferred. A similar

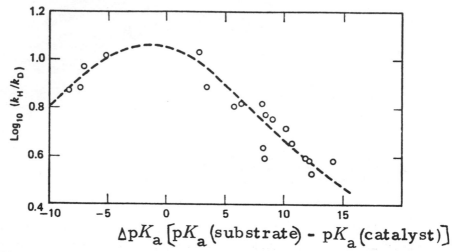

Fig. 5.7.   Relation between kinetic isotope effect and pK difference of the reacting systems in the bromination of organic substrates. From R. P. Bell and D. M. Goodall, *Proc. R. Soc. London,* **A294,** 273 (1966).

maximum is seen in proton abstraction from a series of phenonium ions of varying reactivity in the aromatic hydrogen exchange reaction.

We wish to point out that the preceding discussion paints only a general picture of the use of isotope effects and does not enumerate the problems involved in interpretation of results. However, it is our intention only to make the student aware of such tools.

5.5.   THE RELATION OF GENERAL TO SPECIFIC CATALYSIS

Let us consider the differences between general and specific catalysis, using as an example general acid and specific hydronium ion catalysis. We wish to consider the question of why some reactions are general acid-catalyzed whereas other reactions are specific hydronium ion–catalyzed.

One obvious function of general catalysis is to avoid the formation of high energy intermediates and, at the same time, contribute to the polarization necessary for catalysis. For example, large concentrations of hydronium or hydroxide ion are obviously unstable in neutral solution. However, large concentrations of general acids (or general bases) are stable in neutral solution and can effect, to some extent, the same kind of electronic perturbation that the hydronium or hydroxide ion can.[13]

A relationship between the inherent reactivity of a system and the kind of catalysis observed in the system can be advanced in several ways. Proton transfer must constitute the slow step of a reaction when the rest of the rate processes of the reaction are fast. In general acid catalysis, this phenomenon is seen in slow proton transfers to or from a carbon atom. But when proton transfers involve oxygen or nitrogen atoms rather than carbon atoms, other aspects of the rate processes must be considered. For example, the hydrolysis of acetals usually involves specific hydronium ion catalysis, whereas the hydrolysis of orthoesters involves general acid catalysis. This difference can be rationalized by saying that the rate-determining step in acetal hydrolysis is a more difficult process than in orthoester hydrolysis, since the carbonium ion produced in the latter reaction is more resonance-stabilized than in the former reaction. The reaction in which general acid catalysis is seen may thus be described either as one in which the other rate processes are fast enough so that proton transfer becomes rate-determining, or alternatively, as a facile reaction where a strong acid, hydronium ion, is not necessary and a weak or general acid, suffices.

A somewhat different approach to this problem was taken by Bell, who calculated the appearance of general acid catalysis as a function of the Brönsted α of the reaction.[14] He set up a hypothetical catalysis by acetic acid in a buffer containing 0.1 M acetic acid and 0.1 M acetate ion. Table 5.2 shows the relative contributions to the total velocity made by hydronium ion, water, and acetic acid, assuming three different values of α. When $\alpha = 1.0$, the k_{cat}'s of hydronium ion, water, and acetic acid are 99.8, 5×10^{-12}, and 0.2 respectively; thus the first is a strong catalyst, the second a weak one, and the third a medium catalyst. However, when α is equal to 0.1, the predominant catalyst is water.

Earlier it was pointed out that Brönsted α is a function of the ease of reaction. As seen before, when α is equal to 1.0, the overriding reaction is the hydronium ion-catalyzed reaction. It seems likely that general acid (base) catalysis will only be detectable for intermediate values of $\alpha(\beta)$.

In a multistep reaction consisting of one proton transfer step plus other steps, general catalysis will prevail if the proton transfer step is rate-determining, while specific catalysis will prevail if one of the other steps is rate-determining. Thus one may convert a reaction from specific to general catalysis by speed-

Table 5.2. The Effect of α on the Occurrence of General Acid Catalysis or Specific Hydronium Ion Catalysis[a]

Percent of Catalysis by Various General Acids		
by $H_3O^{\oplus}$	by H_2O	by CH_3COOH
α		
0.1	0.002	98	2
0.5	3.6	0.01	96.4
1.0	99.8	5×10^{-12}	0.2

Source: R. P. Bell, *Acid-Base Catalysis*, Clarendon Press, Oxford. 1941, p. 94.
[a]A hypothetical system containing 0.1 M acetic acid and 0.1 M acetate ion.

ing up those steps not involving proton transfer while in the latter case the reverse would be true. There are a number of avenues for effecting such a transition. One possibility is to increase the concentration of a reagent that is involved in one of the steps not involving proton transfer. For example, in the aldol condensation, an increase in acetaldehyde concentration leads to an increase in the rate of the condensation step so that the proton transfer step becomes rate-determining. Another possibility is to increase the temperature of the reaction. If the activation energy of the steps not involving proton transfer is higher than the activation energy of the proton transfer step, a transition will occur at higher temperature from specific to general catalysis. Entropic factors can also be significant. In enzymatic catalysis, the bond making and breaking steps other than proton transfer may be entropically accelerated to such an extent that proton transfer often becomes rate-determining.[15]

As there is no absolute distinction between specific and general acid catalysis, general catalysis should occur at any time that the concentration of the general acid becomes sufficiently elevated with respect to the concentration of hydronium ion. For the ethylene oxide ring opening reaction of epichlorohydrin with sodium iodide in the presence of acetic acid, the transition from specific hydronium ion to general acid catalysis was observed by increasing the concentration of acid.[13] In the hydrolysis of ethyl orthoformate, specific hydronium ion catalysis is found in aqueous acetic acid buffer, but when the medium is changed to an aqueous dioxane solution containing acetic acid buffer, general acid catalysis by acetic acid occurs. The aqueous dioxane solution was effective since the ratio $[CH_3CO_2H]/[H_3O^\oplus]$ is approximately 1000 times higher in aqueous dioxane than in water (because of the change in pK_a of acetic acid).[16]

5.6. MECHANISMS OF GENERAL ACID AND GENERAL BASE CATALYSIS

General acid–base catalysis, a widespread phenomenon in organic chemistry (see Tables 5.3–5.5) can be represented by the following kinetic schemes. The simplest general acid catalysis involves a rate-determining proton transfer followed by a fast decomposition step.

$$S + HA \underset{\longleftarrow}{\overset{slow}{\rightleftharpoons}} SH^\oplus + A^\ominus \xrightarrow{fast} \text{Products} \qquad (5.26)$$

$$\text{rate} = k[S][HA]$$

Rate-determining proton transfer may also occur after a covalent change involving heavy (heavier than hydrogen) atoms [Eq. (5.27)].

$$S \overset{fast}{\rightleftharpoons} S'$$

$$S' + HA \underset{\longleftarrow}{\overset{slow}{\rightleftharpoons}} S'H^\oplus \xrightarrow{fast} \text{Products} \qquad (5.27)$$

$$\text{rate} = k[S'][HA] = k_{obs}[S][HA]$$

where

$$k_{obs} = kK_{eq}$$

The significant point about each of these equations is that the rate-determining step is solely a proton transfer.

Proton transfer may occur simultaneously with covalent changes involving heavy atoms. Two possibilities exist: (1) The proton transfer and covalent changes occur directly [Eq. (5.28)]; or (2) the rate-determining proton transfer and covalent changes occur only after a preliminary prototropic preequilibrium [Eq. (5.29)].

$$S + HA \overset{slow}{\rightleftarrows} SH^{\oplus} + A^{\ominus} \overset{fast}{\longrightarrow} \text{Products}$$

$$\text{rate} = k[S][HA] \tag{5.28}$$

$$S + HA \overset{fast}{\rightleftarrows} SH^{\oplus} + A^{\ominus} \overset{slow}{\longrightarrow} \text{Products} \tag{5.29}$$

$$\text{rate} = k[SH^{\oplus}][A^{\ominus}] = k_{obs}[S][HA]$$

where

$$k_{obs} = kK_{eq}$$

A second substrate molecule, however, is often involved; if such is the case, prior complexation between two of the three reactants can occur before any of the steps listed in Eqs. (5.26)–(5.29).

The general acid catalyses of Eqs. (5.26)–(5.29) have their counterparts in the general base catalyses of Eqs. (5.30)–(5.33). Equations (5.29) and (5.30) consist solely of rate-determining proton transfers with or without prior covalent change involving heavy atoms.

$$SH + B \overset{slow}{\rightleftarrows} S^{\ominus} + BH^{\oplus} \overset{fast}{\longrightarrow} \text{Products}$$

$$\text{rate} = k[SH][B] \tag{5.30}$$

$$SH \overset{fast}{\rightleftarrows} S'H$$

$$S'H + B \overset{slow}{\rightleftarrows} S'^{\ominus} + BH^{\oplus} \overset{fast}{\longrightarrow} \text{Products} \tag{5.31}$$

$$\text{rate} = k[S'H][B] = k_{obs}[SH][B]$$

where

$$k_{obs} = k K_{eq}$$

Equations (5.32) and (5.33) involve a rate-determining step that includes both proton transfer and a covalent change involving heavy atoms. They differ by the prior prototropic equilibrium in Eq. (5.33)

$$\text{SH} + \text{B} \underset{}{\overset{slow}{\rightleftharpoons}} \text{S'}^{\ominus} + \text{BH}^{\oplus} \overset{fast}{\longrightarrow} \text{Products} \qquad (5.32)$$

$$\text{rate} = k[\text{SH}][\text{B}]$$

$$\text{SH} + \text{B} \underset{}{\overset{fast}{\rightleftharpoons}} \text{S}^{\ominus} + \text{BH}^{\oplus} \overset{slow}{\longrightarrow} \text{Products} \qquad (5.33)$$

$$\text{rate} = k[\text{S}^{\ominus}][\text{BH}^{\oplus}] = k_{obs}[\text{SH}][\text{B}]$$

where

$$k_{obs} = k K_{eq}$$

In Tables 5.3–5.5, two varieties of general catalysis appear: (1) reactions involving proton transfer to or from carbon atoms; and (2) reactions involving proton transfer to or from nitrogen, oxygen, and sulfur atoms. Chapter 2 indicates that the first group can have solely a rate-determining proton transfer, while the second group probably cannot.

Equations (5.26)–(5.29) contain many mechanistic ambiguities. In all general acid mechanisms except Eq. (5.29), the general acid catalyst acts as a proton donor. However, Eq. (5.29) is mechanistically a hydronium ion–general base or nucleophilic combination.

Mechanistic ambiguities in general base catalyses are even more prevalent. All general base mechanisms given except Eq. (5.29) can be interpreted either as general base reactions as shown or, alternatively, as nucleophilic reactions. The sole exception, Eq. (5.29), is mechanistically a general acid–hydroxide ion combination, since the equilibrium $\text{B} + \text{H}_2\text{O} \rightleftharpoons \text{BH}^{\oplus} + \text{OH}^{\ominus}$ leads to the kinetic equivalence of $[\text{B}][\text{H}_2\text{O}]$ and $[\text{BH}^{\oplus}][\text{OH}^{\ominus}]$ in aqueous solution.

Establishing criteria for these pathways is a major problem in analyzing the mechanisms of general acid and general base catalysis. There are essentially three kinds of ambiguities involved:

1. Is the catalyst acting as a proton transfer agent or a nucleophile?
2. In which step of a multistep process is the catalyst acting?
3. What is the position of the proton in the important transition state?

We will discuss these problems in Sections 5.7–5.8.

Table 5.3. General Acid Catalysis Where an Extensive Series Exists

Proton transfer to carbon
   Diazoacetic ester decomposition
   Nitromethane ion + HA
   Nitroethane ion + HA
   Azodicarbonate decomposition
   Acetylacetone enolate ion + HA
   Hydration of p-methoxy-α-methylstyrene
   Cleavage of C–Hg bond
   Hydration of mesityl oxide and crotonaldehyde
   Addition of H_2O to acetylenic thiolesters
   Aromatic hydrogen exchange
Proton transfer from carbon
   Acetone iodination
   Keto-enol transformation
   Bromination of nitromethane
   Bromination of acetoacetic ester
Carbonyl addition and reverse
   Dihydroxyacetone depolymerization
   Dehydration of acetaldehyde hydrate
   Schiff base formation
   Hydrolysis of Schiff bases
   Semicarbazone formation
   Carbonyl compound + anilines
   Carbonyl compound + H_2O, ROH
   Carbonyl compound + thiourea
   Oxime formation
   Bisulfite addition
   Nitrone formation
   Thiol addition to carbonyl
   Phenylhydrazone formation
   Hydrolysis of ortho esters
   Hydrolysis of N-arylglucosamines
   Hydration of 1-benzyl-1,4-dihydronicotinamide
   Triose condensation
   Reactions in ice
Reactions of carboxylic acid and phosphoric acid derivatives
   Esterification of acetic acid by methanol
   Phosphate hydrolysis
   Hydroxylaminolysis of hydroxamic acid
   Hydrolysis of piperazine-2,5-dione
   Esterification
   Amide hydrolysis
   Hydroxylaminolysis of amides
Miscellaneous
   Cleavage of C–B bond
   Cleavage of C–Sn bond
   Hydrolysis of $NaBH_4$
   Nucleophilic aromatic substitution and elimination

Source: Reference 11.

Table 5.4. General Base Catalysis Where an Extensive Series Exists

Proton transfer from carbon
  Bromination of ketones
  Hydrogen exchange of isobutyraldehyde
  Nitramide decomposition (from nitrogen, not carbon atoms)
  Halogenation, isomerization, and deuterium exchange of many organic substances
    containing an acidic hydrogen atom, and the racemization of such substances when
    the acidic hydrogen is on a asymmetric center
  Aldol condensation
  Condensation of glyceraldehyde and dihydroxyacetone
  Transamination
Carbonyl addition
  Hydrolysis of Schiff base
  Cyanohydrin formation
  Semicarbazone formation
  Phenylhydrazone formation
  Solvolysis of nitrostyrene
  Condensation of formaldehyde and tetrahydrofolic acid
Aromatic substitution
  Aromatic nucleophilic substitution
  Aromatic electrophilic substitution
Nucleophilic reactions, mainly of carboxylic acid derivatives
  Solvolysis of tetrabenzyl pyrophosphate
  Ammonolysis of phenyl acetate
  Hydrolysis of phosphoric carbonic anhydride
  Hydration of carbon dioxide
  Hydrolysis of Leuch's anhydride
  Hydrolysis of ethyl trifluorothiolacetate
  Hydrolysis of lactones and amides
  Aminolysis of esters and thiol esters
  Hydrolysis of glycine ester–cobalt complex
  Ethanolysis of ethyl trifluoroacetate
  Hydrolysis of γ-phenyltetronic acid enol esters
  Hydrolysis of phenyl acetates
  Hydrolysis of anhydrides
  Hydroxylaminolysis of esters
  Hydrolysis of acyl cyanides
  Hydrolysis of acetyl fluoride
  Nucleophilic reactions of N-acetylimidazole
  Hydrolysis of 1-acetyl-3-methylimidazolium ion
  Ester hydrolysis
  Aminolysis of phenyl acetates
  Imidazolysis of esters
  Aminolysis of acetyl phosphate
  Aminolysis of imido esters
  Hydrolysis of methyl ethylene phosphate
  Hydrolysis of trifluoroacetanilide
  Hydrolysis of N-methyltrifluoroacetanilide
  Hydrazinolysis of esters

Table 5.4. (*continued*)

Hydrolysis of aspirin anion
S–N acyl transfer
Cleavage of carbon–carbon bonds
Hydrolysis of *N*-acylimidazoles
Hydrolysis of amidines
Miscellaneous
    Cyclization of 4-chlorobutanol
    Hydrolysis of 4(5)-hydroxymethylimidazole
    Silicon–oxygen bond cleavage
    Addition of $HSiCl_3$ to acetylene

Source: Reference 11.

Table 5.5. General Acid–Base Catalysis Where an Extensive Series Exists

Reaction
    Hydrolysis of diisopropyl phosphorofluoridate
    Aminolysis of thiol esters
    Mutarotation of glucose
    Hydration of acetaldehyde
    Enolization of ketones
    Amide hydrogen exchange

Source: Reference 11.

5.7. GENERAL BASE VERSUS NUCLEOPHILIC CATALYSIS

The question of whether a catalyst acts as a proton transfer agent or a nucleophile is a common one. This ambiguity is particularly prevalent in reactions of carboxylic acid derivatives, because of the common occurrence of both general base and nucleophilic catalyses in these reactions. To illustrate the problem and its resolution, we will consider methods used to differentiate general base from nucleophilic catalysis in these reactions.[11]

   1. Catalysis by the leaving group of a reaction is evidence of general base rather than nucleophilic catalysis. The reaction shown in Eq. (5.34) cannot be explained by nucleophilic attack by B on the carboxylic acid derivative since such a reaction could only regenerate the starting material, but not lead to chemical transformation.

$$\underset{\text{RCB}}{\overset{\text{O}}{\overset{\|}{}}} + H_2O \overset{B}{\longrightarrow} \underset{\text{RCOH}}{\overset{\text{O}}{\overset{\|}{}}} + HB \qquad (5.34)$$

Therefore, the function of B must be that of a general base. Examples of catalysis by the leaving group of a reaction include the hydrolysis of acetic anhydride by acetate ion, the hydrolysis and thiolation of *N*-acetylimidazole by imidazole, and the hydrolysis of acetyl fluoride by fluoride ion. This mechanistic criterion is probably the most rigorous one for distinguishing between general base and nucleophilic catalysis.

2.   Catalysis by a second mole of an attacking nucleophile is evidence that general base catalysis occurs. The first mole of attacking nucleophile may be involved either as a nucleophilic catalyst or a nucleophile; this can be ascertained by product analysis. A second mole of nucleophile cannot act either as nucleophilic catalyst or nucleophile, but can act as general base catalyst.

However, there are several complications to this interpretation. Kinetic dependence on a second mole of nucleophile is usually not seen until very high concentrations of nucleophile are introduced into the system, leading to a changing medium and all the attendant errors that may obscure the true kinetics.

In addition, there may be preequilibria involving the nucleophile.

$$BH + OH^{\ominus} \rightleftharpoons B^{\ominus} + H_2O$$

$$BH + BH \rightleftharpoons B^{\ominus} + BH_2^{\oplus} \qquad (5.35)$$

Such reactions suggest the possibility that $B^{\ominus}$ rather than BH is the reactive species or, alternatively, that $B^{\ominus}$ and $BH_2^{\oplus}$ are the true reactants rather than 2 moles of BH. The combination of ions can be ruled out if the calculated rate constant for the reaction of the substrate with $B^{\ominus}$ and $BH_2^{\oplus}$ is larger than that of a diffusion-controlled process.

3.   Observation of a transient intermediate that can be identified as the intermediate of nucleophilic catalysis by chemical characterization and adherence of the kinetics of the overall process to a nucleophilic catalysis is proof that the catalyst is acting as a nucleophile rather than a general base. The observation may be a direct observation of the appearance and disappearance of the transient, or it may be a trapping experiment in which the reaction mixture is treated with a reagent to trap the hypothetical intermediate as an isolable species. For example, in the imidazole-catalyzed hydrolysis of p-nitrophenyl acetate, the formation and decomposition of N-acetylimidazole can be observed spectrophotometrically at 243 nm, the λ_{max} of this species, yielding data that rigorously fit a two-step process of nucleophilic catalysis.[17]

Conversely, failure to detect an intermediate that should be observable by independent evidence (either direct observation or the results of trapping experiments) can be used as evidence for general base catalysis. For example, the acetate ion–catalyzed hydrolysis of p-chlorophenyl acetate must proceed via general base catalysis since the presumed intermediate of nucleophilic catalysis, acetic anhydride, is not trapped by aniline, a proven trapping agent for this species.

4.   A Brönsted plot for a series of nucleophiles or bases can be characteristic of either a general base-catalyzed reaction or a nucleophile-catalyzed reaction. A Brönsted plot for a family of general bases follows, with few ex-

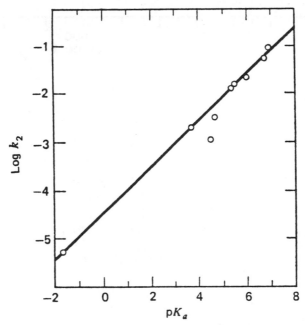

Fig. 5.8. Brönsted plot of the catalytic constants for general base-catalyzed hydrolysis of ethyl dichloroacetate. No statistical corrections have been made. From W. P. Jencks and J. Carriuolo, *J. Am. Chem. Soc.*, **83**, 1747 (1961). Copyright © 1961 by the American Chemical Society. Reproduced by permission of the copyright owner.

ceptions, a single line, independent of the structural characteristics of the bases making up the plot. On the other hand, a Brönsted-type plot of a series of nucleophiles of differing steric requirements and nucleophilic atoms can show considerable deviations from a single line. Plots of these two types are shown in Fig. 5.8, a Brönsted plot for true general base catalysis, the hydrolysis of ethyl dichloroacetate, and in Fig. 5.9, a Brönsted-type plot for the nucleophilic reactions of *p*-nitrophenyl acetate, only some of which are catalytic.[†] To take a particular example, pyridine is equivalent to acetate ion in basicity and falls at the same point on a Brönsted plot for general base catalysis. However, pyridine is at least a thousand times more reactive as a nucleophile than acetate ion, a difference common to nitrogen and oxyanion nucleophiles of the same basicity, and on a Brönsted-type plot of nucleophilic reactions, there is a wide discrepancy in these points. Another pair of nitrogen and oxyanion nucleophiles of identical basicity, imidazole and monohydrogen phosphate ion, also shows this behavior in general base and nucleophilic catalyses. Thus general base catalysis is characterized by Brönsted plots such as that in Fig. 5.8 showing identical catalytic rate constants for bases of identical strength. Conversely, nucleophilic catalysis is characterized by a Brönsted-type plot such as Fig. 5.9

[†]The catalytic reactions show scattering no different from the noncatalytic reactions.

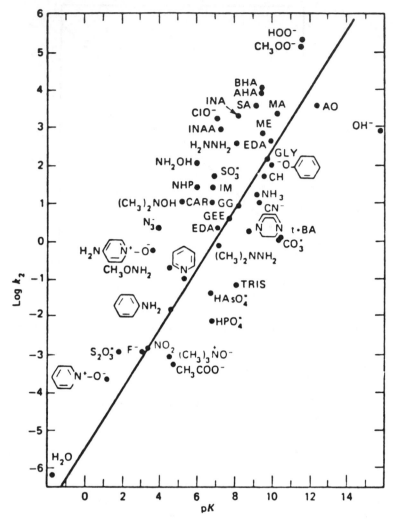

Fig. 5.9. Rates of nucleophilic reactions with *p*-nitrophenyl acetate in aqueous solution at 25° C plotted against the basicity of the attacking reagent. Abbreviations: AHA, acetohydroxamic acid; AO, acetoxime; BHA, *n*-butyryl-hydroxamic acid; CAR, carnosine; CH, chloral hydrate at 30° C; EDA, ethylenediamine; GEE, glycine ethyl ester; GG, glycylglycine; GLY, glycine; IM, imidazole; INA, isonitrosoacetone; INAA, isonitrosoacetylacetone; MA, sodium mercaptoacetate; ME, mercaptoethanol; NHP, *N*-hydroxyphthalimide; SA, salicylaldoxime; *t*-BA, *t*-butylamine; TRIS, tris-(hydroxymethyl)aminomethane. No statistical corrections have been made. From W. P. Jercks and J. Carriuolo, *J. Am. Chem. Soc.*, **82**, 1779 (1960). Copyright © 1960 by the American Chemical Society. Reproduced by permission of the copyright owner.

and by large differences in catalytic rate constants for (N and O) pairs of catalysts of equal basicity.[†]

The slope of a Brönsted plot for nucleophilic catalysis is usually higher than that for general base catalysis.[14] In the examples given here, the reaction of p-nitrophenyl acetate with nucleophiles has a slope of 0.68, while the hydrolysis of ethyl dichloroacetate with general bases has a slope of 0.3. The Brönsted plot for general base catalysis, furthermore, cannot have a slope greater than one (see, however, Section 5.3). In a Brönsted-type plot for nucleophilic catalysis, however, no such dictum applies.

5. Steric hindrance is of less importance in general base catalysis than in nucleophilic catalysis, a corollary of the arguments concerning Brönsted plots.[14] A priori this should be so, since the former involves proton abstraction while the latter involves attack at a carbon atom, the steric requirements of the former being less than the latter. Steric hindrance of nucleophilic catalysis is particularly evident in the pyridine-catalyzed hydrolysis of acetic anhydride; compounds such as 2-picoline (2-methylpyridine), 2,6-lutidine, (2,6-dimethylpyridine) and 2,4,6-collidine (2,4,6-trimethylpyridine) do not catalyze hydrolysis at all, whereas pyridine does so efficiently. Moreover, 2,6-lutidine is a general base catalyst in the solvolysis of a pyrophosphate ester.

6. A nucleophile exhibiting an α-effect, that is, enhanced nucleophilicity due to the juxtaposition of two atoms containing unshared electron pairs, will be an exceptionally good nucleophilic catalyst, but only an ordinary general base catalyst, another corollary of the arguments concerning Brönsted plots. For example, hydroxide ion is a stronger base than hydroperoxide ion, but the latter is a stronger nucleophile by as much as 10^4 in certain nucleophilic displacement reactions.[18] Thus a comparison of the reactivity of these two substances with a given substrate can be used to distinguish between nucleophilic and general base reactions, including catalytic processes. This differentiation is based on the fact that the α-effect is not seen in proton transfer reactions.

7. The common ion effect can be used as a method of distinguishing general basic and nucleophilic catalyses. In the pyridine-catalyzed hydrolysis of acetic anhydride,[19] for example [see Eq. (5.36)], the addition of acetate ion significantly decelerates reaction, indicating nucleophilic catalysis by pyridine. Such decelerations require the reversible formation of an intermediate.

$$(5.36)$$

[†]The thousandfold difference in nucleophilicity between pyridine and acetate ion may in fact be greater since, as mentioned above, acetate ion serves as a general base catalyst toward p-nitrophenyl acetate, whereas pyridine serves as a nucleophilic catalyst. This change in mechanism is contrary to the tacit assumption of a constant mechanism made above.

the subsequent hydrolysis of which is rate-determining. Likewise, the imidazole-catalyzed hydrolysis of trifluoroethyl acetate is inhibited by added trifluoro-ethanol, but the acetate ion-catalyzed hydrolysis of acetylimidazole is not inhibited by imidazole. A corollary of this criterion is the incorporation of a radioactive (or other) tracer into a reactant using a common ion. This result, accomplished in the pyridine-catalyzed hydrolysis of acetic anhydride, also points to nucleophilic rather than general base catalysis.

8.   Deuterium oxide solvent isotope effects can, with caution, be used to distinguish general base and nucleophilic catalyses. For example, the imidazole-catalyzed hydrolysis of p-nitrophenyl acetate has a solvent isotope effect k_{H_2O}/k_{D_2O} of 1.0.[11] Conversely, the imidazole-catalyzed hydrolysis of ethyl dichloroacetate has a solvent isotope effect of 3.0. The previous discussion presents other criteria indicating that imidazole acts as a nucleophilic catalyst in the former reaction, but as a general base catalyst in the latter. The deuterium oxide solvent isotope effects are consistent here with the other criteria. No other isotope effect would be expected in a nucleophilic reaction, but a sizable effect would be expected in a general base reaction involving a rate-determining proton transfer because of the difference in zero-point energy of vibration of proton and deuteron bonds (see Section 5.4). It must be emphasized, however, that differences in solvation between the ground and transition states can result in sizable secondary solvent isotope effects in reactions not involving a rate-determining proton transfer. Furthermore, some general catalyses involving proton transfer show little or no primary isotope effect (see below). Therefore, this criterion must be used in conjunction with other arguments.

9.   Product analysis can sometimes be helpful in distinguishing between general base and nucleophilic catalysis. A nucleophilic catalysis by imidazole yields an unstable quaternary ammonium ion as hydrolytic intermediate. However, a nucleophilic analog of imidazole, aniline, in a comparable nucleophilic reaction will give an anilide as a stable product rather than an intermediate. Thus the latter nucleophile can be used in conjunction with the former catalyst to verify the catalytic mechanism. Conversely, if imidazole acts as a general base yielding a hydrolytic product, aniline should likewise yield a hydrolytic product. The limitation in these arguments is that one must ascertain that both imidazole and aniline react by the same mechanism, a requirement difficult to fulfill.

5.8.   SPECIFICATION OF GENERAL CATALYSIS IN STEPWISE PROCESSES

Another mechanistic problem in a multistep reaction is that of specifying the step in which general catalysis occurs. There are many multistep reactions subject to general acid–base catalysis. The more prominent ones include:

1.   Additions to carbonyl and imine groups.[20]

2. Nucleophilic reactions of carboxylic acid derivatives.[21]
3. Aromatic substitution reactions.[22]

While the experimental basis for a multistep pathway will not be explicitly given unless it directly touches on the catalysis, we will attempt to provide mechanistic arguments as to the step catalyzed in these reactions.

In illustrating the principles discussed earlier, addition to carbonyls is an excellent example to consider since they represent such a large fraction of the processes of biological interest. However, we are not suggesting that the results of the analysis below apply to all carbonyl additions but rather that the analysis itself is typical of mechanistic arguments. Consider the reaction of a semicarbazide with a carbonyl compound.

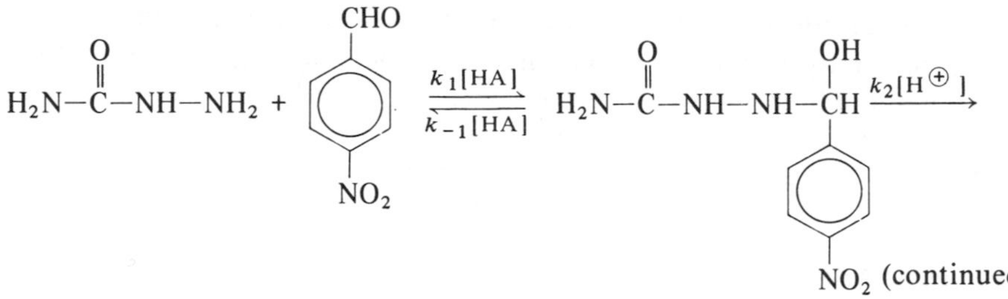

$$H_2N-\overset{O}{\overset{\|}{C}}-NH-NH_2 + \underset{NO_2}{\overset{CHO}{\bigcirc}} \underset{k_{-1}[HA]}{\overset{k_1[HA]}{\rightleftharpoons}} H_2N-\overset{O}{\overset{\|}{C}}-NH-NH-\underset{\underset{NO_2}{\bigcirc}}{\overset{OH}{\overset{|}{CH}}} \xrightarrow{k_2[H^{\oplus}]} \text{(continued)}$$

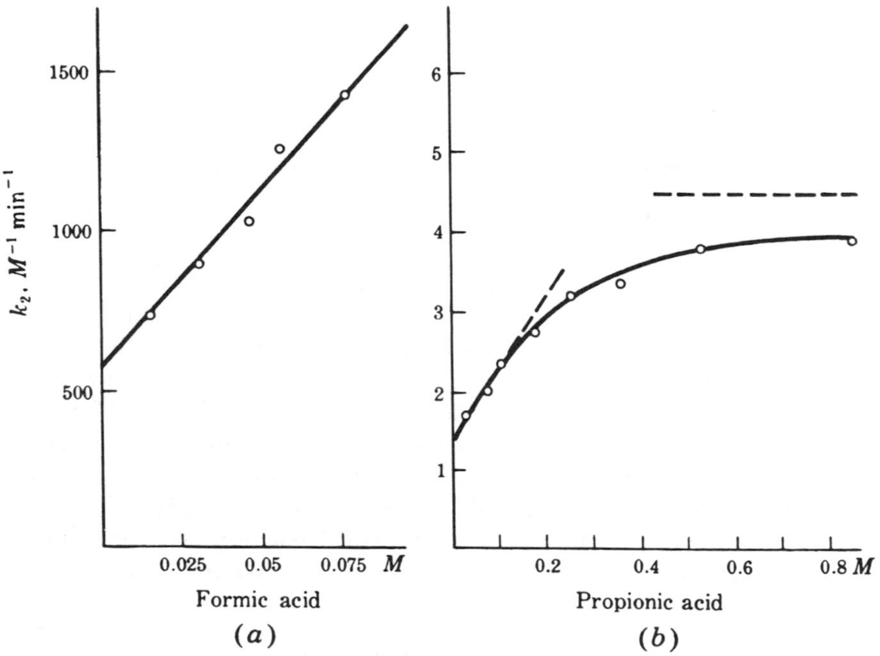

Fig. 5.10. (a) General acid catalysis by formic acid of p-nitrobenzaldehyde semicarbazone formation at pH 3.27. (b) General acid catalysis by propionic acid of acetophenone semicarbazone formation at pH 4.10. The horizontal dashed line is the rate of the carbinolamine dehydration step at this pH (Ref. 23).

$$\overset{\text{O}}{\overset{\|}{H_2N\ \ C-NHNH}}=C\overset{H}{\diagdown} \quad + H_2O$$

(with nitrobenzene ring, NO_2)

When the reaction is carried out at pH 3.27, the attack of the amine on the carbonyl compound is rate-determining and general acid-catalyzed. This is shown nicely in Fig. 5.10. When the pH is held constant, the reaction rate increases linearly as the formic acid concentration increases. However, with propionic acid at a higher pH of 4.10, the rate first increases with increasing acid but as one continues to add acid, the rate then levels off (Fig. 5.10).[23]

The reason for this seemingly peculiar behavior is that the catalyst increases the rate of the first step so much it becomes even faster than the second step. The overall rate then approaches the rate of dehydration. This "changeover in mechanism" is not an uncommon phenomenon.

Symmetrical Stepwise Processes

The most important example of symmetrical catalytic partitioning of a tetrahedral intermediate is exemplified by

$$\overset{\text{O}}{\overset{\|}{RCX}} + YH \underset{BH^{\oplus}}{\overset{B}{\rightleftharpoons}} \overset{O^{\ominus}}{\underset{Y}{\overset{|}{RC-X}}} \underset{B}{\overset{BH^{\oplus}}{\rightleftharpoons}} \overset{\text{O}}{\overset{\|}{RCY}} + HX, \qquad (5.37)$$

the general base-catalyzed ethanolysis of ethyl trifluoroacetate. This symmetrical reaction must show symmetrical partitioning of its tetrahedral intermediate both catalytically and otherwise. From the symmetry of the reaction together with the requirement that the microscopic reverse of general acid catalysis must be general base catalysis Eq. (5.37) can be written. Other reactions approaching this symmetrical situation should conform to this scheme also. One of these is the general base-catalyzed hydrolysis of ethyl trifluoroacetate, which shows a pH-independent kinetic profile from pH 1–5.

An interesting property of such symmetrical reactions is that although both the formation and decomposition of the tetrahedral intermediate are catalyzed, the overall reaction is only first order in catalyst. This conclusion follows from the fact that only one of the steps is rate-limiting (otherwise an inequality of rate constants occurs). If the reaction proceeded with the same catalysis without an intermediate, the reaction would be second order in catalyst, since then both the addition and decomposition processes would occur simultaneously. First-order dependence of a reaction on a general base, however, cannot serve as a criterion for a stepwise rather than a one-step process involving a single general base. Both reactions show the same kinetic behavior.[22]

5.9.   ENFORCED GENERAL ACID–BASE CATALYSIS

The most common mechanism of catalysis of chemical or biochemical reactions involves one or more proton transfers. The same sort of action by enzymes takes place, involving facilitation by general acids or general bases.

The sudden large change in the pK_a of reacting groups gives the impetus for general acid–base catalysis of carbonyl addition and other like reactions. There is an increase in acidity of the amine of at least 22 pK_a units, and an increase in basicity of 13 pK_a units resulting in the general catalysis of carbonyl addition and related reactions involving the formation of unstable intermediates and transition states in which catalysis can accelerate either by trapping of intermediates or by stabilizing or bypassing the transition states.

$$
\begin{array}{cccc}
pK_a \sim 30 & pK_a \sim -4 & pK_a \sim 8 & pK_a \sim 9 \\
\end{array}
$$

$$
\underset{\underset{H}{|}}{\overset{\overset{H}{|}}{R-N}} + \underset{}{\overset{}{\diagup}}C{=}O \rightleftharpoons R\overset{\oplus}{\underset{\underset{H}{|}}{\overset{\overset{H}{|}}{N}}}-\underset{|}{\overset{|}{C}}-O^{\ominus} \tag{5.38}
$$

It has not been clear why some reactions of this kind are subject to general acid–base catalysis while others are not or what the driving force and maximum rate advantage for such catalyses are. Some reactions in which catalysis *must* occur because of the properties of unstable intermediates have been reviewed.[24] These reactions provide partial answers to the preceding questions and some insight into the mechanism of general acid–base catalysis.

Class (e-s) and (n-s) Reactions

General acid–base catalysis of these reactions can be represented by class **e** reactions [Eq. (5.39)], in which the catalyst donates a proton to the electrophilic reagent in one direction and removes it in the reverse direction, or class **n** reactions [Eq. (5.40)], in which the catalyst facilitates proton transfer from or to

$$
N^{\ominus} + \underset{}{\overset{}{\diagup}}C{=}X + HA \rightleftharpoons \overset{(z+1)}{N}\underset{|}{\overset{|}{-}}\underset{|}{\overset{|}{C}}-X-H + A^{\ominus} \tag{5.39}
$$

$$B + H-N + \;\; {\overset{\diagdown}{\diagup}}C{=}X \overset{\textcircled{2}}{\rightleftharpoons} BH^{\oplus} + N-\overset{|}{\underset{|}{C}}-X\underset{\bigcirc}{(z-1)} \tag{5.40}$$

the nucleophilic reagent H—N. In class **e-s** and *n-s* reactions (**s** refers to a *slow* proton transfer), the pK_a of the initially formed tetrahedral addition intermediate is such that the intermediate is not immediately trapped by proton transfer to or from the solvent; that is, in **e-s** reactions, the pK_a of $N\textcircled{\scriptsize z+1}$—$R_2C$—XH is below 14, and in **n-s** reactions, the pK_a of $NH\textcircled{\scriptsize z+1}$—$R_2C$—X is above 0.

TRAPPING BY SIMPLE PROTON TRANSFER

When a nucleophile is expelled faster than the initially formed tetrahedral addition compound is trapped by solvent-mediated proton transfer there is catalysis by general acids or bases. Such reactions, in which proton transfer is required to form a stable product, are therefore subject to *enforced* general acid–base catalysis. The addition of thiol anions to the carbonyl group [Eq. (5.41)] as an example of a class **e-s** reaction. The rate at which the anionic

$$RS^{\ominus} + {\overset{\diagdown}{\diagup}}C{=}O \underset{k_{-1}}{\overset{k_1}{\rightleftharpoons}} RS-\overset{|}{\underset{|}{C}}-O^{\ominus} \underset{k_{-h}}{\overset{k_h(HOH)}{\rightleftharpoons}} RS-\overset{|}{\underset{|}{C}}-OH + OH^{\ominus} \tag{5.41}$$
$$T^{\ominus}$$

tetrahedral addition intermediate, $T^{\ominus}$ is trapped by solvent mediated proton transfer depends on its basicity and is given by $k_h = k_{-h}K_w/K_a$, in which K_a is the acid dissociation constant of the hemithioacetal and k_{-h} is the diffusion-controlled abstraction of a proton by hydroxide ion (approximately $10^{10} \, M^{-1}$ sec^{-1}. The addition of the weakly basic methyl mercaptoacetate anion to acetaldehyde is subject to general acid catalysis because the initially formed unstable tetrahedral intermediate breaks down to reactants at a rate ($k_{-1} \simeq 2 \times 10^8$ sec^{-1}) that is comparable to the rate at which it is trapped by proton abstraction from water ($k_h \simeq 2 \times 10^8$ sec^{-1} determined by its pK_a of 12.4). Therefore, diffusion-controlled protonation of $T^{\ominus}$ upon encounter with a low concentration of a buffer acid provides an additional route to product by trapping the anionic intermediate and thereby increases the observed rate. The more basic ethanethiol anion is expelled more slowly ($k_{-1} = 5 \times 10^8$ sec^{-1}), so that every time the anionic intermediate is formed it abstracts a proton from water faster than it reverts to reactants ($k_h = 7 \times 10^8$ sec^{-1}) and no catalysis is observed.

If the attacking nucleophile contains a proton that becomes acidic when the addition compound is formed, the solvent-mediated proton transfer is likely to occur through a "proton switch" mechanism with a rate constant k_s of approximately 10^6–10^8 sec^{-1}. Thus the zwitterionic intermediate, $T^{\oplus}$, that is formed from the attack of methoxyamine on *p*-chlorobenzaldehyde [Eq. (5.42)] abstracts a proton from solvent slowly ($k_h \simeq 3 \times 10^4$ sec^{-1}) because of its relatively low basicity, but is instead trapped by a proton switch, probably through

$$HN + \underset{\substack{\diagup \diagdown}}{\overset{O}{\underset{\|}{C}}} \underset{k_{-1}}{\overset{k_1}{\rightleftharpoons}} \underset{T^{\oplus}}{\overset{H}{\underset{-N-C-}{\overset{\oplus}{\underset{|}{|}}\,\overset{O^{\ominus}}{\underset{|}{|}}}}} \overset{\substack{-k_s \\ \rightleftharpoons \\ k_{-s}}}{\longrightarrow} \underset{\substack{k_A[HA] \\ \swarrow \nwarrow \\ k_{-A}[A^{\ominus}]}}{} \qquad \pm H^{\oplus} \quad (5.42)$$

$$\overset{HO}{\underset{>N-C-}{\overset{|}{\underset{|}{}}}}$$

$$\overset{H\ OH}{\underset{-N-C-}{\overset{\oplus}{\underset{|}{|}}\,\overset{|}{\underset{|}{}}}}$$

solvent molecules, with a rate constant of $k_s = 6 \times 10^6$ sec^{-1}. Since the intermediate reverts to reactants faster than this, with $k_{-1} = 3 \times 10^8$ sec^{-1}, protonation by buffer acids gives catalysis by trapping. The more stable intermediate that is formed from trimethylamine, a more basic amine, and formaldehyde, a more reactive aldehyde, breaks down to reactants more slowly ($k_{-1} = 3.4 \times 10^3$ sec^{-1}) and shows no catalysis by trapping because every molecule of intermediate formed abstracts a proton from water.

A group of reactions for which the observed catalysis has been attributed to trapping by an **e-s** mechanism or by the corresponding **n-s** mechanism, in which the addition intermediate is trapped by proton removal after encounter with a buffer base, is listed in Table 5.6. An important characteristic of such reactions is

Table 5.6. Examples of Reactions for Which Catalysis Has Been Attributed to Trapping

Reactants	Reactions
RNH_2 and	
Aldehydes	Carbinolamine formation
Imines	Transimination
Cyanic acid	Urea synthesis
Esters	Ester aminolysis
Thiol esters	Ester aminolysis
Acetylimidazole and acetyltriazole	Amide aminolysis
Formamidine	Methenyl–THFA model
Trinitrobenzenes	Nucleophilic aromatic substitution
Triaryl carbonium ions	Addition to carbon
$RS^{\ominus}$ and aldehydes	Hemithioacetal formation
ROH and thiol esters	Ester alcoholysis
ROH and amides	Amide alcoholysis
$C^{\ominus}$ and aldehydes	Aldol condensation

Source: Refrence 24, with permission.

that their properties, such as structure–reactivity correlations and isotope effects, depend only on the equilibrium for formation of the tetrahedral addition intermediate and the rate constant for proton transfer, not the rate of attack of the nucleophile. Examples of the experimental manifestations of this catalysis follow.

The Brönsted slopes α and β for catalysis by relatively strong acids and bases are close to zero because trapping of the intermediate is diffusion-limited and independent of the pK_a of the catalyst when the proton transfer is strongly favorable energetically. This is seen in the catalysis of methoxyamine addition to p-methoxybenzaldehyde by amines of pK_a 7 (see Fig. 6.1). As the pK_a of the catalyzing acid is increased so that the proton transfer becomes energetically unfavorable, the slope of the Brönsted plot as it approaches a limiting value of $\alpha = 1.0$ following an "Eigen curve" (Fig. 2.2), for simple proton transfer reactions; here, the rate of the proton-transfer step approaches the diffusion-controlled limit in the reverse sense. The breakpoint in the curve ($\Delta pK_a = 0$) occurs when the pK_a of the catalyst is equal to that of the protonated intermediate and the observed break, at $pK_a = 8.6$, is close to the estimated pK_a of the intermediate of 9.1. Both the proton and the hydroxide ion exhibit positive deviations of 10–50-fold from such curves because of their faster rates of diffusion-limited proton transfer. In spite of its simplicity, the demonstration that a Brönsted plot follows an Eigen curve is not a trivial matter, largely because of differences in the behavior of different catalysts.

As the buffer concentration increases, the rate of addition of methyl mercaptoacetate to acetaldehyde [Eq. (5.41)] first increases, but then levels off as all of the molecules of the anionic addition intermediate are trapped by encounter with HA and the observed rate is limited by the rate of the uncatalyzed addition step k_1. Such a change in rate-limiting step will generally be detectable if the rate constant k_{-1} is in the region of $10^9 \ sec^{-1}$ or less; it provides unequivocal evidence for a two-step reaction and an intermediate if it can be shown that the leveling off is not caused by association of the catalyst or by salt or solvent effects. If the rate constant for the diffusion-controlled trapping step can be estimated, the observation of a change in rate-limiting step provides a "clock" that permits the calculation of k_{-1} and the equilibrium constant $K_1 = k_1/k_{-1}$. For example, when the change in rate-limiting step and leveling off is half-complete, the intermediate breaks down to reactants and products at equal rates; i.e., $k_{-1} = k_{HA}[HA]$.

The rate constants for hydroxide ion–catalyzed expulsion of acidic thiols from hemithioacetals [Eq. (5.41), right to left] are 0.8–$1.0 \times 10^{10} \ M^{-1} \ sec^{-1}$, in the range calculated for a diffusion-limited reaction. The rate constant estimated from equilibrium and rate constants for the addition of methyl mercaptoacetate to acetaldehyde and for catalysis of other carbonyl addition reactions are also in the range of diffusion-limited trapping. The rate of urea formation from cyanic acid exhibits general acid and base catalysis with a dependence on the basicity of the attacking amine for weakly basic amines ($\beta_{nuc} = 0.8$), but no buffer catalysis for strongly basic amines ($\beta_{nuc} = 0.3$). Weakly basic amines are expelled rapidly from the zwitterionic tetrahedral addition intermediate $T^{\oplus}$ to give reactants ($k_{-1} > k_s$), so that the rate is increased by buffer catalyzed trapping, but strongly basic amines are expelled more slowly, so that the intermediate yields products ($k_{-1} < k_s$) and the rate-limiting step changes to the uncatalyzed formation of the intermediate.

The values of the rate constants for catalysis by strong acids and bases found for urea formation from cyanic acid and aniline are similar to those that would be expected if both types of catalysis involved diffusion-limiting encounter of the catalyst with the same zwitterionic tetrahedral addition intermediate $T^{\oplus}$. Bicarbonate ion, which can act as a bifunctional acid–base catalyst, shows a thirty-fold positive deviation from the Eigen curve for e-s catalysis by monofunctional general acids of the intramolecular aminolysis of S-acetyl-mercaptoethylamine. Such bifunctional catalysts can trap a dipolar intermediate through a proton-switch mechanism [Eq. (5.43)], which gives bifunctional an advantage over monofunctional catalysts.

$$(5.43)$$

The kinetics of this reaction requires that it proceed through at least three sequential steps. Since only two can represent formation and breakdown of a tetrahedral addition intermediate, one must represent a proton transfer. The

$$(5.44)$$

rate-limiting step for diffusion-limited acid catalysis of thiol ester aminolysis $[k_a - k_{-a}$, Eq. (5.44)] corresponds to a product-determining step when 2-methyl-Δ^2-thiazoline (**5.3**) undergoes hydrolysis through the same tetrahedral addition intermediates to thiol ester and amide. This step shows a decrease with increasing viscosity in 0–60% glycerol that is expected for a diffusion-limited reaction.

Similarly, hydrolysis of phenyl acetimidates (**5.4**) generates the same tetrahedral addition intermediates that are formed in the acid-catalyzed aminolysis of

$$CH_3-C\underset{NR}{\overset{OAr}{<}}$$

5.4

phenyl esters. Eigen-type Brönsted plots are observed for general acid catalysis of the partitioning of these intermediates to give ester and amine. This suggests that the product-limiting step of imidate hydrolysis and the rate-limiting step of general acid catalyzed ester aminolysis is a stepwise proton transfer. The partitioning of the intermediate in the absence of catalyst is essentially unchanged by electron-withdrawing substituents on the phenol, as is expected if proton transfer and not phenolate expulsion controls the partitioning. Solvent deuterium oxide isotope effects on these reactions are usually determined by secondary isotope effects on any pre-equilibrium steps and a small effect on the rate of diffusion. The small isotope effect $(k_{H_2O}/k_{D_2O}) = 1.25$ for the hydroxide-ion-catalyzed breakdown of the hemithioacetal of acetaldehyde and thioacetic acid is in the expected range for a diffusion-limited reaction.

THE PREASSOCIATION MECHANISM

When the breakdown of the tetrahedral intermediate is faster than the separation of the intermediate and catalyst, the reaction *must* proceed through a preassociation mechanism.

HYDROGEN BONDING

When the lifetime of an intermediate is sufficiently short that the reaction proceeds through a preassociation mechanism with the catalyst in the correct position for subsequent proton transfer, an **e-s** reaction *must* exhibit catalysis by hydrogen bonding to HA with $\alpha > 0$ when the carbonyl oxygen atom in the transition state is basic enough to perturb the H–A bond.

One can then ask whether buffer catalysis is observed *only* for reactions in which the brief lifetime of the intermediate provides an additional advantage.

The answer is no, because catalysis has been observed for the breakdown of the bisulfite addition compound of p-methoxybenzaldehyde, for which proton transfer is faster than breakdown of the intermediate dianion. The catalytic constants are correlated with the equilibrium constants for association of sulfite dianion with acids and cations. This may mean that the catalysis involves stabilization by hydrogen bonding to the dianionic transition state. A possible mechanism is shown in Eq. (5.45).

$$\begin{array}{c} | \\ -C-OH \\ | \\ SO_3^{\ominus} \end{array} \rightleftharpoons \left[\begin{array}{c} B \\ \overset{|}{\underset{\underset{\delta}{|}{O}}{\overset{\delta^-}{C}}} \cdots H \\ O-S-O \cdots H \\ \overset{\delta^-}{O} \end{array} \overset{\oplus}{\underset{\ominus}{}} O \cdot HB \right]^{\ddagger} \overset{B}{\rightleftharpoons} \begin{array}{c} \diagdown \\ C=O + HSO_3^{\ominus} \\ \diagup \end{array} \qquad (5.45)$$

CONCERTED CATALYSIS

If the lifetime of an intermediate in a catalyzed reaction becomes shorter than that for a preassociation mechanism, the "intermediate" no longer exists and the reaction *must* proceed through a concerted mechanism.

Proton Transfers to or from Carbon.   A proton may be transferred from a general acid to a carbanion, an olefin, or an aromatic compound in a rate-determining process. Some of these reactions are shown in Table 5.1. Since general base catalysis is the microscopic reverse of general acid catalysis, reversals of the aforementioned reactions show general base catalysis. Examples include the removal of a proton from ketones, esters, aldehydes, and nitro compounds.

The bromination of acetone, first investigated by Lapworth in 1904,[31] has been of special importance in the theories of physical organic chemistry. The fact that the base-catalyzed bromination is first-order in acetone, first-order in base, but zero-order in bromine led Lapworth to the idea of a stepwise process. In this reaction, bromine appears not in the rate-determining step but rather in the fast product-determining step.

$$CH_3COCH_3 \underset{k_{-1}[BH^{\oplus}]}{\overset{k_1[B]}{\rightleftharpoons}} CH_3COCH_2^{\ominus} \overset{k_2[Br_2]}{\longrightarrow} \text{Product} \qquad (5.46)$$

The rate constant expression for this reaction is

$$k_{obs} = \frac{k_1 k_2 [Br_2][B]}{k_{-1}[BH^{\oplus}] + k_2[Br_2]} \qquad (5.47)$$

When $k_2[Br_2] > k_{-1}[BH^{\oplus}]$, $k_{obs} = k_1[B]$. At high halogen concentration, this formulation predicts that the rate of bromination should equal the rate of iodination, which it does. This scheme also predicts that halogenation, deuterium exchange, and racemization of ketones should have identical rates, which has likewise been shown to be true. These data mean that the rate-determining step of base-catalyzed halogenation must involve a proton transfer; in other words, this is a general base-catalyzed reaction.

The halogenation of ketones is also subject to general acid catalysis [Eq. (5.48)].

$$CH_3COCH_3 \underset{k_{-1}[A^{\ominus}]}{\overset{k_1[HA]}{\rightleftharpoons}} CH_3\overset{\overset{\displaystyle OH^{\oplus}}{\|}}{C}-CH_3 \underset{k_{-2}[HA]}{\overset{k_2[A^{\ominus}]}{\rightleftharpoons}} CH_3\overset{\overset{\displaystyle OH}{|}}{C}=CH_2 \overset{k_3[X_2]}{\longrightarrow} \text{Product} \qquad (5.48)$$

The mechanism is complicated by the preequilibrium protonation of the ketone; otherwise it is analogous to general base catalysis.[25]

The mechanisms for these general acid- and base-catalyzed halogenation reactions are somewhat ambiguous because kinetic dependence on a general base may be mechanistically described either in terms of the action of a general base [Eq. (5.46)] or of the action of hydroxide ion and a conjugate general acid [Eq. (5.49)]. These two mechanisms are kinetically indistinguishable.

$$CH_3COCH_3 \underset{k_{-1}}{\overset{k_1[HB]}{\rightleftharpoons}} CH_3COCH_3 \cdot HB \overset{OH^{\ominus}}{\longrightarrow} CH_3\overset{\overset{\displaystyle O^{\ominus}}{|}}{C}=CH_2 \qquad (5.49)$$

The kinetic dependence on a general acid may be mechanistically described either in terms of general acid (concerted) catalysis [Eq. (5.48)] or of hydronium ion and conjugate general base (stepwise) catalysis [Eq. (5.50)]

$$CH_3COCH_3 \underset{k_{-1}}{\overset{k_1[HA]}{\rightleftharpoons}} CH_3COCH_3 \cdot HA \overset{H_2O}{\longrightarrow} CH_3\overset{\overset{\displaystyle OH}{|}}{C}=CH_2 \qquad (5.50)$$

This ambiguity (of the position of the proton in the transition state) is met in essentially all general acid–base-catalyzed reactions. There are a number of ways in which this problem can be handled.

One of the methods of differentiating two kinetically indistinguishable mechanisms is to study an intermediate species that is completely protonated. For example, in the racemization of D-α-phenyl-isocaprophenone in concentrated sulfuric acid, a decrease in rate constant occurs in going from 85 to 94% sulfuric acid solution.[25] This decrease occurs at an acidity at which the ketone is essentially completely converted to its conjugate acid. It must therefore reflect the abstraction of a proton from this conjugate acid by the base, specifying Eq. (5.48) rather than Eq. (5.50) as the mechanism of the acid-catalyzed enolization of a ketone.

Solvent deuterium isotope effects tend to confirm the description of Eq. (5.48) for the catalysis of enolization of α-phenylisocaprophenone by acetic acid. For this reaction, the solvent isotope effect $(k_{HOAc}/k_{DOAc}) = 1.09$ at 100°C, which is very close to unity. Mechanism (5.48) predicts a solvent isotope effect of around 1. On the other hand, mechanism (5.50) predicts a solvent isotope effect of the order of 3 since it involves a rate-determining proton transfer. Therefore, the most reasonable explanation for the catalysis of enolization by acetic acid is mechanism (5.48), which is consistent with the preceding argument.

The mechanisms (5.49) and (5.50) are consistent with the susceptibility of enolization of various ketones to general base and general acid catalysis. For example, the enolization of acetone in aqueous solution is strongly accelerated by general acids, while the enolization of bromoacetone and of pyruvic acid are insensitive to acid catalysis. The insensitivity of the latter compounds is related to a lower pK_a of the carbonyl groups in those compounds with electronegative substituents.

Proton Transfers to or from Electronegative Atoms. Although proton transfers to or from carbon can quite clearly be slower than other bond-making or bond-breaking processes, proton transfers to or from electronegative atoms such as oxygen, nitrogen, or sulfur, whose rate constants are often diffusion-controlled are in general faster than other bond-making and bond-breaking processes (Chapter 4). The description of general acid–base catalysis involving proton transfers to and from such atoms is therefore one of considerable interest. Several typical examples of these reactions will be discussed from the viewpoint of the relation of the proton transfers to other processes occurring in these reactions. Two general mechanisms may be advanced for these processes, either a solely rate-determining proton transfer or a rate-determining proton transfer concerted with heavy atom bond-making/breaking.

The hydrolysis of orthoesters exhibits a simple version of general acid catalysis. This reaction, long the object of study, has been described as:

1. A fast proton transfer followed by a rate-determining carbon–oxygen cleavage.

$$RC(OR)_3 \underset{H_2O}{\overset{H_3O^\oplus}{\rightleftharpoons}} \overset{\displaystyle \overset{H}{\underset{|}{O^\oplus R}}}{RC(OR)_2} \xrightarrow[-ROH]{slow} RC^\oplus(OR)_2 \xrightarrow[H_2O]{fast} \overset{\displaystyle \overset{O}{\|}}{RCOR} + ROH \quad (5.51)$$

2. A fast proton transfer followed by participation of the conjugate base of the catalyst in a subsequent rate-determining cleavage.

$$RC(OR)_3 \underset{H_2O}{\overset{H_3O^\oplus}{\rightleftharpoons}} \overset{\displaystyle \overset{H}{\underset{|}{O^\oplus R}}}{RC(OR)_2} \xrightarrow[\underset{-ROH}{slow}]{+A^\ominus} RC^\oplus(OR)_2 \longrightarrow Products \quad (5.52)$$

3.   A slow proton transfer reaction concerted with carbon–oxygen cleavage.

$$A\!-\!H + \overset{\displaystyle R}{\underset{\displaystyle O}{|}}\!-\!CR(OR)_2 \longrightarrow [A\text{-}\text{-}\text{-}\overset{\delta^{\ominus}}{H}\text{-}\text{-}\text{-}\overset{\displaystyle R}{\underset{\displaystyle O}{|}}\text{-}\text{-}\text{-}\overset{\delta^{\oplus}}{CR(OR)_2}] \longrightarrow$$
$$A^{\ominus} + ROH + RC^{\oplus}(OR)_2 \longrightarrow \text{Products} \qquad (5.53)$$

4.   A slow proton transfer reaction followed by a fast carbon–oxygen cleavage,

$$RC(OR)_3 + H_3O^{\oplus} \underset{}{\overset{\text{slow}}{\rightleftharpoons}} \overset{\displaystyle \overset{H}{O^{\oplus}}R}{\underset{}{RC(OR)_2}} + H_2O \qquad (5.54)$$
$$\xrightarrow{\text{fast}} \text{Products}$$

Of the various mechanisms, only Eq. (5.54) consists solely of a rate-determining proton transfer while Eqs. (5.52) and (5.53) consist of a rate-determining step involving both proton transfer and bond cleavage.

The most probable mechanism of the hydrolysis of orthoesters is mechanism (5.53). Orthoester hydrolysis is insensitive to structural changes that would normally lead to appreciable carbonium ion stabilization. A change from an orthoformate to an orthocarbonate results in a decrease in rate constant, even though a carbonium ion should be appreciably stabilized in this transformation. In addition, the Hammett rho constant of the hydronium ion-catalyzed hydrolysis of four substituted methyl orthobenzoates in aqueous methanol correlates with σ rather than $\sigma^{\oplus}$. These substituent effects certainly cannot be explained in terms of either Eq. (5.51) or Eq. (5.52), since the substituent effects imply that the transition state is not close to the carbonium ion at all.

Furthermore, the forward rate constant for triethyl orthoacetate [Eq. (5.54)] is estimated to be 10^4 sec^{-1} using the estimated ionization constant of this substrate, and the maximal rate constant for the reverse process, 10^{11} M^{-1} sec^{-1}. However, it is difficult to conceive of a carbon–oxygen bond cleavage being as fast as 10^4 sec^{-1}. Therefore, the mechanism represented in (5.54) seems unlikely.

General acid–base catalysis is widespread in addition to the carbonyl group and reverse reactions. Both general acid catalysis and general base catalysis are found in the addition of weak nucleophilic reagents to the carbonyl group. As before, general acid–base catalysis of these reactions can involve either a rate-determining proton transfer unaccompanied by any bond-making to carbon or a concerted process in which proton transfer occurs at the same time or very nearly the same time as bond formation to carbon.

If a proton transfer alone is the rate-determining step, two mechanisms for the reaction may be written (although they are improbable as discussed on the next page).

$$HX + R_2C=O \xrightarrow{\text{fast}} HX^{\oplus}-CR_2-O^{\ominus} \xrightarrow[A^{\ominus}]{HA} $$

slow

$$H-X^{\oplus}-CR_2OH \xrightarrow{\text{fast}} H^{\oplus} + XCR_2OH \qquad (5.55)$$

$$HX + R_2C=O + H^{\oplus} \xrightarrow{\text{fast}} HX^{\oplus}-CR_2OH \xrightarrow[\text{HA}]{A^{\ominus}} XCR_2OH \qquad (5.56)$$

slow

Mechanisms (5.55) and (5.56) are kinetically indistinguishable because of the equilibrium $H_2O + HA \rightleftharpoons H_3O^{\oplus} + A^{\ominus}$, as discussed earlier. In the forward direction, these reactions involve a general acid proton donation or a general base proton removal in the rate-determining step. The reverse reactions are described in the opposite manner. Since these reactions require proton transfer from oxygen to be slower than bond-making to a carbon atom, they are unlikely. If $HX^{\oplus}-C-O^{\ominus}$ is a stronger base than $A^{\ominus}$, the free energy of proton transfer to the intermediate is negative and therefore the slow step of reaction (5.55) is diffusion-controlled. In most instances, this inequality in base strength will hold (although it is not mandatory). When it does, it is difficult to conceive of a preequilibrium addition of HX to the carbonyl group that can be faster than diffusion-controlled. Therefore, this mechanistic possibility cannot be taken seriously. However, when $HX^{\oplus}-C-O^{\ominus}$ is a weaker base than $A^{\ominus}$, this dictum no longer applies.

Several pieces of experimental evidence make a solely rate-determining proton transfer improbable. One is that ^{13}C kinetic isotope effects have been found in the formation of semicarbazones, hydrazones, and 2,4-dinitrophenylhydrazones from carbonyl compounds containing ^{13}C in the carbonyl group.[20] Since an isotope effect can occur only if the carbonyl carbon atom participates in the rate-determining step, this fact rules out Eq. (5.55) or (5.56). Furthermore, if the free energy of proton transfer of reactions (5.55) or (5.56) is negative, no dependence of the rate of the reaction on the pK_a of the general catalyst should be observed, a result contrary to experiment. Thus one must exclude mechanisms (5.55) or (5.56) for general acid- (and general base-) catalyzed reactions when the free energy of the proton transfer is negative.

More reasonable mechanisms of these general acid and base catalyses involve concerted addition of HX and a proton transfer, as described by eqs. (5.57)–(5.60).

$$HX + \overset{CH_3}{\underset{CH_3}{>}}=O + HA \xrightarrow{\text{slow}} H-X^{\oplus}-\overset{CH_3}{\underset{CH_3}{\overset{|}{\underset{|}{C}}}}-O-H + A^{\ominus} \xrightarrow{\text{fast}}$$

$$\qquad\qquad (5.57)$$

$$H^{\oplus} + A^{\ominus} + X-\overset{CH_3}{\underset{CH_3}{\overset{|}{\underset{|}{C}}}}-OH$$

$$A^{\ominus} + HX + \underset{CH_3}{\overset{CH_3}{>}}=O^{\oplus}H \;\overset{slow}{\rightleftharpoons}\; A-H + X-\underset{CH_3}{\overset{CH_3}{\underset{|}{\overset{|}{C}}}}-OH$$

$$\Updownarrow$$

$$A^{\ominus} + H^{\oplus} + HX + \underset{CH_3}{\overset{CH_3}{>}}=O \tag{5.58}$$

$$A^{\ominus} + HX + \underset{CH_3}{\overset{CH_3}{>}}=O \;\overset{slow}{\rightleftharpoons}\; A-H + X-\underset{CH_3}{\overset{CH_3}{\underset{|}{\overset{|}{C}}}}-O^{\ominus} \;\overset{fast}{\rightleftharpoons}\; A^{\ominus} + X-\underset{CH_3}{\overset{CH_3}{\underset{|}{\overset{|}{C}}}}-OH \tag{5.59}$$

$$A^{\ominus} + HX + \underset{CH_3}{\overset{CH_3}{>}}=O \;\overset{fast}{\rightleftharpoons}\; X^{\ominus} + \underset{CH_3}{\overset{CH_3}{>}}=O + HA \;\overset{slow}{\rightleftharpoons}\; X-\underset{CH_3}{\overset{CH_3}{\underset{|}{\overset{|}{C}}}}-OH + A^{\ominus} \tag{5.60}$$

Mechanism (5.55) may be described in the forward direction as a general acid-catalyzed reaction and as a hydronium ion–general base-catalyzed reaction in the reverse direction. Mechanism (5.58) can be described in the forward direction as a hydronium ion–general base-catalyzed reaction and as a general acid-catalyzed reaction in the reverse direction. Mechanisms (5.57) and (5.59) are kinetically indistinguishable for reasons given earlier and are equivalent to one another in a mirror image sense. In a similar fashion, mechanism (5.59) involves a general base catalysis in the forward direction and general acid catalysis in the reverse direction, while mechanism (5.60) involves general acid catalysis in the forward direction and general base catalysis in the reverse direction. Again, mechanisms (5.59) and (5.60) are kinetically indistinguishable.

Several methods for distinguishing mechanistic pairs such as Eqs. (5.57) and (5.58) or Eqs. (5.59) and (5.60) exist.[11]

1. A substrate containing an alkyl group can be substituted for one containing a hydrogen atom. The position of the alkyl group is, of course, fixed by synthesis and cannot equilibrate with the medium as a proton can; thus the ambiguous position of the proton can be defined by the unambiguous position of the alkyl group if analogy between the alkyl (usually methyl) group and the proton is correct. If the alkyl-substituted substrate reacts at the same rate as the protonated compound, it can be inferred that the proton is in the same position as the alkyl group. The hydrolyses of acetals, ketals, and glycosides proceed via specific acid catalysis involving a preequilibrium addition of a proton followed by a rate-determining heterolysis. The reverse of these reactions is a rate-determining addition of alcohol to the carbonyl group according to Eq. (5.58), except that an alkyl group rather than a proton is on the carbonyl oxygen atom. Since the hydrolysis of acetals shows no general acid

catalysis, the reverse reaction must show no general base catalysis. However, catalysis is well known for the addition of water and alcohols to carbonyl compounds. Thus the hydrolysis of acetals is not a model for carbonyl hydration, and Eq. (5.58) may be ruled out. By process of elimination, Eq. (5.57) can be considered the correct description for the addition of alcohols and semicarbazides to carbonyl compounds.

2. The reaction can be examined under conditions of pH in which the substrate is completely protonated (or deprotonated). Obviously, under such conditions, a catalyst cannot be used to protonate (or deprotonate) the substrate.

3. A mechanism can be rejected if the calculated rate constant for a given step of the reaction, usually the rate-determining step, is greater than the rate constant for a diffusion-controlled reaction.

4. The sensitivity of a reaction to general acid catalysis changes as the nucleophilicity of the attacking reagent increases. A change in the opposite direction is predicted when a general acid-catalyzed reaction is viewed as such or is viewed as the kinetically equivalent general base–hydronium ion reaction because the general acid is now a general base. Thus one can distinguish the individuals of a pair from one another.

5.10. THE POSITION OF THE PROTON IN THE TRANSITION STATE OF GENERAL CATALYZED REACTIONS

General catalysis in the overall sense involves the transfer of a proton, preceded by hydrogen bonding (Chapter 2). On this basis, the efficiency of a general catalysis may depend on: (1) the equilibrium constant of formation of the hydrogen-bonded complex; (2) the degree of bond polarization induced by complex formation; and (3) the rate of proton transfer. Evaluation of such factors depends on the intimate description of the proton transfer occurring in general catalysis.

5.11. THE MARCUS THEORY OF PROTON TRANSFER

The considerable curvature found in some Brönsted plots can be accounted for by the use of Marcus' theory of atom transfer.[26] The Marcus theory is described here. Marcus' formalism for a proton-transfer reaction is given in Eqs. (5.61)–(5.63). In them ΔF^*_{HA} is the free energy of activation for the reaction catalyzed by HA, $\Delta F^{\circ}_R{}'$ is the standard free energy of reaction within the reaction complex, λ is the free energy of activation within the reaction complex, and W^r is the work required to form the reaction complex.

$$\Delta F^*_{HA} = W^r + \frac{\left(1 + \dfrac{\Delta F^{\circ}_R{}'}{\lambda}\right)^2 \lambda}{4} \tag{5.61}$$

$$\left(1 > \frac{\Delta F_R^{o\,\prime}}{\lambda} > -1\right)$$

$$\Delta F_{HA}^* = W^r \tag{5.62}$$

$$\left(-1 > \frac{\Delta F_R^{o\,\prime}}{\lambda}\right)$$

$$\Delta F_{HA}^* = W^r + \Delta F_R^{o\,\prime} \tag{5.63}$$

$$\left(\frac{\Delta F_R^{o\,\prime}}{\lambda} > 1\right)$$

The reaction complex has sometimes been called an encounter complex, but it is plain that considerably more than a simple encounter is required to generate a structure in which proton transfer will take place. Since the work required to form the reaction complex is W^r, operationally it is that part of ΔF_{HA}^* which is insensitive to changes in pK_{aHA}. For proton transfers from a homogeneous series of oxygen acids to a single substrate, it has been suggested that W^r is made up of the entropy of localization and the free energy of desolvation of the acid. The free energy for desolvation of tertiary amines shows little systematic relation to their basicity. It is reasonable to assume that the desolvation free energy of the ammonium salts will be unrelated to their acidity since that of the amines has been attributed mostly to van der Waals forces, which would also operate on the solvated ions. Even hydrogen-bond energies have been shown to be only weakly related to acid strengths. Thus, to the extent that W^r is attributable to these causes, it should be approximately constant for a series of closely related acids.

Quantitatively, W^r is the parameter that governs the ultimate rate which will be achieved by strongly spontaneous variants of the reaction in question, as shown by Eqs. (5.61) and (5.62), and λ is the parameter that governs the curvature of the Brönsted plot. By twice differentiating ΔF_{HA}^* with respect to ΔF_{HA}^o, it can be shown that $\delta \Delta F_{HA}^* / \delta \Delta F_{HA}^o$ is $1/(2\lambda)$. Thus the Marcus formalism can be regarded as a particular quantification of the "Hammond postulate."[27]

The Marcus theory shows the exact relation between rate and equilibrium constants for proton transfer reactions and thus gives a mathematical explanation for the curvature in many Brönsted plots, indicating that rate constants are systematically, though not linearly, related to equilibrium constants in many reactions and that these relations can be used to separate from the overall energy of activation that part required in preliminary steps. This separation now appears to show that much, perhaps most, of the heavy atom and solvent reorganization that accompanies a proton transfer precedes or follows the rate-determining step. There can be no other explanation for the large values of W^r consistently observed. This observation should considerably simplify the construction of the required potential surface for the rate-determining step.[28]

5.12. SUSCEPTIBILITY TO GENERAL ACID–BASE CATALYSIS

General acid–base catalysis is a widespread but not universal phenomenon in organic reactions. Earlier, consideration was given to the occurrence of general or specific catalysis in a given reaction. Let us now try to specify the influence of the structure of the reagents in a related family of reactions on the susceptibility of the reaction to general catalysis. That is, we wish to determine the effect of the reactivity of the substrate or nucleophile on the Brönsted α or β values, a measure of the susceptibility of the reaction to general catalysis.[29]

First, the relation between the reactivity of the substrate and α (β) value is described. Let us consider a system of i substrates (characterized by σ_i) and two general acid catalysts (characterized by pK_{a_1} and pK_{a_2}). By combining the Hammett and Brönsted equations for a family of general acid-catalyzed reactions, one can obtain

$$\frac{\sigma_i}{\alpha_0 - \alpha_i} = \frac{pK_{a_2} - pK_{a_1}}{\rho_2 - \rho_1} = \text{constant} \tag{5.64}$$

where α is a Brönsted coefficient,[1] and σ refers to the Hammett equation.[30] The left-hand side of Eq. (5.64) depends only on the nature of the substrates, whereas the right-hand side depends only on the nature of the general acid catalysts. For a family of base-catalyzed reactions, one can obtain the expression

$$\frac{\sigma_i}{\beta_i - \beta_0} = \frac{pK_{a_1} - pK_{a_2}}{\rho_1 - \rho_2} = \text{constant} \tag{5.65}$$

where the symbolism is similar. Equation (5.64) indicates that electron withdrawing substituents on the substrate will result in a smaller value of α for general acid catalysis; that is, as the reactivity of the substrate increases, the value of α will linearly decrease. Equation (5.65) is a similar relationship in which increased reactivity of the substrate results in a lowered sensitivity to general base catalysis, that is, a decreased value of β.

In a similar manner, a relationship between the Brönsted parameters, α or β, and the reactivity of a nucleophile participating in a general acid or base reaction can be derived from the Brönsted and Swain–Scott equations. Here a system of n nucleophiles (characterized by n_k) and two general acid catalysts (characterized by pK_{a_1} and pK_{a_2}) is considered. For general acid catalysis, the relationship is

$$\frac{pK_{a_2} - pK_{a_1}}{S_2 - S_1} = \frac{n_k}{\alpha_k - \alpha_0} = \text{constant} \tag{5.66}$$

and for general base catalysis, it is

$$\frac{pK_{a_2} - pK_{a_1}}{S_2 - S_1} = \frac{n_k}{\beta_0 - \beta_k} = \text{constant} \qquad (5.67)$$

in which n is a measure of nucleophilic reactivity, often proportional to the basicity of the nucleophile, and S is a measure of the sensitivity to nucleophilic reactivity. Equation (5.66) states that as the nucleophilicity (basicity) of the attacking reagent increases, the sensitivity to the strength of the catalyzing acid α linearly decreases. Similar statements can be made for Eq. (5.67).

The conclusion from Eq. (5.66) is experimentally confirmed in the general acid-catalyzed addition of nucleophiles to carbonyl compounds, as shown in Table 5.4. Clearly, as the pK_a of the nucleophile increases, α decreases.

The preceding correlations indicate that general catalysis will be seen to a greater extent with substrates and nucleophiles of low reactivity.

REFERENCES

1. J. N. Brönsted and K. Pederson, *Z. Phys. Chem.*, **108**, 185 (1924). This paper is worthy of careful perusal even at this date.

2. J. N. Brönsted, *Chem. Rev.*, **5**, 322 (1928).

3. R. P. Bell, *The Proton in Chemistry*, Cornell University Press, Ithaca, New York, 1959.

4. M. Eigen, *Angew. Chem. Int. Ed.*, **3**, 1 (1964).

5. F. G. Bordwell and W. J. Boyle, Jr., *J. Am. Chem. Soc.*, **93**, 511, 512 (1971).

6. F. G. Bordwell, personal communication; *J. Org. Chem.*, **43**, 3107 (1978).

7. A. J. Kresge, *Acc. Chem. Res.*, **8**, 354 (1975).

8. A. J. Kresge, *Pure Appl. Chem.*, **7**, 259 (1969).

9. A. I. Hassid, M. M. Kreevoy, and T. M. Liang, *Faraday Symp. Chem. Soc.*, **10**, 69 (1975).

10. R. P. Bell, in *Correlation Analysis in Chemistry*, Plenum Press, N. B. Chapman and B. Shorter, Eds., New York, 1980, pp. 55–84.

11. M. L. Bender, *Mechanisms of Homogeneous Catalysis from Protons to Proteins*, Wiley-Interscience, New York, 1971.

12. R. L. Schowen, in *Isotope Effects on Enzyme-Catalyzed Reactions*, W. W. Cleland, M. H. O'Leary, and D. B. Northrop, Eds., University Press, Baltimore, MD, 1977, pp. 64–69.

13. C. G. Swain, *J. Am. Chem. Soc.*, **74**, 4108 (1952).

14. R. P. Bell, *Acid-Base Catalysis*, Clarendon Press, Oxford, 1941.

15. M. L. Bender and F. J. Kezdy, in *Proton-Transfer Reactions*, E. F. Caldin and V. Gold, Eds., Chapman and Hall Pub. Co., London, 1975, pp. 385–401.

16. R. H. DeWolfe and R. M. Roberts, *J. Am. Chem. Soc.*, **76**, 4379 (1954).

17. M. L. Bender and B. W. Turnquest, *J. Am. Chem. Soc.*, **79**, 1652 (1957).

18. R. G. Pearson, in *Mechanisms of Inorganic Reactions*, R. F. Gould, Ed., *Advances in Chemistry Series*, Vol. 49, A.C.S. Publications, Washington, DC, 1971, pp. 21–29.

19. V. Gold and E. G. Jefferson, *J. Chem. Soc.*, **1953**, 1409.

20. W. P. Jencks, *Prog. Phys. Org. Chem.*, **2**, 63 (1964).

21. M. L. Bender, *Chem. Rev.*, **60**, 53 (1960).

22. T. C. Bruice and S. J. Benkovic, *Bioorganic Mechanisms*, Vol. I., W. A. Benjamin, New York, 1966.

23. E. H. Cordes and W. P. Jencks, *J. Am. Chem. Soc.*, **84,** 4319 (1962).

24. W. P. Jencks, *Acc. Chem. Res.*, **9,** 425 (1976).

25. D. J. Cram, C. A. Kingsbury, and B. Rickborn, *J. Am. Chem. Soc.*, **83,** 3688 (1961).

26. R. A. Marcus, *J. Chem. Phys.*, **24,** 966 (1956); *Discuss. Faraday Soc.*, **29,** 21 (1960); *J. Am. Chem. Soc.*, **91,** 7224 (1969).

27. J. R. Murdock, *J. Am. Chem. Soc.*, **94,** 4410 (1972).

28. M. M. Kreevoy and S.-W. Oh, *J. Am. Chem. Soc.*, **95,** 4805 (1953).

29. W. P. Jencks, *Catalysis in Chemistry and Enzymology*, McGraw-Hill Book Co., New York, 1969.

30. L. Zucker and L. P. Hammett, *J. Am. Chem. Soc.*, **61,** 2785 (1939).

31. L. P. Hammett, *Physical Organic Chemistry*, McGraw-Hill Book Co., Inc., New York, 1940.

6 General Acid–Base Catalysis: Enzymatic Reactions

6.1. ENZYMATIC GENERAL ACIDS (BASES)

The formation of covalent bonds between enzymes and substrates proceeds so efficiently relative to nonenzymatic reactions that one is forced to conclude that either the reacting nucleophiles or electrophiles or both are somehow activated. This activation was found to occur through the numerous general acid and base functions common to all enzymes. These functions are listed in Table 6.1.

As with intramolecular reactions, the effective molarity of these acids (bases) is very much higher than corresponding intermolecular catalysts. Furthermore, reactions occurring in the enzyme active site have the added advantage of having the reacting groups properly aligned. Both the high effective general acid or general base concentrations and the proper stereochemistry contribute to reaction accelerations. The first indication of the importance of proton-transfer reactions in enzyme catalysis came from the observation that the rate of most enzyme-catalyzed reactions displays a relatively simple sigmoidal or bell-shaped pH dependence. Thus enzymatic reactions require a small number of acids (bases) in a definite state of ionization. Later mechanistic studies indeed confirmed that in many cases these acids and bases—usually identifiable from the pK_a values of the pH-rate constant profile—act as proton donors and acceptors in the rate-limiting step of the catalytic process (Table 6.1). Since in biological systems enzymatic reactions occur almost invariably around neu-

Table 6.1. Prototropic Groups of Enzymes

Amino Acid Side Chain on Protein	Acid Function	Base Function	pK_a
N-terminal	$\alpha\text{-NH}_3^{\oplus}$	$\alpha\text{-NH}_2$	7.8
C-terminal	$\alpha\text{-CO}_2\text{H}$	$\alpha\text{-CO}_2^{\ominus}$	3.8
Aspartic acid $\Big\}$ Glutamic acid	$\beta,\gamma\text{-CO}_2\text{H}$	$\beta,\gamma\text{-CO}_2^{\ominus}$	4.4 4.6
Histidine	Imidazolium ion	Imidazole	7.0
Cysteine	—SH	$\text{S}^{\ominus}$	8.7
Tyrosine	$-\text{C}_6\text{H}_4\text{OH}$	$\text{C}_6\text{H}_4\text{O}^{\ominus}$	9.6
Lysine	$\epsilon\text{-NH}_3^{\oplus}$	NH_2	10.4
Serine $\Big\}$ Threonine	$\beta\text{-OH}$	$\beta\text{-O}^{\ominus}$	13
Arginine	$-\text{NH(C}=\text{NH}_2)\text{NH}_2^{\oplus}$	$-\text{NH(C}=\text{NH)NH}_2$	12.5
Peptide	$\text{R·CO·NHR}'$	$\text{R·CO·N}^{\ominus}\text{R}'$	14.8

Source: M. L. Bender and F. J. Kezdy, in *Proton Transfer Reactions*, E. F. Caldin and V. Gold, Eds., Chapman and Hall, London, 1975, Chapter 12.

trality, where hydronium and hydroxide ion concentrations are at a minimum, it is not surprising that enzymes make extensive use of general acid and base catalysis.

6.2. GENERAL ACID (BASE) CATALYSIS BY ENZYMATIC NUCLEOPHILES

The near neutral pH of the reaction medium imposes a severe limitation on the pK_a of the acidic (basic) functions that can act as an efficient general acid (general base). This is a direct consequence of the inverse relationship between the catalytic efficiency and the ionizability of the catalyst. In the following derivation, this relationship will be demonstrated for the case of general base catalysis only;[1] the same considerations can also be used for general acid catalysis. The catalytic rate constants (k_b) for general base catalysis obey the equation (Brönsted and Pedersen)[†] (Chapter 5).

$$k_B = G_B \left(\frac{1}{K_a}\right)^{\beta} \tag{6.1}$$

or

$$\log k_B = \beta p K_a + A$$

where $0 < \beta < 1$.

[†] The limitations of these equations are delineated in Chapter 5.

The pK_a of the catalyst also determines its degree of ionization and hence the concentration of the catalytically active species $[A^\ominus]$, according to the equation

$$[A^\ominus] = \frac{[HA]_T}{1 + \dfrac{[H_3O^\oplus]}{K_a}} \tag{6.2}$$

Thus the experimentally observable, pH-dependent catalytic rate constant $(k_{obs} = k_B \cdot [A^\ominus])$ at a given concentration of the catalyst $[C_0]$ at any pH can be expressed by

$$\log k_{obs} = A + \log[HA]_T + \beta pK_a - \log\left(1 + \frac{[H_3O^\oplus]}{K_a}\right) \tag{6.3}$$

or

$$k_{obs} = [HA]_T \frac{K_a^{-\beta}}{1 + \dfrac{[H_3O^\oplus]}{K_a}} \tag{6.4}$$

For any given pH and β, there is only one value of $K_a (= K_{max})$ for which k_{obs} is at a maximum. K_{max} can be found by differentiating Eq. (6.4) with respect to K_a and solving the resulting equation for the value of K_a that yields $(dk_{obs}/dK_a) = 0$. In this way, we find

$$K_{max} = [H_3O^\oplus] \frac{1-\beta}{\beta} \qquad \text{for } \beta < 1 \tag{6.5}$$

That is, the optimal general base catalysis occurs when the pK_a value of the catalyst is as close as possible to the pH value of the reaction medium. These considerations are of value for $0.2 < \beta < 0.8$, the range usually observed for general base catalysis.

Figure 6.1 further illustrates the occurrence of maximal catalysis when $pK_a \simeq pH$. The theoretical $\log k_{obs}$ versus pK_a curves shown were calculated from Eq. (6.3) with pH = 7 and A = −3. When β varies from 0.3 to 0.7, a shift of less than 0.5 pK units occurs in the value of pK_{max} on either side of the vertical line representing $pK_a = pH$.

In light of the preceding considerations, the approximately neutral pH of biological media should restrict general acid–general base catalysis in enzymatic reactions to the side chains of amino acids possessing pK_a values between 4 and 10. This was indeed found to be the case; the groups most often occurring in enzymes as general acid–general base catalysts are the imidazole group of histidine, and the carboxylate ions of aspartic or glutamic acids. In some enzymes, such as acetoacetate decarboxylase, the ϵ-amino group of lysine acts as

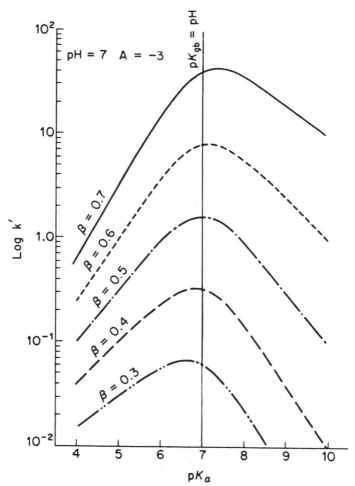

Fig. 6.1.   The dependence of the rate constant of a general base-catalyzed reaction at pH $= 7$ on the pK_{gb} of the catalyst, according to Eq. (6.3) (Ref. 1).

a general acid–general base catalyst, but in these cases, the pK_a value of the catalytic amino group is lowered from the expected pK_a of 10.8 to a value close to 7, presumably because of some factor in the environment.

As a corollary, enzymes acting in acidic media, for example, acidic proteases, would be expected to use general acid base catalysts having pK_a values in the acidic range. Carboxylates or imidazoles of slightly lower than usual pK_a's were indeed found to act as catalysts in these enzymes.

The neutrality of the reaction medium and the pK_a value of the conjugate acid of the catalyzing base imposes some restrictions on the choice of the nucleophile too. At present, no quantitative theory exists relating the reactivity of the nucleophile to the efficiency of general base catalysis for the same nucleophile. It is well established, however, that *increasing the reactivity of the nucleophile* (n_i)

decreases the sensitivity of the nucleophilic reaction to general base catalysis (β_i), according to the equation

$$n_i C = \beta_0 - \beta_i \qquad (6.6)$$

where C and β_0 are constants.[2]

On the other hand, n_i is approximately proportional to the pK_N value of the protonated nucleophile:

$$n_i = \gamma pK_N \qquad (6.7)$$

although the proportionality factor γ decreases toward the region of high pK_N values. On the basis of Eqs. (6.3), (6.6) and (6.7), one is then able to predict that toward a given electrophilic substrate, and for the same general base catalyst, an increase in the pK_N value of the nucleophile will first result in an increase in the rate of the reaction. At high pK_N values, however, the decrease of β_i, combined with a decreasing value of γ, renders the general base-catalyzed reaction less and less efficient. Thus, for a given substrate and general base, a pK_N value must exist that results in maximal catalytic rates.

Experimentally, it was found that for serine proteases, which act in the pH range of 7–8, the general base catalyst is imidazole with $pK_a = 7$. The serine alcohol has an estimated $pK_N = 12.5$, and with this value, we calculate $\Delta pK_a = pK_a \text{(acceptor)} - pK_a \text{(donor)} = -5.5$ pH units. In the same way, for acidic sulfhydryl proteases ($pK_N = 8.5$), the general base has $pK_a = 3.5$, and thus again $\Delta pK_a = -5$ pH units. It would thus appear that $\Delta pK_a = -5$ is a suitable value for maximizing the rates of general base-catalyzed enzymatic amide hydrolyses. That this choice of nucleophile and general base is not excessively sensitive to the nature of the leaving group of the substrate is shown by the fact that serine proteases are excellent catalysts of the hydrolysis of a variety of acyl derivatives with a wide range of leaving groups.

The preceding analysis concerning the acidity of nucleophiles and acid–base catalysts in enzymes shows that for the enzymatic catalysis of a given type of reaction not all acidic functions of the protein are appropriate. In fact, the choice in each pH range is limited to perhaps not more than two different amino-acid side chains. For this reason, the mechanistic diversity found in a class of enzymes catalyzing the same type of reaction is extremely limited. In dwelling on this aspect of structure-function relationships in enzymatic reactions, our intention was not to show the power of teleological argument. Rather, we hoped to illustrate the prominent place occupied by acid–base catalysis in enzymatic processes and the lesser importance of most other mechanistic details with respect to proton transfer.[3-5]

A survey of reactions catalyzed by enzymes reveals that probably all types of proton transfers observed in organic reactions also occur in enzymatic reactions. The transfer can occur to or from oxygen, nitrogen, sulfur, or carbon atoms and it may consist of a simple acid–base reaction—as in the case of epimerases—or it may be intimately coupled with other bond-making, bond-breaking, or electron-transfer processes. Donors or acceptors can be linked covalently to the peptide chain of the enzyme, or they can be situated on substrate or cofactor molecules. Finally, hydrogen can also be transferred enzymatically by mechanisms involving hydride ions or hydrogen atoms.

Rather than attempting an exhaustive review of all possible enzymatic proton transfers, which are manifold, we would like to present a few well-documented mechanisms involving different types of prototropic reactions. Of these, enzymatic amide and ester hydrolyses will be treated the most thoroughly, since at the present time they are the most extensively studied and their corresponding nonenzymatic model systems provide the most solid foundation for discussion.

Excellent reviews exist on several important aspects of enzymatic proton transfers. For the enzymology of carbon-acid reactions the reader is referred to the chapter by Rose in *The Enzymes;*[6] an interesting discussion of proton transfers in biological redox reactions can be found in Hamilton's review.[7]

6.3. ENZYMATIC PROTON TRANSFER: SERINE PROTEASES

Chymotrypsin is the most studied member of the serine protease family of enzymes. The enzyme-catalyzed hydrolytic reaction has been shown to occur in at least three kinetically distinguishable steps [Eq. (6.8)]. The first of these consists of a very fast, diffusion-controlled formation of a noncovalent enzyme-substrate complex followed by an acylation step. In the latter, the acyl group (Ac) of the substrate becomes covalently attached to a serine alcohol of the active site of the enzyme with the concomitant release of the amine of an amide substrate, the first product (P_1); then the enzyme intermediate is hydrolyzed by water, thus regenerating the free enzyme and releasing the carboxylic acid, the second product (P_2):

$$E + S \rightleftharpoons E \cdot S \xrightarrow{\text{acylation}} E - Ac \xrightarrow{\text{deacylation}} E + P_2 \qquad (6.8)$$
$$+P_1$$

From the experimental point of view, the deacylation step is easiest to study, since no binding step is associated with it. The mechanistic information most pertinent to a description of the deacylation step includes the following:[8]

1. The acyl-enzyme is an ester that was shown by both chemical and physical methods to result from the attachment of the acyl moiety of the substrate to Ser-195 of the enzyme.
2. A base of $pK_a = 7$ is required for the reaction.
3. This base is the imidazole ring of His-57.
4. The deacylation is a nucleophilic reaction and a series of substituents in the acyl group yields a Hammett ρ-constant of $+1.6$.

5. The reaction is first order with respect to the nucleophile, as determined from the kinetics of the methanolysis of the acyl-enzyme.

6. The nucleophile reacts in the protonated form with the acyl-enzyme, as evidenced by the pH dependence of the reaction of acetyl, isopropoxyphosphonyl-, and dimethylphosphoryl-chymotrypsins with the nucleophiles isonitrosoacetone, glycine-hydroxamic acid and phenylacetohydroxamic acid. All of these reactions exhibit bell-shaped curves depending on two groups, one with $pK_a = 7$ (general base catalyst) and the other with $pK_a = 8.9$ (binding of S to E).

7. The reactivity of amine and alcohol nucleophiles depend only slightly on their basicities.

8. No intermediate was observed in the deacylation step, although indirect evidence suggests the existence of at least one. While experimental evidence for the formation of an acyl-enzyme is old,[8] evidence for the formation of a tetrahedral intermediate is recent. This evidence is spectroscopic in nature (for the carboxylesterase-catalyzed reaction of benzil) and is shown in Fig. 6.2.[9]

9. A kinetic solvent isotope effect (k_{H_2O}/k_{D_2O}) of 2.5–3 is associated with the deacylation step.

10. X-ray crystallographic data show that the amino acid His-57 is hydrogen bonded to the carboxylate group of the amino acid Asp-102. At the same time, deacylation pH-rate profiles show that the reaction is insensitive to the ionization of any carboxylic acid in the pH range 2–9, but dependent on an ionization of a base of $pK_a \simeq 7$.

The components of the transition state of the deacylation reaction *must* then include the acyl-serine ester, an unprotonated imidazole group immobilized by a

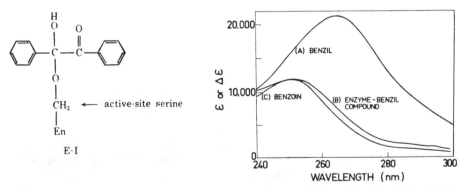

Fig. 6.2. Absorption spectra at 25°C in 0.05 M Tris–HCl buffer, 0.15 M in KCl, pH 7.5, [acetonitrile] = 2.4% v/v: (A) [benzil], 1.018 × 10⁻⁵ M; (B) [E] > [I], [enzyme] = 1.325 × 10⁻⁵ M, [benzil] = 1.018 × 10⁻⁵ M versus [enzyme] = 1.325 × 10⁻⁵ M, ~99.7% of benzil is bound at the active site (from K_1 = 1.0 × 10⁻⁸ M); (C) [benzoin], 1.069 × 10⁻⁴ M, in 0.05 M phosphate buffer, pH 7.5, [acetonitrile] = 0.83% v/v. Duplicate spectra agreed within 1% at λ_{max}.

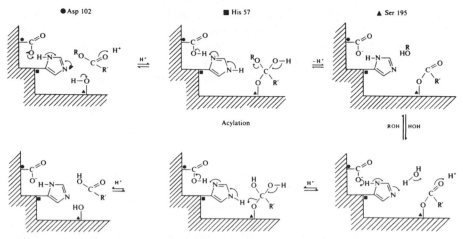

Fig. 6.3.   The mechanism of α-chymotrypsin-catalyzed hydrolysis of an ester substrate. From M. L. Bender, *Mechanisms of Homogeneous Catalysis from Protons to Proteins*, Wiley-Interscience, New York. 1971, Chapter 16.

hydrogen bond and a molecule of water as the acyl acceptor. A rate-determining proton transfer, which must shift a proton from the water molecule (or from another nucleophile) to the imidazole base, is presumably concerted with attack of the nucleophile on the acyl carbon, and perhaps the shift of the proton from imidazole to carboxylate ion in another hydrogen bond. The reaction scheme shown in Fig. 6.3 takes into account all the experimental data enumerated earlier. The mechanism shown involves the formation of a tetrahedral inter-mediate. In spite of considerable effort expended, the existence of this intermediate is still based only on indications from structure-activity relation-ships and spectroscopic evidence with the enzyme carboxylesterase[9] (Fig. 6.2), since isotopic oxygen exchange studies in the deacylations of cinnamoylcarbonyl-^{18}O-chymotrypsin and *p*-nitrobenzoyl-carbonyl-^{18}O-chymotrypsin yielded nega-tive results.

The mechanism depicted in Fig. 6.3 utilizes the unique ability of imidazole to serve simultaneously as proton donor and acceptor. The concerted nucleophilic attack and proton transfers enhance the kinetic efficacy of the catalyst. At the same time, all transition states are neutral in agreement with experimental data showing that the reaction is insensitive to variations in ionic strength and dielectric constant. Finally, the mechanism possesses a high degree of symmetry: The serine nucleophile and the alcohol or amine leaving group of the *acylation step* are mechanistically equivalent to the water nucleophile and the serine leaving group of the *deacylation step*.

6.4.   ENZYMATIC PROTON TRANSFER: SULFHYDRYL PROTEASES

Sulfhydryl proteases, such as papain, ficin, and bromelain, follow essentially the mechanism outlined in Fig. 6.3: the acyl acceptor is the sulfhydryl group of a cysteine moiety at the active site, as shown by chemical inhibition studies, by the pH dependence of the catalytic reaction ($pK_a = 8.4$ appears in acylation but not in deacylation) and by spectroscopic identification of a thiol ester in the acyl-enzyme. Papain-catalyzed reactions show again a substantial deuterium oxide kinetic solvent isotope effect, implying a rate-determining proton transfer. The chemical nature of the general base catalyst was discussed earlier. Since thiol esters react more readily with amines than do oxygen esters, papain is a better catalyst for transpeptidation than is chymotrypsin.

6.5. ENZYMATIC PROTON TRANSFER: DECARBOXYLASES

Enzymatic proton transfers to and from nitrogen atoms will be illustrated with the mechanism of acetoacetate decarboxylase, one of the many enzymes that form a Schiff base (imine) as a catalytic intermediate. Acetoacetate decarboxylase (AAD) from *Clostridium acetobutylicum* catalyzes the formation of acetone from acetoacetic acid. Chemical and kinetic evidence demonstrates the presence of an essential ϵ-amino group of lysine at the active site. In the presence of the substrate, a Schiff base salt of acetone has been trapped by borohydride reduction. By using radioactive $3\text{-}^{14}C$-acetoacetate, the peptide containing the label was isolated after acid hydrolysis and shown to contain N^{ϵ}-isopropyl-lysine. The enzyme also catalyzes the de-deuteration of acetone-d_6 in water and the carbonyl–oxygen exchange between acetone or acetoacetate and water. The pH-rate profile of the enzymatic reaction is bell-shaped, indicating the requirement for a base of $pK_a = 5.5$ and an acid of $pK_a = 6.5$. The base was identified as the ϵ-amino group of lysine acting as the nucleophile in Schiff-base formation, while the acid of $pK_a = 6.5$ acts as a general acid catalyst in the protonation of the enamine intermediate. The reaction mechanism scheme shown in Fig. 6.4 is consistent with all observations.

Oxaloacetate decarboxylase[10] from *Micrococcus lysodeicticus* catalyzes the decarboxylation of the β-ketoacid oxaloacetate with the same stoichiometry as acetoacetate decarboxylase.[8] The former, however, requires a $Mn^{\oplus 2}$ ion for activity and is insensitive to the action of sodium borohydride. This duality of mechanism is not unlike the one observed for enzymatic aldol condensation, where enzymes of Class I react by forming Schiff-base intermediates, whereas enzymes of Class II show metal-ion requirements. Oxaloacetate decarboxylase from cod also catalyzes the reduction by borohydride of the enzymatic reaction product pyruvate. This is evidenced by the accumulation of D-lactate in the presence of enzyme, reducing agent, and manganous ions. It has been proposed that both reduction and decarboxylation occur by way of an enzyme-metal–ion-substrate complex (see Section 9.2.2) in which the metal ion acts as an electron sink, thereby stabilizing the enolate ion formed in the decarboxylation reaction:

$$\text{E--Mn}^{\oplus 2}\underset{\underset{\ominus}{O}}{\overset{\overset{\ominus}{O}}{\Big\langle}}\begin{array}{l}\text{CO}\\ |\\ \text{C=CH}_2\end{array}$$

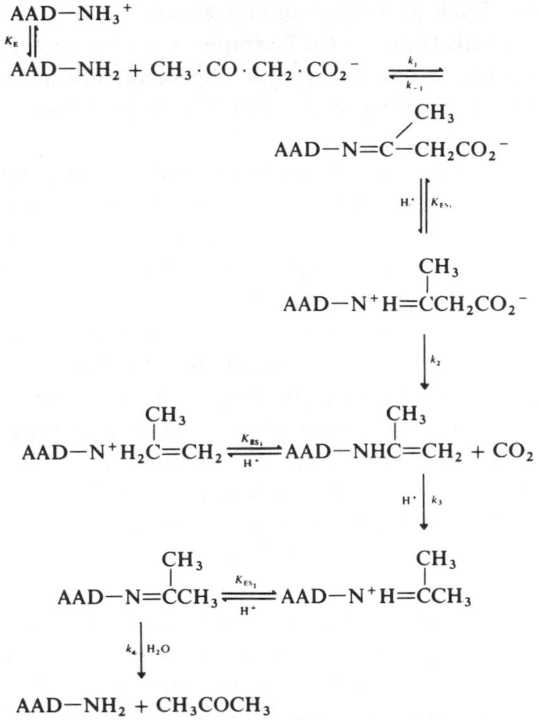

$$\text{AAD}-\text{NH}_3{}^+$$
$$K_E \, \| \|$$
$$\text{AAD}-\text{NH}_2 + \text{CH}_3 \cdot \text{CO} \cdot \text{CH}_2 \cdot \text{CO}_2{}^- \quad \underset{k_{-i}}{\overset{k_i}{\rightleftharpoons}}$$

$$\text{AAD}-\text{N}=\overset{\displaystyle \overset{\text{CH}_3}{\diagup}}{\text{C}}-\text{CH}_2\text{CO}_2{}^-$$

$$\text{H}^\cdot \, \| \| \, K_{ES_i}$$

$$\text{AAD}-\text{N}^+\text{H}=\overset{\displaystyle \overset{\text{CH}_3}{|}}{\text{C}}\text{CH}_2\text{CO}_2{}^-$$

$$\Big\downarrow k_2$$

$$\text{AAD}-\text{N}^+\text{H}_2\overset{\displaystyle \overset{\text{CH}_3}{|}}{\text{C}}=\text{CH}_2 \underset{\text{H}^\cdot}{\overset{K_{ES_i}}{\rightleftharpoons}} \text{AAD}-\text{NH}\overset{\displaystyle \overset{\text{CH}_3}{|}}{\text{C}}=\text{CH}_2 + \text{CO}_2$$

$$\text{H}^\cdot \, \Big| k_3$$

$$\text{AAD}-\text{N}=\overset{\displaystyle \overset{\text{CH}_3}{|}}{\text{C}}\text{CH}_3 \underset{\text{H}^+}{\overset{K_{ES_i}}{\rightleftharpoons}} \text{AAD}-\text{N}^+\text{H}=\overset{\displaystyle \overset{\text{CH}_3}{|}}{\text{C}}\text{CH}_3$$

$$k_4 \, \Big| \text{H}_2\text{O}$$

$$\text{AAD}-\text{NH}_2 + \text{CH}_3\text{COCH}_3$$

Fig. 6.4. Mechanism of action of acetoacetate decarboxylase (AAD). Ref. 1

The duality of hydrolytic enzyme mechanisms by chymotrypsin, and carboxy-peptidase and that of decarboxylation by metal-ion and Schiff base enzymes, essentially reflects the sensitivity of carbonyl functions toward acid or base catalysis. Whereas the protein side chains provide adequate *nucleophiles* and *bases* for catalytic activity, the superiority of metal ions as *acidic* catalysts in comparison with protons is amply demonstrated by metalloenzymes. A further reason for the occurrence of numerous metalloenzymes might be sought in the multidentate nature of metal complexes. The precise stereochemical positioning of several reaction components in the same complex provides an easy optimization of proximity and orientation effects.

6.6.   ENZYMATIC PROTON TRANSFER: METALLOENZYMES

A large number of enzymatic reactions require metal ions as one of the essential components of the catalytic process. A variety of metal ions can be involved, ranging from the alkali metals $Na^\oplus$ and $K^\oplus$ through the divalent ions $Ca^{2\oplus}$, $Mg^{2\oplus}$, $Zn^{2\oplus}$, $Fe^{2\oplus}$, $Cu^{2\oplus}$, $Co^{2\oplus}$, $Ni^{2\oplus}$ to the rarely occurring vanadium and molybdenum ions of higher valence. Their role can be purely structural, but most often they are tightly bound to the active site where they participate in the catalytic reactions. Such participation can be limited to stereospecific ligand formation with the substrate, as for example, with the phosphate group of a substrate. In the case of redox enzymes, the metal ion acts as an electron-transfer agent by undergoing a reversible change between two states of oxidation.

Most often, however, as in metal ion catalysis of organic reactions, the role of the metal ion directly participating in the enzymatic reaction is generally that of a Lewis acid, a substitute for a proton, but more efficient, especially at neutrality (a superacid—Chapter 9). These reactions always involve a simultaneous proton transfer between substrate and enzyme and it is in this sense that one can discuss metal ion catalyzed enzymatic proton transfers. As an example, the mechanism of carboxypeptidase-catalyzed reactions will be considered.

Carboxypeptidase is a metalloenzyme containing one atom of zinc per protein molecule. It catalyzes the hydrolysis of C-terminal peptide bonds in proteins and oligopeptides and that of carboxyl esters of α-hydroxyacids. The rate constant of the reaction shows a kinetic solvent isotope effect of 2 for the ester substrate O-(*trans*-cinnamoyl)-L-β-phenyllactate, but only 1.33 ± 0.15 for the peptide N-(N-benzoylglycyl)-L-phenylalaninate.[11] The three-dimensional structure of the enzyme, determined by X-ray crystallography, shows a globular protein containing a zinc ion coordinated with two histidines. The active site also contains a carboxylate ion (Glu-270), a phenolic group (Tyr-248), and a guanidino group (Arg-145); the latter was shown to form a salt bond with the free carboxylate ion of the substrate. The simplest mechanism accounting for all present structural and kinetic information is shown in Fig. 6.5.[11,12]

Thus the reaction proceeds through the transient formation of an acyl-anhydride intermediate with simultaneous release of the terminal carboxylic

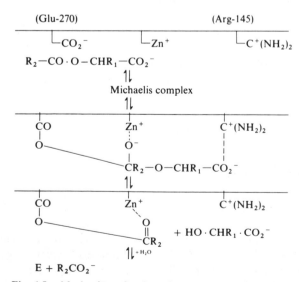

Fig. 6.5. Mechanism of action of carboxypeptidase. Ref. 1

acid. In a second step, the anhydride then rapidly reacts with water. It is possible that the phenol of Tyr-248 acts as a general base in the deacylation step by abstracting a proton from water and perhaps even as a general acid in the first step by donating a proton to the leaving group. An alternative mechanism in which Glu-270 acts as a general base is also possible.[13] No definite experimental proof is available on this point. It is clear, however, that the role of the metal ion is limited to polarizing the carbonyl function as a general acid and at the same time to contribute to the rigid orientation of the peptide or ester bond so as to facilitate the formation of the tetrahedral intermediate indicated in the scheme. Such a polarization is necessary in view of the weak nucleophilicity of the attacking carboxylate ion. The metal ion does not appear to play any direct role in the proton transfer or to interact with any other proton donor or acceptor participating in the reaction.

Metalloenzymes will be considered in more detail in Chapter 9.

6.7. ENZYMATIC PROTON TRANSFER: MANDELIC ACID RACEMASE

On the basis of presently available information, mandelic acid racemase might display the simplest enzymatic proton-transfer mechanism. The enzyme, isolated from *Pseudomonas putida*, catalyzes the epimerization reaction[14] without a cofactor requirement.

$$
\begin{array}{ccc}
CO_2^{\ominus} & & CO_2^{\ominus} \\
| & & | \\
H-C-OH & \rightleftharpoons HO-C-H \\
| & & | \\
C_6H_5 & & C_6H_5
\end{array}
$$

Deuterium exchange with the solvent occurs during the reaction; but at a rather slow rate. A symmetrical reaction intermediate must exist, since the rate of incorporation of tritium from the solvent into D-mandelate as the substrate yields equimolar amounts of D- and L- product. Thus the data are consistent with the formation of an α-carbanion intermediate with an enzymatic base acting as proton acceptor. The proton transfer has to be rate-limiting, as indicated by the approximately five-fold primary isotope effect for deuterium. In the enzyme-substrate complex, the epimerization occurs with a rate constant of the order of 10^3 sec^{-1}, indicating the upper limit of an enzymatic proton transfer.

6.8. ENZYMATIC PROTON TRANSFER: LYSOZYME

The enzyme lysozyme hydrolyzes the $\beta(1 \rightarrow 4)$ glycosidic linkages of mucopolysaccharides which are constituents of the cell wall of certain microorganisms.[15] In addition to the natural substrate, which is composed of N-acetylmuramic acid and *N*-acetylglucosamine, the enzyme also hydrolyzes chitin (poly-*N*-acetyl-*β*-glucosamine) and its degradation products and a number of *β*-aryl di-*N*-chitobiosides. The complete three-dimensional structure of the enzyme shows a globular protein, roughly the shape of a butterfly, with one-half containing a good deal of helix, and the other mostly *β*-pleated sheet. The active site is situated between these two halves and has been described as a cleft or groove adapted to the long-chain polymer natural substrate.

The catalytically active functional groups have been suggested to be the carboxyl groups of aspartic acid-52 and glutamic acid-35. The mechanism of lysozyme action, the enzyme whose structure is the best defined, is still somewhat uncertain. In total, three mechanisms have been proposed for lysozyme action:

1.   The neighboring *N*-acetylamino group of the substrate acts as a nucleophile and aspartic acid as a proton donor.
2.   The carboxylate anion of glutamic acid-36 further assists the *N*-acetylamino group by acting as a proton acceptor.
3.   The carboxylate anion of aspartic acid-52 acts as a nucleophile, and the carboxylic acid of glutamic acid-35 acts as a proton donor.

Lysozyme catalyzes the hydrolysis of *p*-nitrophenyl 2-acetamido-4-*O*-(2-acetamido-2-deoxy-*β*-D-glucopyranosyl)-2-deoxyglucopyranoside with a value of k 20 times greater than *p*-nitrophenyl-4-*O*-(2-acetamido-2-deoxy-*β*-D-glucopyranosyl)-*β*-D-glucopyranoside.[16] These results suggest acetamido participation.

A concerted proton donor acceptor or a combined proton donor–nucleophilic catalysis is favored, however, on the basis of kinetic results with small substrates. Transglycosylation was demonstrated in lysozyme-catalyzed reactions. This reaction has been used to synthesize *p*-nitrophenyl *β*-D-glucoside and a corresponding oligomer. Since lysozyme hydrolyzes these glycosidic bonds and since they contain no acetylamino group, anchimeric assistance by the *N*-acetylamino group can be ruled out as a necessary pathway for the enzyme-catalyzed cleavage of glycosidic bonds. The lysozyme active site, while not completely defined, must certainly be large, for the enzyme can interact with a hexasaccharide and there is strong kinetic dependence on the length of the saccharide. This implies the importance of nonproductive complexes (complexes of enzyme and substrate that do not lead to reaction). Evidence for oxocarbonium ion formation and participation of glutamic acid-35 is particularly convincing. Retention of configuration at the anomeric center means that if an oxocarbonium ion is formed, it must be constrained sterically by the enzyme.

6.9.   ENZYMATIC PROTON TRANSFER: STEREOCHEMISTRY

The use of isotopes in carbon-acid substrates is an invaluable tool for the determination of the stereochemistry of the enzymatic proton transfer. In contrast to organic reactions, stereospecific proton transfers are the rule rather than the exception in enzymatic reactions owing to the inherently asymmetric nature of the protein surface. An example is the pair of isotopic exchange reactions between dihydroxyacetone phosphate and tritiated water catalyzed by the enzymes aldolase and triose phosphate isomerase. In these two cases, a different α-hydrogen of the ketone is exchanged with water, leading to two discrete monotritiated derivatives **6.1** (dihydroxyacetone phosphate labeled by the enzyme isomerase) and **6.2** (dihydroxyacetone phosphate labeled by the enzyme aldolase).[17]

H   H

T—C—OH   HO—C—T

C=O   C=O

$CH_2OPO_3H^-$ $CH_2OPO_3H^-$

6.1   **6.2**

REFERENCES

1.   M. L. Bender and F. J. Kezdy, in *Proton Transfer Reactions*, E. Caldin and V. Gold, F. R. S., Eds., Chapman and Hall, London, 1975, Chapter 12. Most of this chapter has been taken from here with permission.

2.   W. P. Jencks, *Catalysis in Chemistry and Enzymology*, McGraw-Hill, New York, 1969; W. P. Jencks and M. Gilchrist, *J. Am. Chem. Soc.*, **90**, 2622 (1968).

3.   For a review, see W. P. Jencks, *Cold Spring Harbor Symp. Quant. Biol.*, **36**, 1 (1971).

4.   M. L. Bender and F. J. Kezdy, *Ann. Rev. Biochem.* **34**, 49 (1965).

5.   G. P. Hess, in *The Enzymes*, P. D. Boyer, Ed., Vol. 3, 3rd ed., Academic Press, New York, 1971, p. 231.

6.   I. A. Rose, in *The Enzymes*, P. D. Boyer, Ed., Vol. 2, 3rd ed., Academic Press, New York, 1970. p. 281.

7.   G. A. Hamilton, *Prog. Bioorg. Chem.*, **1**, 83 (1971); M. W. Hunkapiller, S. H. Smallcombe, D. R. Whitaker, and J. H. Richards, *Biochemistry*, **12**, 4732 (1973); E. C. Lucas and A. Williams, *Biochemistry*, **8**, 5125 (1969); J. Drenth, J. N. Jansonius, R. Koekoek, H. M. Swen, and B. G. Wolthers, *Nature*, **218**, 929 (1968); M. L. Bender and L. J. Brubacher, *J. Am. Chem. Soc.*, **88**, 5880 (1966).

8.   M. L. Bender, G. E. Clement, C. R. Gunter, and F. J. Kezdy, *J. Am. Chem. Soc.*, **86**, 3697 (1964).

9.   M. C. Berndt, J. De Jersey, and B. Zerner, *J. Am. Chem. Soc.*, **99**, 8332 (1977).

10.   D. Herbert, *Methods Enzymol.*, **1**, 753 (1955).

11.   E. T. Kaiser and B. L. Kaiser, *Acc. Chem. Res.*, **5**, 219 (1972).

12.   J. E. Coleman, *Prog. Bioorg. Chem.*, **1**, 159 (1971).

13.   R. Breslow and D. Wernick, *J. Am. Chem. Soc.*, **98**, 259 (1976).

14.   D. Hegeman, Y. Rosenberg, and G. L. Kenyon, *Biochemistry*, **9**, 4036 (1970).

15.   D. C. Phillips, *Sci. Am.*, **215**, No. 5, 78 (1966).

16.   G. Lowe and G. Sheppard, *J. Chem. Soc. Chem. Commun.*, **1968**, 529.

17.   S. V. Reider and I. A. Rose, *J. Biol. Chem.*, **234**, 1007 (1959).

7 | Nucleophilic and Electrophilic Catalysis

7.1. INTRODUCTION

Nucleophilic and electrophilic catalyses occur via several partial reactions, each of which is the same as a noncatalytic reaction.[1,2] What, then, is the difference between a nucleophilic (or electrophilic) reaction, and a nucleophilic (or electrophilic) catalyzed reaction?

In order for nucleophilic (N) or electrophilic catalysis to occur at a significant rate, several requirements must be met. (1) The catalyst, whether it is a nucleophile or an electrophile, must react with the substrate faster than it does with the final acceptor [Eq. (7.1)]. (2) The intermediate must react with the final acceptor or otherwise decompose faster than does the original substrate.[3]

$$S Q + N \longrightarrow S—N \xrightarrow{[X]} PX + N + Q \qquad (7.1)$$

That is, the catalyst must be an unusually effective nucleophile (electrophile), and the intermediate must be unusually susceptible to nucleophilic (electrophilic) attack or other decomposition. (3) The equilibrium between the catalyst and substrate yielding the intermediate must not be as favorable as between the substrate and the eventual acceptor.

Nucleophilic and electrophilic catalyses have been recognized for many years, although these names are of more recent origin. Examples of nucleophilic and

130

electrophilic catalyses are numerous; we will attempt to assess their importance and to delineate some selected mechanisms. Enzymatic examples of these catalyses will be discussed in Chapter 8.

7.2.   NUCLEOPHILIC CATALYSIS

7.2.1.   Nucleophilicity

Nucleophilic catalysis ultimately depends on nucleophilicity and for this reason, we briefly consider what makes a good nucleophile. Unfortunately, there is no simple correlation between structure and nucleophilicity, but we will examine some of the factors involved. Edwards has suggested two components of nucleophilicity: one related to polarizability or oxidizability and the other related to classical basicity [Eq. (7.2)].[4]

$$\log \frac{k}{k_0} = \alpha P + \beta H \qquad (7.2)$$

Here P is defined in terms of the relative molar refractivity, a measure of polarizability [Eq. (7.3)],

$$P = \log \frac{R_N}{R_{H_2O}} \qquad (7.3)$$

where R = refractive index and N = nucleophile, and H [Eq. (7.4)] is a function of the basicity,

$$H = pK_a + 1.74 \qquad (7.4)$$

Equation (7.2) correlates a number of disparate nucleophiles in several families of reactions.

However, if attention is confined to a single nucleophilic atom, a rough correlation between basicity and nucleophilicity is noted [Eq. (7.4)]. In fact, within a limited range of structural variation, a quantitative logarithmic correlation between basicity and nucleophilicity holds, as seen in the family of amines in Fig. 5.9. Such a quantitative relationship may be viewed as a manifestation of either a Brönsted or a Hammett relation, both of which are linear free-energy correlations.

When the nucleophilic atom is altered, all correlations between basicity and nucleophilicity vanish. For example, sodium butylmercaptide is about as basic as sodium phenoxide but is 10^3 times as nucleophilic toward saturated carbon; phenoxide and bromide ions are of comparable nucleophilicity toward saturated carbon, but differ by a factor of about 10^{17} in basicity. Even when atoms of the same column of the periodic table are compared, there is often no correlation between basicity and nucleophilicity. For example, in water, one finds the orders of decreasing nucleophilicity toward saturated carbon of:

$$I^{\ominus} > Br^{\ominus} > Cl^{\ominus} > F^{\ominus}$$

$$RSe^{\ominus} > RS^{\ominus} > RO^{\ominus} \qquad (7.5)$$

The best nucleophiles in these series form the weakest bonds to carbon, indicating no correlation between thermodynamics and kinetics in these reactions. It has been suggested many times that the electronic charge cloud of the nucleophile must be distorted by the electrophilic center of the substrate in the transition state. Thus such polarizability may explain the preceding sequences, although orbital overlap, steric hindrance, and solvation energies may also need to be considered. Solvation is of considerable importance in determining the nucleophilicity of anions. In fact, the order of nucleophilicity of the halide ions can be reversed by changing the solvent (Chapter 3).

Another factor determining nucleophilicity is seen in nucleophiles possessing an adjacent electronegative atom containing one or more pairs of unshared electrons. Examples of such nucleophiles are hydroxylamine, hydrazine, hydroperoxide ion, hypochlorite ion, oximate ion, and hydroxamate ion. These nucleophiles show exceptional nucleophilicity not accounted for on the grounds of either polarizability or basicity. The reactivity of these nucleophiles has been attributed to the stabilization of the transition state of the nucleophilic reaction by the neighboring lone pair electrons. For example, when the hypochlorite anion reacts with an electrophile, the charge on the oxygen is diminished. The chlorine helps to stabilize this situation by resonance donation of its nonbonding pairs. This phenomenon has been termed the "alpha effect."[5]

The order of nucleophilic (electrophilic) reactivity depends not only on the nucleophile (electrophile), but also on the nature of the substrate. Analysis of the vast amount of data on substitution reactions has indicated that for some substrates the rates are sensitive chiefly to the ordinary proton *basicity* of the nucleophile, while for other substrates the rates are sensitive chiefly to the *polarizability* of the nucleophile. The properties of the electrophilic center determine which type of behavior is followed. If the properties of the electrophilic center are those that make it a *hard* (nonpolarizable) acid, then *basicity* is the dominant factor. If the electrophilic site is a *soft* (polarizable) center, then *polarizability* is the most important factor in the rates. For example, electrophilic centers such as $RCO^{\oplus}$, $H^{\oplus}$, $RSO_2^{\oplus}$, $(RO)_2PO^{\oplus}$, and $(RO)_2B^{\oplus}$ react rapidly with nucleophiles that are strongly basic to the proton and not very *polarizable*, such as $OH^{\ominus}$ and $F^{\ominus}$. Other electrophilic centers such as $RCH_2^{\oplus}$, $R_2P^{\oplus}$, $RS^{\oplus}$, $Br^{\oplus}$, $R_2N^{\oplus}$, $RO^{\oplus}$, and $Pt^{2\oplus}$ react rapidly with highly *polarizable* nucleophiles such as $I^{\ominus}$ and $R_3P^{\ominus}$. These facts have been generalized in terms of the principle of "*hard and soft acids and bases*":[6] *Hard* electrophilic centers (acids) react rapidly with *hard* nucleophiles (bases), and *soft* electrophilic centers (acids) react rapidly with *soft* nucleophiles (bases). The rule refers to both S_N2 and E-2 mechanisms and may also refer to enzyme mechanisms of this kind.

We can explore the relative nucleophilicity toward the carbonyl carbon atom, the tetrahedral phosphorus atom, and the saturated carbon atom by means of the data in Tables 7.1 and 7.2. Table 7.1 indicates that the important component of the nucleophile with respect to p-nitrophenyl acetate is its basicity. There are some obvious exceptions to this statement pointing out the fact that those nucleophiles containing adjacent electronegative atoms with unshared electron pairs are exceptionally good nucleophiles (α-effect). Table 7.1 also points out the fact (Chapter 6) that nitrogen nucleophiles are considerably more reactive than oxygen nucleophiles of the same pK_a. The behavior of nucleophiles toward a tetrahedral phosphorus atom, shown in Table 7.1, is similar. The reactive centers of both p-nitrophenyl acetate and the phosphonofluoridate can thus be described as *hard* acids or electrophiles.

On the other hand, the data of Table 7.2, pertaining to reactions at saturated carbon, show a different order of nucleophilicity. The most important nucleophilic component appears to be *polarizability* rather than *basicity*, although both contribute something to the overall nucleophilicity. The importance of *polarizability* in reactions at saturated carbon is also seen from a comparison of the reactivities of diethyl malonate and ethoxide ions toward ethyl bromide. The former reaction is two times faster than the latter, although ethoxide ion is some 500 times more basic than diethyl malonate ion.

Table 7.1. Nucleophilic Reactivity Toward p-Nitrophenyl Acetate and Isopropyl Methylphosphonofluoridate[a]

Nucleophile	pK_a of Conjugate Acid	p-Nitrophenyl Acetate	Isopropyl Methyl-phosphonofluoridate
$HOO^{\ominus}$	11.5	2×10^5	1.0×10^5
Acetoximate	12.4	3.6×10^3	—
Salicylaldoximate	9.2	3.2×10^3	1.5×10^3
$OH^{\ominus}$	15.7	9×10^2	1.6×10^3
$\phi O^{\ominus}$	10.0	1×10^2	34
NH_2OH	6	1×10^2	1.3
$OCl^{\ominus}$	7.2	1.6×10^3	7×10^2
$CO_3^{2\ominus}$	10.4	1.0	75
NH_3	9.2	16	—
$CN^{\ominus}$	10.4	11	—
$\phi S^{\ominus}$	6.4	—	7.4×10^{-3}
ϕNH_2	4.6	1.5×10^{-2}	—
C_5H_5N	5.4	0.1	—
$NO_2^{\ominus}$	3.4	1.3×10^{-3}	—
$CH_3CO_2^{\ominus}$	4.8	5×10^{-4}	—
$F^{\ominus}$	3.1	1×10^{-3}	Very reactive
$S_2O_3^{2\ominus}$	1.9	1×10^{-3}	Unreactive
H_2O	−1.7	6×10^{-7}	1×10^{-6}

Source: J. O. Edwards and R. G. Pearson, *J. Am. Chem. Soc.*, **84**, 16 (1962). © 1962 by the American Chemical Society. Reprinted by permission of the copyright owner.
[a] Rate constants in M^{-1} min^{-1}.

Table 7.2.   Nucleophilic
Reactivity Toward
Methyl Chloride

Nucleophile	$k(M^{-1} sec^{-1})$
$SO_3^{\ominus 2}$	2.3×10^{-4}
$S_2O_3^{\ominus 2}$	1.7×10^{-4}
$SC(NH_2)_2$	2.5×10^{-5}
$I^{\ominus}$	1.2×10^{-5}
$CN^{\ominus}$	1×10^{-5}
$SCN^{\ominus}$	3.2×10^{-5}
$NO_2^{\ominus}$	1.8×10^{-5}
$OH^{\ominus}$	1.2×10^{-6}
$N_3^{\ominus}$	8×10^{-7}
$Br^{\ominus}$	5×10^{-7}
NH_3	2.2×10^{-7}
$Cl^{\ominus}$	1.1×10^{-7}
C_5H_5N	9×10^{-8}
H_2O	1×10^{-10}

Source: J. O. Edwards and R. G.
Pearson, *J. Am. Chem. Soc.*, **84**, 16
(1962). © 1962 by the American
Chemical Society. Reprinted by
permission of the copyright owner.

Toward aromatic carbon, the following order of nucleophilicity is seen:

$$C_6H_5S^{\ominus}, CH_3O^{\ominus} > C_5H_{10}NH > C_6H_5O^{\ominus} >$$
$$N_2H_4 > OH^{\ominus} > C_6H_5NH_2 > Cl^{\ominus} > CH_3OH$$

There is a large spread in these rate constants. Methoxide ion is 10^4 times faster than aniline, which in turn is 10^9 times faster than methanol. Nucleophilicity toward aromatic carbon apparently involves both *polarizability* and *basicity*.

The *softness–hardness* principle can be used as a guide for selecting catalysts in substitution reactions. The rules are simply that if the leaving group is a *hard* base, then a *hard* electrophile is used as catalyst (or vice versa). Likewise, when the electrophilic center is *hard*, a *hard* nucleophile is used as catalyst (or vice versa).[3-8]

7.2.2.   Mechanisms of Nucleophilic Catalysis

Catalysis by a base or a nucleophile can be interpreted mechanistically either as general base or nucleophilic catalysis (Chapter 6). The principal criteria for nucleophilic catalysis include the following: (1) observation of a transient intermediate whose structure and kinetics indicate that it is an intermediate in the pathway of reaction; (2) trapping of the intermediate as an isolable species; (3)

dependence of relative catalytic power on nucleophilicity rather than basicity, characterized by the nucleophilic reactivity orders discussed earlier, considerable steric hindrance, and importance of nucleophiles exhibiting the alpha effect; (4) occurrence of a common ion effect by the leaving group of the substrate (this amounts to trapping the intermediate as the reactant); (5) no primary deuterium oxide solvent isotope effect; and (6) product analysis of noncatalytic nucleophiles analogous to alleged catalytic nucleophiles.[9]

7.2.3. Halide Ions as Catalysts

An example of nucleophilic catalysis is the bromide ion–catalyzed hydrolysis of methyl iodide:

$$CH_3I + Br^{\ominus} \longrightarrow CH_3Br + I^{\ominus} \qquad (7.6)$$

$$CH_3Br + H_2O \longrightarrow CH_3OH + Br^{\ominus} \qquad (7.7)$$

In nucleophilic substitution reactions, methyl iodide reacts with bromide ion faster than it does with water; furthermore, the intermediate methyl bromide reacts with water faster than does the starting material. Thus the principal requirements of nucleophilic catalysis are met, although none of the rigorous mechanistic criteria have been demonstrated.[10] It might seem anomalous that bromide ion could be both a good nucleophile and leaving group at the same time, but, considering microscopic reversibility, this behavior can occur when factors other than basicity are involved. Nucleophilicity and leaving group ability depend on different aspects of structure (polarizability vs. bond strength) and thus one suitably constituted group may contain both properties. The free-energy–reaction coordinate diagram for the catalyzed process is compared in Fig. 7.1 with the corresponding uncatalyzed reaction. The catalyzed reaction bears the relationship to the uncatalyzed reaction predicted by the simplest theory of catalysis: the pathway is changed to one of greater complexity and the height of any individual peak in the catalyzed reaction is less than the height of the single peak in the uncatalyzed reaction.

The fluoride ion-catalyzed hydrolysis of ethyl chloroformate proceeds through a similar nucleophilic catalysis, since the chloroformate may be readily transformed to the fluoroformate in nonaqueous medium and the hydrolysis of the latter is about 30 times faster than that of the former.

$$C_2H_5O\overset{\overset{\displaystyle O}{\|}}{C}Cl \underset{}{\overset{+F^{\ominus}}{\rightleftharpoons}} C_2H_5O\overset{\overset{\displaystyle O}{\|}}{C}F \overset{H_2O}{\longrightarrow} C_2H_5OH + CO_2 + HF$$

Halide ion catalysis is also seen in acid-catalyzed reactions. For example, the perchloric acid-catalyzed hydrolyses of methyl p-toluenesulfinate and sulfite esters are accelerated by the addition of sodium chloride, and even more so by the addition of sodium bromide. These reactions, which occur with sulfur–oxygen

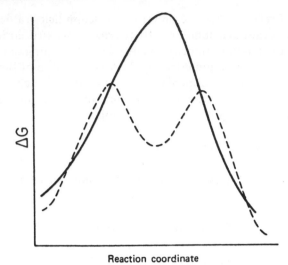

Fig. 7.1. The hydrolysis of methyl iodide. Solid line, uncatalyzed; dashed line, catalyzed by bromide ion.

bond cleavage can be explained most readily by postulating that catalysis occurs by the conversion of the ester into the readily hydrolyzable sulfinyl chloride or bromide **(7.1)**. More effective catalysis by bromide ion than chloride ion is consistent with this mechanism since the former is a better nucleophile (and HBr is a stronger acid). Enhanced catalysis by halogen acids over perchloric acid in the hydrolysis of glucose-6-phosphate may be explained in a similar manner. This reaction, which follows an A-2 (acid-catalyzed, second-order) mechanism with carbon–oxygen cleavage, probably proceeds through nucleophilic attack by halide ion on the anomeric carbon atom, leading to a glucosyl halide that rapidly hydrolyzes to products.

$$CH_3-O-\overset{\displaystyle O}{\overset{\displaystyle \|}{S}}-X$$

7.1

7.2.4. Oxyanions As Catalysts

The saponification of an ester in aqueous base, although frequently described as nucleophilic catalysis, is really a nucleophile-promoted hydrolysis, since the catalyst is not regenerated.[9]

$$R-\overset{\displaystyle O}{\overset{\displaystyle \|}{C}}-OR' + OH^{\ominus} \rightleftharpoons R-\overset{\displaystyle O^{\ominus}}{\underset{\displaystyle OH}{\overset{\displaystyle |}{C}}}-OR' \longrightarrow R-CO_2^{\ominus} + R'OH \quad (7.8)$$

The mechanism of this reaction, the addition of hydroxide ion to the carbonyl group to form an unstable intermediate that collapses to give products, is of the form of nucleophilic catalysis even though the catalyst is not regenerated. However, the alkoxide ion–catalyzed alcoholysis of an amide is true nucleophilic catalysis, following the same mechanism as the saponification of an ester.

$$
R-\overset{\overset{\displaystyle O}{\|}}{C}-NH_2 + R'O^{\ominus} \longrightarrow R-\underset{\underset{\displaystyle OR'}{|}}{\overset{\overset{\displaystyle O}{\|}}{C}}-NH_2 \dashrightarrow R-\overset{\overset{\displaystyle O}{\|}}{C}-OR' + NH_3
$$

$$
R-\overset{\overset{\displaystyle O}{\|}}{C}-OR' \overset{OH^{\ominus}}{\longrightarrow} R-\overset{\overset{\displaystyle O}{\|}}{C}-O^{\ominus} + R'O^{\ominus} \tag{7.9}
$$

Since (1) alcohol is a better nucleophile than water, (2) alkoxide ion is a better nucleophile than hydroxide ion (toward all carboxylic acid derivatives), and (3) esters are unstable hydrolytically compared to amides, either alcohol or alkoxide ion should serve as a nucleophilic catalyst for the hydrolysis of an amide.

Carboxylate ions catalyze the hydrolysis of several carboxylic acid derivatives. As mentioned in Chapter 5, some of these reactions proceed via general base catalysis while others proceed via nucleophilic catalysis. The formate ion-catalyzed hydrolysis of acetic anhydride is an example of nucleophilic catalysis [Eq. (7.10)].

$$
CH_3\overset{\overset{\displaystyle O}{\|}}{C}O\overset{\overset{\displaystyle O}{\|}}{C}CH_3 + HCO_2^{\ominus} \rightleftharpoons CH_3\overset{\overset{\displaystyle O}{\|}}{C}O\overset{\overset{\displaystyle O}{\|}}{C}H + CH_3CO_2^{\ominus} \overset{H_2O}{\longrightarrow} H\overset{\overset{\displaystyle O}{\|}}{C}OH + CH_3\overset{\overset{\displaystyle O}{\|}}{C}OH \tag{7.10}
$$

The postulated intermediate in Eq. (7.10), acetic formic anhydride, is more susceptible to hydrolysis than the reactant acetic anhydride. Furthermore, catalysis by formate ion is greater than that by acetate ion, contrary to the order of their base strengths. On these grounds, formate ion catalysis conforms to Eq. (7.10).

In the acetate ion-catalyzed hydrolysis of 2,4-dinitrophenyl esters, evidence for nucleophilic catalysis has been demonstrated both by an isotopic tracer experiment [Eq. (7.11)] and by experiments with aniline as a trapping agent.

Nucleophilic catalysis requires incorporation of oxygen-18 into the benzoic acid product since the fission of the intermediate acetic benzoic anhydride should take place preferentially at the acetyl carbonyl carbon atom, based on the known reactivities of acetyl and benzoyl derivatives. Experimentally, 75% of the oxygen-18 derived from one atom of the original acetate ion is found in the benzoic acid product, indicating that the major pathway of this catalysis is indeed a nucleophilic one. The trapping by aniline of the anhydride intermediate from the acetate ion–catalyzed hydrolysis of 2,4-dinitrophenyl acetate accounts for essentially 100% of the reaction product, indicating that this reaction proceeds only by nucleophilic catalysis.[11]

Phosphate ion is a catalyst for a number of hydrolyses including those of p-nitrophenyl acetate, esters of thiocholine, N-acetylimidazole, chloramphenicol, acetic anhydride, methyl acetate, tetraethyl pyrophosphate, and the reaction of dialkyl sulfides with iodine. The kinetics of some of these reactions indicate that phosphate ion, probably in the form of monohydrogen phosphate, participates in the reaction. In the hydrolysis of p-nitrophenyl acetate, the catalytic constant for monohydrogen phosphate is approximately 1000 times less than that for imidazole, implying a nucleophilic order as suggested in Chapter 6.

In the hydrolysis of tetraethyl pyrophosphate, both trapping and isotopic tracer experiments indicate nucleophilic catalysis with the intermediacy of $(C_2H_5O)_2P(=O)OPO_3^{2\ominus}$. This should more readily decompose to metaphosphate ion than the neutral reactant would [Eq. (7.12)], and eventually lead to the regeneration of the catalytic orthophosphate ion, HPO_4^{-2}.

$$
\underset{\substack{\|\\O}}{(EtO)_2P}-O-\underset{\substack{\|\\O}}{P(OEt)_2} \overset{HPO_4^{2\ominus}}{\rightleftharpoons} \underset{\substack{\|\\O}}{(EtO)_2P}-O-\underset{\substack{\|\\O\\|\\O^\ominus}}{P}-O^\ominus + \underset{\substack{\|\\O}}{(EtO)_2P}-O^\ominus + H^\oplus
$$

$$
\underset{\substack{\|\\O}}{(EtO)_2P}-O-\underset{\substack{\|\\O\\|\\O^\ominus}}{P}-O^\ominus \longrightarrow \underset{\substack{\|\\O}}{(EtO)_2P}-O^\ominus + PO_3^\ominus \tag{7.12}
$$

$$
PO_3^\ominus + H_2O \rightleftharpoons H_2PO_4^\ominus \rightleftharpoons H^\oplus + HPO_4^{2\ominus}
$$

Several nucleophiles exhibiting the alpha effect serve as catalysts for carboxylic and phosphoric acid derivatives. These catalysts include nitrite, hypochlorite, and oximate ions. Nitrite ion catalyzes the hydrolyses of acetic anhydride [Eq. (7.13)] and bis-(dimethylamino)-phosphorochloridate [Eq. (7.14)]

$$
\underset{\substack{\|\\O}}{CH_3C}\underset{\substack{\|\\O}}{OCCH_3} + NO_2{}^\ominus \longrightarrow \underset{\substack{\|\\O}}{CH_3C}ONO + CH_3CO_2{}^\ominus
$$

$$CH_3CONO - \left\{ \begin{array}{l} \xrightarrow{H_2O} \quad CH_3\overset{\displaystyle O}{\overset{\|}{C}}OH + NO_2{}^{\ominus} \\ \\ \xrightarrow{ NH_2 } \end{array} \right.$$

(7.13)

$$\begin{array}{c} \text{N=N} \\ \text{(naphthalene ring)} \quad \text{(naphthalene ring)} + CH_3CO_2H \\ NH_2 \end{array}$$

$$\begin{array}{ccc} \underset{Me_2N}{\overset{Me_2N}{\diagdown}}P\underset{Cl}{\overset{O}{\diagup}} & \xrightarrow[-Cl^{\ominus}]{+ONO^{\ominus}} & \underset{Me_2N}{\overset{Me_2N}{\diagdown}}P\underset{ONO}{\overset{O}{\diagup}} & \xrightarrow[-ONO^{\ominus}]{+H_2O} & \underset{Me_2N}{\overset{Me_2N}{\diagdown}}P\underset{OH}{\overset{O}{\diagup}} \end{array}$$

(7.14)

Evidence for the postulated intermediate, acetyl nitrite [Eq. (7.13)] has been obtained by the addition of α-naphthylamine to the system. This diverts acetyl nitrite from its usual hydrolytic path to form 4-amino-1,1′-azonaphthalene.

Hypochlorite ion is an efficient catalyst for the hydrolysis of isopropyl methylphosphonofluoridate (Sarin). This reaction shows first-order dependence on both substrate and hypochlorite ion, but the latter is not consumed. Anions such as chloride ion have no effect. On this basis, mechanism (7.15) can be written:

$$\begin{array}{ccc} \underset{RO}{\overset{RO}{\diagdown}}P\underset{F}{\overset{O}{\diagup}} & \xrightarrow[-F^{\ominus}]{+OCl^{\ominus}} & \underset{RO}{\overset{RO}{\diagdown}}P\underset{OCl}{\overset{O}{\diagup}} & \xrightarrow[-OCl^{\ominus}]{+H_2O} & \underset{RO}{\overset{RO}{\diagdown}}P\underset{OH}{\overset{O}{\diagup}} \end{array}$$

(7.15)

In the second step, cleavage of the chlorine–oxygen bond may take place. Hypochlorite ion shows approximately the same nucleophilicity towards Sarin as hydroxide ion, even though its basicity is eight powers of ten less than that of hydroxide ion. This fact, together with the instability of acyl hypochlorites, leads to a facile nucleophilic catalysis.

7.2.5. Tertiary Amines As Catalysts

The synthetic utility of pyridine in acylation reactions has long been recognized. The acylations of amines, alcohols, and phenols with an acyl chloride or anhydride in pyridine solution or with catalytic amounts of pyridine in inert solvents are well-known synthetic procedures. These catalyses can be explained on the same basis as the pyridine-catalyzed hydrolysis of acetic anhydride, a reaction intensively investigated from many points of view. A number of generalizations can be made about pyridine catalysis. (1) When pyridines are substituted so that they do not sterically inhibit proton abstraction, these bases show a linear relationship between log k_{obs} and pK_a. However, sterically hindered compounds such as 2-picoline, 2,6-lutidine, and 2,4,6-collidine are not catalysts at all. (2) The addition of acetate ion significantly decelerates the reaction. (3) Added hydroxylamine traps an intermediate as a hydroxamic acid. (4) The postulated intermediate, acetylpyridinium chloride, was isolated from the reaction of acetyl chloride and pyridine under anhydrous conditions. These data are consistent with Eq. (7.16).

$$CH_3\overset{O}{\overset{\|}{C}}O\overset{O}{\overset{\|}{C}}CH_3 + \underset{N}{\bigodot} \rightleftharpoons CH_3\overset{O}{\overset{\|}{C}}N\underset{\oplus}{\bigodot} + {}^{\ominus}O\overset{O}{\overset{\|}{C}}CH_3$$

$$\downarrow H_2O \qquad\qquad (7.16)$$

$$\underset{N}{\bigodot} + H^{\oplus} + HO\overset{O}{\overset{\|}{C}}CH_3$$

The intermediate acylammonium ion is probably the most reactive acylating agent known. It is not to be confused with the protonated form of an amide, where protonation occurs on oxygen with resultant conservation of resonance stabilization.

The use of pyridines as acylation catalysts has been substantially improved by the introduction of alkyl groups in the 4-position as in the case of 4-dimethyl-amino pyridine (DMAP) and 4-pyrolidinopyridine (PPX). Unfortunately, these catalysts are only of value in aprotic nonpolar solvents. Because of their enhanced nucleophilicity, these bases form high concentrations of N-acyl-pyridinium salts on reaction with acylating agents in these apolar solvents. These N-acyl- 4-dialkylaminopyridinium salts are loosely bound ion pairs. Attack of the nucleophile on the activated acylium ion is further facilitated by the general base catalysis of the counterion. These nucleophilic catalysts are thousands of times more effective than pyridine itself. We will describe further such multiple catalyses in Chapter 11.

The hydrolysis of acetyl phosphate is catalyzed by tertiary amines, and in the reactions catalyzed by pyridine, 4-methylpyridine, triethylene-diamine, and trimethylamine, the reaction occurs with P—O cleavage. This is nucleophilic catalysis since an intermediate phosphorylated tertiary amine was trapped by fluoride ion to give phosphonofluoridate.

$$CH_3\overset{O}{\overset{\|}{C}}O-\overset{O^{\ominus}}{\underset{O^{\ominus}}{\overset{|}{P}}}=O \xrightarrow[-CH_3CO_2^{\ominus}]{+NR_3} \overset{{}^{\ominus}O}{\underset{{}^{\ominus}O}{\overset{\|}{O}}}=\overset{\oplus}{\underset{}{P}}-NR_3 \quad \begin{array}{l} \xrightarrow{H_2O} H_2PO_4^{\ominus} \\[2ex] \xrightarrow{F^{\ominus}} HOPO_2F^{\ominus} \end{array} \quad (7.17)$$

7.2.6. Imidazole As Catalyst

The tertiary amine, imidazole, has achieved prominence in nucleophilic catalysis because:

1. As part of the side chain of the amino acid histidine, it is a constituent of essentially every enzyme.

2. It is an amine nucleophile of pK_a 7.0 and therefore it can operate efficiently at neutrality both as a base and nucleophile.

3. Acylimidazoles and phosphorylimidazoles, the initial products of nucleophilic attack on acyl and phosphoryl derivatives, are reactive species.

Nucleophilic reactions of many carboxylic acid derivatives are catalyzed by imidazole.[12] The imidazole-catalyzed hydrolysis of p-nitrophenyl acetate is illustrative of the process. Kinetic experiments with substrate concentration much larger than the imidazole concentration show a catalytic process depending on the unionized form of the nucleophile. Complementary experiments with the imidazole concentration much larger than the substrate concentration also show the same kinetic dependency; in addition, the formation and decomposition of an unstable intermediate was followed spectroscopically. The intermediate can be isolated from the reaction in nonaqueous medium and shown to be N-acetylimidazole. Furthermore, the kinetics of the overall reaction are completely accounted for by the formation and decomposition of N-acetylimidazole. On this basis, the mechanism must be described by the nucleophilic catalysis of Eq. (7.18), where the first step is ordinarily rate-determining.[12]

Nucleophilic catalysis by imidazole combines the properties of a good nucleophile with an intermediate that is quite unstable. In nucleophilicity toward p-nitrophenyl acetate, imidazole is approximately 10^5 times better than acetate ion. However, in the hydrolysis of the intermediates formed by nucleophilic attack, N-acetylimidazole is only tenfold slower than acetic anhydride, which hydrolyzes quite easily. Thus the requirements of an efficient nucleophilic catalysis—high nucleophilicity and an unstable intermediate—are met.

The reason for the simultaneous occurrence of these two properties lies in the unique nature of the intermediate that is subject to nucleophilic attack by neutral nucleophiles such as water through the ready protonation of the N-acylimidazole ($pK_a \sim 5$) to a species resembling an acyltrialkylammonium ion, which is markedly susceptible to nucleophilic attack. A superficial way to account for the nucleophilic instability of N-acylimidazoles is to view these compounds as nitrogen analogs of anhydrides.

Many other imidazole and histidine derivatives serve as catalysts for the hydrolysis of p-nitrophenyl acetate, including N-methylimidazole, benzimidazole derivatives, and histidine-containing peptides, as shown in Table 7.3. Since catalysis by N-methylimidazole is approximately equivalent to that by imidazole, a neutral N-acylimidazole is not a necessary intermediate in this catalysis. The catalytic rate constants of many imidazole derivatives of constant steric requirement are dependent on the pK_a of the catalyst as in the pyridine family. Thus, in comparing catalysis by various imidazole derivatives, the relative basicity of the catalysts must be considered first.

A favorite occupation has been to attempt to synthesize polypeptides showing a greater catalytic activity than imidazole toward p-nitrophenyl acetate.[13] Table 7.3 indicates the limited success so far in this pursuit. The most effective peptide is L-seryl-γ-aminobutyryl-L-histidyl-γ-aminobutyryl-L-aspartic acid, which has a catalytic rate constant toward p-nitrophenyl acetate sevenfold greater than that of imidazole; the pK_a of this peptide is unknown. This peptide exhibits partial stereoselectivity in the hydrolysis of optically active phenylalanine esters. A polymer of ethylenimine containing histidine moieties also has catalytic activity toward p-nitrophenyl acetate (see Chapter 12).

Imidazole may also serve as a nucleophilic catalyst for the hydrolysis of phosphate derivatives. Thus, although hindered 2,6-lutidine acts as a general base catalyst in the solvolysis of tetrabenzyl pyrophosphate, unhindered imidazole and N-methylimidazole serve as nucleophilic catalysts in this reaction. The catalysis results from nucleophilic attack by the amine on phosphorus, forming N-(dibenzylphosphoryl)imidazolium ion by displacing dibenzyl phosphate ion. The kinetics of the overall process with imidazole are in accord with the intermediacy of N-(dibenzylphosphoryl)imidazole.

(7.19)

Many other reactions show catalysis by tertiary amines, phosphines, sulfides, tertiary amine oxides, and phosphine oxides. Some examples are the dialkyl sulfide catalysis of the disulfide–sulfinic acid reaction, the tertiary phosphine-

Table 7.3. Catalysis by Imidazoles of the Hydrolysis of _p_-Nitrophenyl Acetate

Catalyst	pK_a	k (sec^{-1})
Imidazoles		
Imidazole	6.95	20.2
2-Methylimidazole	7.75	2.7
4-Methylimidazole	7.45	25.1
_N_-Methylimidazole	7.05	0.5
4-Bromoimidazole	3.7	0.28
4-Hydroxymethylimidazole	6.45	5.6
4-Nitroimidazole	1.5 (9.1)	35.5
Histidines		
Histidine	6.0	
_N_-Acetylhistidine	7.05	11.2
Histidine methyl ester	5.2	5.6
_N_-Benzoyl-L-histidine methyl ester		
1-Methylhistidine	6.5	
β-Aspartylhistidine	6.9	
Histidylhistidine	6.8	
Histamine	6.0	7.0
Carnosine	6.8	
Anserine	7.0	
Carbobenzoxy-L-histidyl-L-tyrosine ethyl ester	6.25	8.9
8:1 copolymer of alanine + histidine		6.0
Poly-L-histidine		
Bacitracin	~6	<20
Gly-L-His-L-Ser		15
Copoly L-His-L-Ser		9.7
L-Ser-L-His-L-Asp		45
L-Thr-L-Ala-L-Ser-L-His-L-Asp		92
L-Ser-γ-NH$_2$Bu-L-His-γ-NH$_2$Bu-L-Asp		147
Cyclo L-Gly-L-His-L-Ser-Gly-L-His-L-Ser		7
Copoly L-His-L-Asp		~6
Copoly L-His-L-Lys		~6
Copoly L-His-L-Gln		~6
Benzimidazoles		
Benzimidazole	5.4	0.96
2-Methylbenzimidazole	6.1	0.0375
6-Aminobenzimidazole	6.0	2.95
6-Nitrobenzimidazole	3.05	4.8
4-Hydroxybenzimidazole	5.3	2.8
4-Methoxybenzimidazole	5.1	0.31
4-Hydroxy-6-nitrobenzimidazole	3.05	3.75
4-Hydroxy-6-aminobenzimidazole	5.9	6.15
2-Methyl-4-hydroxy-6-nitrobenzimidazole	3.9	1.1
2-Methyl-4-hydroxy-6-aminobenzimidazole	6.65	1.5
4-(2′,4′-Dihydroxyphenyl)imidazole	6.45	9.4

Source: M. L. Bender, _Chem. Rev._, **60**, 53 (1960). © 1960 by the Williams and Wilkins Co. Reprinted by permission of the copyright owner.

catalyzed transesterification of phosphates, phosphonates, and phosphinates with alkyl halides, the phosphine oxide–catalyzed conversion of isocyanates to carbodiimides, the pyridine N-oxide–catalyzed reaction of isocyanates with alcohols, the trimethylamine-catalyzed alcoholysis of N-benzoylphosphoramidate, and the dimethylformamide-catalyzed reactions of phosphorochloridates with alcohols, acids, and amines. Presumably all these reactions involve nucleophilic catalysis with the formation of unstable intermediates. Only in the last reaction has the formation of a reactive intermediate been rigorously demonstrated.

7.2.7.   Catalysis by Primary and Secondary Amines

Many reactions of carbonyl compounds show specific catalysis by ammonia and primary and secondary amines. These reactions include the retrograde aldol reaction of diacetone alcohol, the Michael condensation, the decarboxylation of β-keto acids,[14] the Knoevenagel condensation, semicarbazone formation, and the enolization of acetone. In these reactions, the catalyst and the substrate, in a facile nucleophilic reaction, form an imine or protonated imine that can serve as an intermediate for further reaction. The equilibrium between carbonyl compounds and imines is known to be established rapidly under relatively mild conditions.

   Cations of amines also catalyze some reactions by the formation of imine intermediates.[15] For example, anilinium ions catalyze the formation of semicarbazones from benzaldehydes [Eq. (7.20)]. This catalysis is ten to a thousand-fold more efficient than catalysis by other acids of comparable strength. The following data indicate the presence of an imine intermediate in this catalysis. (1) N-benzylidineanilines in dilute solutions of semicarbazide partition to aldehyde and semicarbazone, the fraction of semicarbazone formed being dependent on the concentration of semicarbazide. (2) The rate of the anilinium ion–catalyzed semicarbazone formation is independent of semicarbazide concentration except in very dilute solutions; in fact, the rate of imine formation accounts quantitatively for the rate of the anilinium ion-catalyzed semicarbazone formation. (3) The rate of anilinium ion-catalyzed semicarbazone formation is identical to the rate of anilinium ion-catalyzed oxime formation. These results taken together strongly suggest that catalysis by anilinium ion must proceed by a nucleophilic pathway.

$$\underset{H}{\overset{\phi}{>}}\!=\!O + \phi NH_3^{\oplus} \underset{\rightleftarrows}{\overset{slow}{\quad}} \underset{H}{\overset{\phi}{>}}\!=\!NH^{\oplus}\!-\!\phi + H_2O$$

$$(7.20)$$

$$\underset{H}{\overset{\phi}{>}}\!=\!NH^{\oplus}\!-\!\phi + NH_2NH\overset{O}{\overset{\|}{C}}NH_2 \longrightarrow \underset{H}{\overset{\phi}{>}}\!=\!NNH\overset{O}{\overset{\|}{C}}NH_2 + \phi NH_3^{\oplus}$$

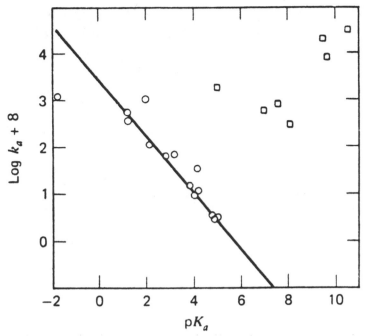

Fig. 7.2. Catalysis of the enolization of acetone by carboxylic acids (O) and ammonium ions (□). The line is an arbitrary line through the carboxylic acid points, giving $\alpha = 0.66$. From M. L. Bender and A. Williams, *J. Am. Chem. Soc.*, **88**, 2504 (1966). Copyright © 1966 by the American Chemical Society. Reproduced by permission of the copyright owner.

Ammonium ion catalysis is also seen in the iodination of acetone and the α-hydrogen exchange of isobutyraldehyde-2-*d*. The deuterium exchange of isobutyraldehyde-2-*d* in the presence of free methylamine, where the aldehyde exists largely as the imine, depends on the methylammonium ion concentration; an important reaction path may therefore involve the rate-controlling removal of deuterium by methylamine from the conjugate acid of the imine. In the iodination of acetone, although amine catalysis adheres to the Brönsted plot defined by other general bases, ammonium ion catalysis is as much as 10^6-fold faster than that predicted on the basis of carboxylic acids of comparable acidity (Fig. 7.2). On this basis, the ammonium ion catalysis of iodination proceeds via imine formation.

The retrograde aldol reaction of diacetone alcohol proceeds with formation of the imine intermediate as shown in Eq. (7.21). Such catalyses are specific to catalysts that are able to form imines, such as primary and secondary amines as well as amino acids, whereas other bases such as tertiary amines or phenoxide ions cannot catalyze the reaction.

$$(CH_3)_2\overset{\overset{\displaystyle OH}{|}}{C}-CH_2-\overset{\overset{\displaystyle O}{||}}{C}-CH_3 \underset{\overset{R_2NH}{\rightleftharpoons}}{} (CH_3)_2\overset{\overset{\displaystyle OH}{|}}{C}-CH_2-\overset{\overset{\displaystyle \overset{\oplus}{N}R_2}{||}}{C}-CH_3 \qquad (7.21)$$

$$(CH_3)_2C{-}CH_2{-}\overset{\overset{\oplus}{N}R_2}{\underset{\|}{C}}{-}CH_3 \qquad \begin{matrix}(7.21)\\(cont.)\end{matrix}$$

$$2\ CH_3\overset{O}{\overset{\|}{C}}CH_3 + HNR_2 \underset{H_2O}{\rightleftharpoons} CH_3\overset{O}{\overset{\|}{C}} + CH_2{=}\overset{NR_2}{\overset{|}{C}}{-}CH_3$$

The intermediate formation of imines is also seen in an aniline-catalyzed Michael condensation. A plausible mechanism is shown in Eq. 7.22.

$$\phi{-}CH{=}CH{-}\overset{O}{\overset{\|}{C}}{-}CH_3 \rightleftharpoons \phi{-}CH{-}CH_2{-}\overset{\overset{\oplus}{N}H\phi}{\underset{\|}{C}}{-}CH_3 \rightleftharpoons \phi{-}CH{=}CH{-}\overset{\overset{\oplus}{N}H\phi}{\underset{\|}{C}}{-}CH_3$$

$$\underset{\overset{\oplus}{N}H_2\phi}{}$$

$$(7.22)$$

Catalyses by primary and secondary amines have several common features: (1) the substrate is a carbonyl compound: (2) the catalysis is almost always proceeds through an imine or immonium ion intermediate formed in a facile reaction under mild conditions; (3) the presence of the imino group in the molecule facilitates subsequent reaction, whether it be a condensation, de-carboxylation, or enolization, by stabilization of the transition state through resonance or electrostatic interactions or both; and (4) the amine catalyst is regenerated by a facile reversion of imine formation. Although the mechanisms have not been rigorously proven, they form a consistent pattern.

7.2.8. Catalysis by Cyanide Ion and Thiazolium Ion

These catalysts are two of the most specific nucleophilic catalysts known, promoting the acyloin condensation, the decarboxylation of α-keto acids,[16,17] and related reactions. The essential features of the acyloin condensation follow.

$$
\text{PhCHO} \underset{}{\overset{CN^{\ominus}}{\rightleftharpoons}} \;
\begin{array}{c} O^{\ominus} \\ | \\ Ph-C-H \\ | \\ CN \end{array}
\underset{B}{\overset{BH^{\oplus}}{\rightleftharpoons}}
\begin{array}{c} OH \\ | \\ Ph-C-H \\ | \\ CN \end{array}
\underset{BH^{\ominus}}{\overset{B}{\rightleftharpoons}}
$$

$$
\left[\;
\begin{array}{c} OH \\ | \\ Ph-C^{\ominus} \\ ||| \\ N \end{array}
\longleftrightarrow
\begin{array}{c} OH \\ | \\ Ph-C \\ || \\ C \\ || \\ N^{\ominus} \end{array}
\;\right] \overset{PhCHO}{\rightleftharpoons}
\tag{7.23}
$$

$$
\begin{array}{c} HO\;\;\;O^{\ominus} \\ |\;\;\;\;\;| \\ Ph-C-CPh \\ |\;\;\;\;\;| \\ NC\;\;\;H \end{array}
\rightleftharpoons
\begin{array}{c} O^{\ominus}\;OH \\ |\;\;\;\;| \\ Ph-C-C-Ph \\ |\;\;\;\;| \\ NC\;\;\;H \end{array}
\overset{-CN^{\ominus}}{\rightleftharpoons}
\begin{array}{c} O\;\;\;OH \\ ||\;\;\;\;| \\ Ph-C-C-Ph \\ |\\ H \end{array}
$$

This condensation must involve the activation of the hydrogen atom attached to a carbonyl group, but a carbanion formed by the removal of such a hydrogen atom has no obvious stabilizing feature in the absence of cyanide ion. However, if cyanide ion is added to benzaldehyde, for example, to form an anion of mandelonitrile, equilibrium between this anion and its conjugate acid produces a species equivalent to the carbanion formed by removal of the hydrogen atom attached to the carbonyl group. This carbanion, stabilized by resonance interaction with the cyano group, can participate in the subsequent condensation. The cyanide ion can then leave the product by reversal of the original carbonyl addition. The unique characteristics of the cyanide ion are its facile, reversible addition to carbonyl groups and its ability to stabilize a carbanionic charge on an adjacent carbon atom.

Cyanide ion catalyzes both the decarboxylation and acyloin condensation of α-keto acids.[16] Catalysis of decarboxylation of an α-keto acid by cyanide ion is shown in Eq. (7.24).[5]

$$
\begin{array}{c} O \\ || \\ CH_3-C-CO_2^{\ominus} \end{array}
\overset{CN^{\ominus}}{\rightleftharpoons}
\begin{array}{c} O^{\ominus} \\ | \\ CH_3-C-CO_2^{\ominus} \\ | \\ CN \end{array}
\underset{B}{\overset{BH^{\oplus}}{\rightleftharpoons}}
\begin{array}{c} OH \\ | \\ CH_3-C-C \\ | \\ CN \;\;\; O \end{array}
\overset{-CO_2}{\rightleftharpoons}
$$

$$
\left[\;
\begin{array}{c} OH \\ | \\ CH_3-C^{\ominus} \\ | \\ C \\ ||| \\ N \end{array}
\longleftrightarrow
\begin{array}{c} OH \\ | \\ CH_3-C \\ | \\ C \\ || \\ N^{\ominus} \end{array}
\;\right]
\underset{B}{\overset{BH^{\oplus}}{\rightleftharpoons}}
\begin{array}{c} OH \\ | \\ CH_3-CH \\ | \\ CN \end{array}
\underset{BH^{\ominus}}{\overset{B}{\rightleftharpoons}}
\begin{array}{c} O^{\ominus} \\ | \\ CH_3-CH \\ | \\ CN \end{array}
\overset{-CN^{\ominus}}{\rightleftharpoons}
\begin{array}{c} O \\ || \\ CH_3CH \end{array}
\tag{7.24}
$$

This catalysis, like the one in Eq. (7.23), depends on the capacity of cyanide ion to add rapidly and reversibly to the carbonyl group to give a species that can stabilize a negative charge.[17] Ordinarily the product of this reaction is an acyloin.

However, when the reaction is carried out in an aqueous dioxane solution of dimedone (5,5′-dimethyldihydroresorcinol), the aldehyde can be trapped as its dimedone derivative.

Although cyanide ion is an efficient catalyst for the decarboxylation of α-keto acids and the acyloin condensation, it is not unique. The coenzyme thiamine catalyzes the acyloin condensation in the absence of enzyme; likewise, other thiazolium ions catalyze this reaction in mildly basic solution. The decarboxylation of α-keto acids is also accelerated (nonenzymatically) by thiamine. It was pointed out[17] that one of the resonance forms of the thiazolium salts, a zwitterion, is analogous to the resonance hybrid of cyanide ion [Eq. (7.25)].

$$(7.25)$$

Both the thiazolium and cyanide ions are strongly stabilized by resonance involving two canonical forms, one of which has a six-electron carbon atom [Eq. (7.25)]. The thiazolium bond zwitterion is a member of a wide class of substances in which this resonance occurs. That the negative charge is concentrated on the suggested carbon is consistent with the observation that one atom of deuterium is incorporated when D_2O reacts with the anion of thiazolium ion at pH 5.[16]

Given that a zwitterion such as that in Eq. (7.25) exists, it is then possible to formulate a mechanism by which it might catalyze the acyloin condensation and the decarboxylation of α-keto acids as cyanide ion does, but without its poisonous character; that is, thiamine pyrophosphate is functionally "biological cyanide." Equation (7.26) illustrates the mechanism of the thiazolium ion catalyzed decarboxylation and acyloin condensation of pyruvic acid.[17,18] Good tracer evidence exists for the intermediate formed between pyruvate and the thiazolium ion. Furthermore, the adduct of acetaldehyde and the thiazolium ion has been isolated from the reactions of the enzyme carboxylase.

$$(7.26)$$

Thiazolium ion   Acyloin of acetaldehyde

The thiazolium ion also catalyzes the transketolase reaction both non-enzymatically and enzymatically. A typical example is the reaction of fructose-6-phosphate with glyceraldehyde-3-phosphate to give a tetrose phosphate and xylulose-5-phosphate. This reaction, given in Eq. (7.27), consists of a retrograde aldol condensation leading to the release of one aldehyde group, followed by the aldol condensation of the acetol adduct of the thiazolium ion with a new aldehyde to form the product.

$$(7.27)$$

$R = -CHOHCHOHCH_2OPO_3H^{\ominus}$
$R' = -CHOHCH_2OPO_3H^{\ominus}$

In this catalysis, the carbanionic intermediate is again resonance-stabilized by the cationic nitrogen which serves as an "electron sink."

7.2.9.   Catalysis by Enzymatic Nucleophiles

There are currently only two established enzymatic nucleophiles (of apo-enzymes): the alcoholic (alkoxide) group of the amino acids serine or threonine and the sulfhydryl (or its conjugate base) group of the amino acid cysteine. The former has a pK_a of 13.2, while the latter has a pK_a of 8.9; the pK_a of the hydroxyl group is abnormally low while that of the sulfhydryl group is not unusual. Both these nucleophiles require the assistance of general bases to be converted to their more reactive anions.

In addition to nucleophilic catalysts of the enzyme protein itself, there are also coenzymic nucleophilic catalysts such as thiamine pyrophosphate and tetra-hydrofolic acid.

7.2.10.   The Transition Between Nucleophilic and General Base Catalysis

We have considered reactions that are subject to nucleophilic or to general base catalysis. Several generalizations can be made as to whether the functionality of interest will act as a nucleophilic or general base catalyst.

1.   Those bases of nucleophiles that are *polarizable*, or *soft*, will tend to act as nucleophilic catalysts. On the other hand, those bases or nucleophiles that are *not polarizable*, but *hard*, will tend to act as general base catalysts.

2.   Since Brönsted plots for nucleophilic catalysis have higher slopes than for general base catalysis, stronger bases will tend to act as nucleophiles, whereas weaker bases will tend to react as general base catalysts. One might then expect a break in a Brönsted plot, weaker bases exhibiting general base catalysis with a lower slope and stronger bases exhibiting nucleophilic catalysis with a higher slope, as shown in Fig. 7.3. The reaction shown in Fig. 7.3 is not entirely hypothetical. It represents in a schematic way the hydrolysis of ethyl dichloro-acetate catalyzed by a series of bases. With bases weaker than ammonia, hydrolysis occurs via general basic catalysis. With bases of the strength of ammonia or higher, a nucleophilic reaction occurs that may lead either to hydrolysis or to other reactions.

3.   The question can also be asked, why is a particular member of a substrate family susceptible to general base catalysis while another is susceptible to nucleophilic catalysis? This dichotomy of mechanism is seen in imidazole catalysis of ester hydrolysis when the structure of the ester is varied. Esters activated in the acyl group and also containing a poor leaving group show general base catalysis by imidazole. On the other hand, esters with a good leaving group are subject to nucleophilic catalysis by imidazole. Likewise, substituted phenyl acetates containing highly electron-withdrawing substituents are subject to nucleophilic catalysis by acetate ion while those with other substituents are subject to general base catalysis by acetate ion. The transition from general base

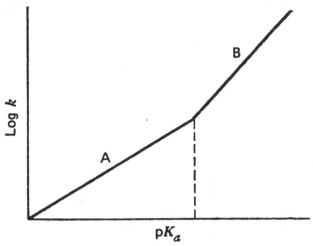

Fig. 7.3.   The Brönsted plot of the general base (A) and nucleophilic (B) catalysis of the hydrolysis of ethyl dichloroacetate. From M. L. Bender, *Mechanisms of Homogeneous Catalysis from Protons to Proteins*, Wiley-Interscience, New York, 1971, p. 177.

to nucleophilic catalysis by imidazole has been analyzed using a structure-reactivity correlation involving the comparison of the rate constants of the imidazole-catalyzed and hydroxide ion–catalyzed reactions of a series of esters, as shown in Fig. 7.4. The hydroxide ion rate constants are assumed to follow a common mechanism for all esters and are used as an empirical measure of structural change rather than a Hammett σ (aromatic) or Taft σ^* (aliphatic) constant. By comparing the rate constants of imidazole catalysis to the rate constants of hydroxide ion catalysis, the structural changes affecting reactivity will automatically be taken into account; thus one can determine structural variations important to the change in catalysis. In this analysis, another factor must be considered in addition to the change in the catalytic mechanism; that is, the rate-determining step of the two-step ester hydrolysis can be changed by structural variation.

The reactions on the upper line of Fig. 7.4 are identified as nucleophilic catalyses on the basis of the demonstration of an *N*-acetylimidazole intermediate. The reactions on the lower line are assumed to be general base-catalyzed reactions. A number of these reactions are discussed in Chapter 6.

At the top of Fig. 7.4 lie a series of acetate derivatives with excellent leaving groups. As we proceed down the upper line of the graph, the leaving groups become progressively poorer. With good leaving groups, essentially every tetrahedral intermediate formed by the addition of imidazole to the carbonyl group is partitioned to products; there will be little effect of structural change of the leaving group on the rate constant and thus the imidazole and hydroxide ion reactions parallel one another well (slope 1.2). As the leaving group becomes poorer, the energy barrier for the decomposition of the intermediate to products (in the imidazole reactions) becomes larger than for the regeneration of the

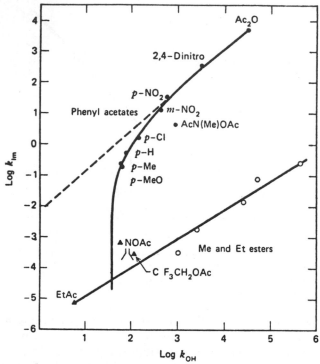

Fig. 7.4. Rates of imidazole-catalyzed ester hydrolysis as a function of the rate of alkaline hydrolysis; nucleophilic reaction of acetates (●); general base catalysis of acetates (▲); general base catalysis of methyl and ethyl esters (O) (ionic strength 1.0; 25°C). Trifluoroethyl acetate rate measured with N-methylimidazole. From J. F. Kirsch and W. P. Jencks, *J. Am. Chem. Soc.*, **86**, 843 (1964). Copyright © 1964 by the American Chemical Society. Reproduced by permission of the copyright owner.

starting material and thus the former becomes rate-determining (Fig. 7.5).[19] Thus a given structural change in the leaving group will be directly reflected in the rate constant. This leads to the downward curvature in the upper line of Fig. 7.4. Finally, as the leaving group becomes still worse, the rate constant of the nucleophilic reaction of imidazole with ester becomes so small that it is not observed; a general base-catalyzed reaction that is less sensitive to the nature of the leaving group takes over, bringing us to the ethyl acetate reaction at the bottom of the sigmoidlike curve. Two breaks are thus seen in this curve, one reflecting the change in rate-determining step from the formation of the tetrahedral intermediate to its decomposition and the second reflecting the transition from nucleophilic to general base catalysis.

Other changes in catalytic mechanism can also occur. While compounds that have very good leaving groups, such as acetic anhydride and p-nitrophenyl acetate have reactions first-order in imidazole, acetate esters of less acidic phenols can have reactions second-order in imidazole (first order in nucleophile and first order in general base) as well as ones first-order in imidazole (as a

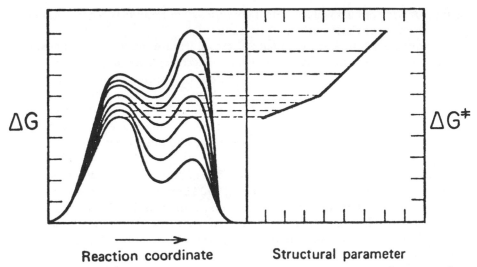

ΔG

Reaction coordinate   Structural parameter

Fig. 7.5. Transition state diagram for a two-step reaction showing how changes in structure that affect principally the second step result in a nonlinear structure-reactivity correlation. From J. F. Kirsch and W. P. Jencks, *J. Am. Chem. Soc.*, **86**, 845 (1964). Copyright © 1964 by the American Chemical Society. Reproduced by permission of the copyright owner.

nucleophile). In comparable reactions of the acetates of acetoxime, trifluoro-ethanol and still worse leaving groups, hydroxide ion is needed to effect nucleophilic catalysis. Finally, ethyl acetate shows no nucleophilic reaction at all.

7.3. ELECTROPHILIC CATALYSIS

Electrophilic catalysis, outside of metal ion catalysis which is treated separately and pyridoxal phosphate, is not of great importance in enzymatic catalysis; therefore, it is treated lightly here. It consists of electrophilic attack by the catalyst on the substrate producing an unstable cationic intermediate that decomposes to give the product and regenerate the catalyst.

7.3.1. Electrophilicity

Although electrophilic catalysts include metal ions, we will discuss metal ion catalysis in Chapter 9 because of its somewhat special character, and concentrate here on reactions catalyzed by electrophilic metal halides, halogens, and carbonyl compounds. No ambiguities exist between electrophilic and general acid catalysts since we arbitrarily define electrophiles as Lewis acids while general acids are defined as proton donors.

Since the order of electrophilicity among electrophiles depends on the nucleophile involved, one must call on the *hard* and *soft* theory for electrophiles. This predicts that *hard* electrophiles will react strongly with *hard* nucleophiles while *soft* electrophiles will react strongly with *soft* nucleophiles. Table 7.4 shows a classification of electrophiles.

The distinguishing features of *hard* electrophiles are small size, high positive oxidation state, and the absence of any outer electrons that are easily excited to higher electronic states. In addition, the hardness of a given acceptor atom is a function of the other groups attached to it. Thus BF_3 is a *hard* electrophile, but BH_3 is a *soft* electrophile, the latter forming complexes such as BH_3CO. *Soft* electrophiles in general have one or more of the following properties: low or zero positive charge, large size, and several easily excited outer electrons. In metals, these outer electrons are d-orbital electrons.

Since hydrogen-bonding molecules are hard, stronger hydrogen bonds are formed to N, O, and F donors than to P, S, and I donors. A similar analysis may be made of the difference in rate constants of proton transfers from oxygen and nitrogen acids versus carbon acids. As mentioned earlier, the *hardness–softness* principle may be used as a guide to selecting electrophilic and nucleophilic catalysts.

7.3.2. Catalysis by Lewis-Acid Metal Halides, Hydrides, and Carbonium Ions

An electrophilic catalysis common to the experience of practically all organic chemists is the use of metal halides of the Lewis-acid variety for accelerating reactions proceeding through carbonium ion intermediates. In these catalyses, assistance in carbonium ion formation is the important factor.

Table 7.4. Classification of Electrophiles[a]

Hard (Class a)	Soft (Class b)
$H^{\oplus}$, $Li^{\oplus}$, $Na^{\oplus}$, $K^{\oplus}$	$Cu^{\oplus}$, $Ag^{\oplus}$, $Au^{\oplus}$, $Tl^{\oplus}$, $Hg^{\oplus}$, $Cs^{\oplus}$
$Be^{\oplus 2}$, $Mg^{\oplus 2}$, $Ca^{\oplus 2}$, $Sr^{\oplus 2}$, $Mn^{\oplus 2}$	$Pd^{\oplus 2}$, $Cd^{\oplus 2}$, $Pt^{\oplus 2}$, $Hg^{\oplus 2}$, $CH_3Hg^{\oplus}$
$Al^{\oplus 3}$, $Sc^{\oplus 3}$, $Ga^{\oplus 3}$, $In^{\oplus 3}$, $La^{\oplus 3}$	$Tl^{\oplus 3}$, $Au^{\oplus 3}$, $Te^{\oplus 4}$, $Pt^{\oplus 4}$
$Cr^{\oplus 3}$, $Co^{\oplus 3}$, $Fe^{\oplus 3}$, $As^{\oplus 3}$, $Ce^{\oplus 3}$	$Tl(CH_3)_3$, BH_3, $Co(CN)_5^{\ominus 2}$
$Si^{\oplus 4}$, $Ti^{\oplus 4}$, $Zr^{\oplus 4}$, $Th^{\oplus 4}$, $Pu^{\oplus 4}$	$RS^{\ominus}$, $RSe^{\ominus}$, $RTe^{\ominus}$
$Ce^{\oplus 4}$, $Ge^{\oplus 4}$, $VO^{\oplus 2}$	$I^{\ominus}$, $Br^{\ominus}$, $HO^{\ominus}$, $RO^{\ominus}$
$UO_2^{\oplus 2}$, $(CH_3)_2Sn^{\oplus 2}$	I_2, Br_2, ICN, and so on
$BeMe_2$, BF_3, BCl_3, $B(OR)_3$	Trinitrobenzene, and so on
$Al(CH_3)_3$, $Ga(CH_3)_3$, $In(CH_3)_3$, AlH_3	Chloranil, quinones, and so on
$RPO_2^{\ominus}$, $ROPO_2^{\ominus}$	Tetracyanoethylene, and so on
$RSO_2^{\ominus}$, $ROSO_2^{\ominus}$, SO_3	O, Cl, Br, I, N
$I^{\oplus 7}$, $I^{\oplus 5}$, $Cl^{\oplus 7}$, $Cr^{\oplus 6}$, $Se^{\oplus 6}$	M^0 (metal atoms)
$RCO^{\oplus}$, CO_2, $NC^{\oplus}$	Bulk metals
HX (hydrogen bonding molecules)	

Source: M. L. Bender, *Mechanisms of Homogeneous Catalysis from Protons to Proteins*, Wiley-Interscience, New York, 1971.
[a]The following are in a borderline class between classes a and b: $Fe^{\oplus 2}$, $Co^{\oplus 2}$, $Ni^{\oplus 2}$, $Cu^{\oplus 2}$, $Zn^{\oplus 2}$, $Pb^{\oplus 2}$, $Sn^{\oplus 2}$, $Sb^{\oplus 3}$, $Bi^{\oplus 3}$, $Rh^{\oplus 3}$, $Ir^{\oplus 3}$, $B(CH_3)_3$, SO_2, $NO^{\oplus}$, $Ru^{\oplus 2}$, $Os^{\oplus 2}$, $R_3C^{\ominus}$.

The alkylation and acylation of aromatic and aliphatic compounds show electrophilic catalyses by metallic halides from every group of the periodic table except group IA. In these Friedel–Crafts reactions, the most common catalysts include halides of aluminum, tin, antimony, iron, zinc, boron, and gallium.

The most probable mechanism of Friedel–Crafts alkylation by BF_3 is shown by

$$RX + BF_3 \rightleftharpoons {}^{\delta\oplus}R\text{--}X\cdots{}^{\delta\ominus}BF_3$$

(7.28)

7.3.3. Catalysis by Halogens and Other Electrophiles

The halogenation of aromatics is catalyzed by many Lewis acids such as metal halides as well as the halogens themselves.[20] The bromination of aromatic compounds by bromine alone and in the presence of iodine follows.

$$k_{obs} = k[Br_2] + k'[Br_2]^2 + k''[Br_2]^3 \tag{7.29}$$

$$k_{obs} = k[Br_2]^2[IBr] + k'[Br_2][IBr]^2 \tag{7.30}$$

The higher powers of bromine and the first and second powers of iodine bromide are halogens acting as a electrophilic catalysts in facilitating cleavage of the bromine–bromine bond of the halogen directly involved in the substitution reaction.

The nitration of activated aromatic compounds is catalyzed by nitrous acid, the rate of nitration being dependent on the concentrations of aromatic compound and of nitrous acid at constant nitric acid concentration. Because this catalysis occurs only with activated aromatics and based on its kinetic dependence, this catalysis probably proceeds through electrophilic attack by $NO^{\oplus}$ forming a nitrosoaromatic intermediate which is then oxidized by nitric acid, yielding the nitroaromatic product and regenerating the nitrous acid catalyst.

7.3.4. Catalysis by Carbonyl Compounds

Since primary and secondary amines are nucleophilic catalysts for many reactions of carbonyl compounds, it may be predicted that carbonyl compounds serve as electrophilic catalysts in reactions of amines. This is clearly demonstrated in the benzaldehyde-catalyzed hydrolysis of ethyl glycinate.

$$NH_2-CH_2-\overset{\overset{\displaystyle O}{\|}}{C}-OEt + H_2O \overset{\phi CHO}{\longrightarrow} NH_2-CH_2-\overset{\overset{\displaystyle O}{\|}}{C}-OH + EtOH$$

The proposed mechanism for this catalysis is given in Eq. (7.31) below.

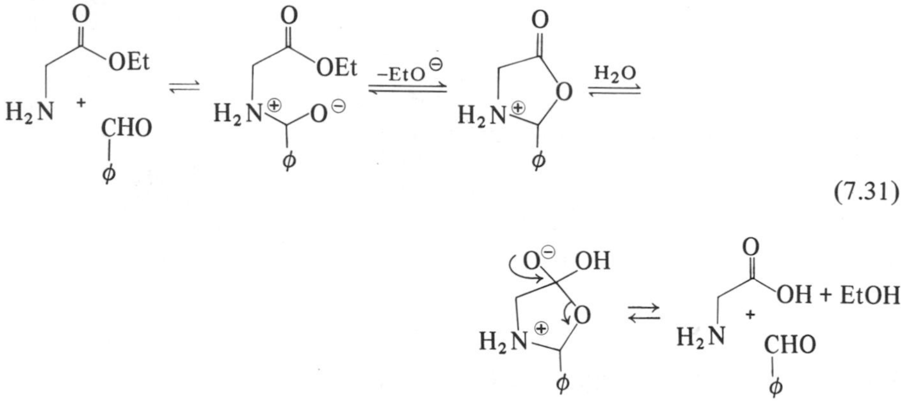

(7.31)

7.3.5. Catalysis by Enzymatic Electrophiles

The most important enzymatic electrophilic catalysis involves the coenzyme pyridoxal. It catalyzes reactions of α-amino acids such as transamination, decarboxylation, racemization, elimination, and condensation. Most of these reactions are catalyzed by pyridoxal alone, albeit less efficiently and less specifically than in the presence of an enzyme.

The nonenzymatic pyridoxal-catalyzed transamination reaction proceeds as shown in Eq. (7.32) in which one α-amino acid reacts with pyridoxal to produce pyridoxamine and an α-keto acid.[21,22]

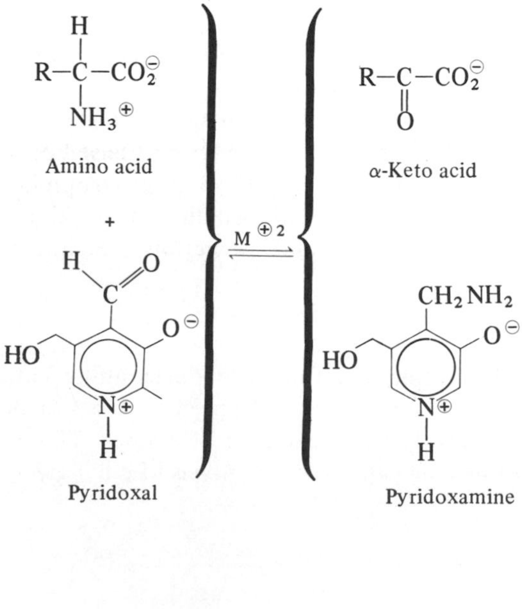

(7.32)

Since the product pyridoxamine can react with another α-keto acid, producing another amino acid and regenerating pyridoxal [the reverse reaction of Eq. (7.32)], the sum of two such reactions constitutes an overall catalysis. This process depends on the unique pi electron system that appears on imine formation. In the conversion of one imine to another, the dihydropyridine moiety serves as an electron source, and in the final step, the second imine is hydrolytically cleaved.

The essential parts of the pyridoxal catalyst are:

1. The aldehyde group for imine formation.
2. The pyridinium nitrogen atom to serve as an electron sink (and source) through a conjugated pi electron system.
3. The *ortho*-hydroxyl group to stabilize the imine.

This mechanistic description is confirmed by the replacement of pyridoxal in these catalyses by both 2-hydroxypyridine-4-carboxaldehyde and 5-deoxypyridoxal indicating that the methyl and hydroxyl groups of pyridine are not required in the catalysis, and by 2-hydroxy-4-nitrobenzaldehyde (but not by 2-hydroxy-3-nitrobenzaldehyde) indicating that any suitably oriented electron sink can replace the pyridinium ion.

Pyridoxal serves as an enzymatic electrophilic catalyst for the racemization of optically active α-amino acids. The reaction is also accelerated by metal ions. As predicted by Eq. (7.32), this reaction accompanies the transamination reaction. In the pyridoxal–alanine–aluminum ion system, racemization predominates at pH 9.6 while transamination is the major reaction at pH 5, although both reactions are competitive at all pH values. The racemization reaction may be viewed simply as the formation of an imine which no longer possesses the original asymmetric center, and reversion to starting materials which in a symmetric (nonenzymatic) environment should lead to a racemic mixture.

Pyridoxal also serves as a catalyst for the decarboxylation of α-amino acids. For example, when α-aminoisobutyric acid is heated with pyridoxal, carbon dioxide, isopropylamine, acetone, and pyridoxamine are produced. This reaction may be interpreted in terms of the cleavage of the α-carbon–carboxylate bond in the imine intermediate rather than the α-carbon–hydrogen bond. Mechanistically, the rest of the process is the same.

In the enzymatic reactions (Chapter 8), pyridoxal also catalyzes eliminations and condensations. Some of these include the conversion of serine and cysteine to pyruvate, internal oxidation–reduction reactions of the pinacol–pinacolone rearrangement type, the formation of tryptophan from serine and indole, the decomposition of threonine to glycine and acetaldehyde, and the conversion of O-phosphohomoserine to threonine. Only partial success has been achieved in effecting these reactions using pyridoxal alone. Vanadium salts in combination with pyridoxal phosphate catalyze a highly specific elimination of hydrogen sulfide from cysteine. The reaction occurs in two stages: ring fission of a thiazolidine intermediate and elimination of hydrogen sulfide. More detailed descriptions of pyridoxal catalyses are made in Chapter 8.

REFERENCES

1. W. Langenbeck, *Die Organischen Katalysatoren*, Springer, Berlin, 1935; *Adv. Enzymol.* **14,** 163 (1952).

2. W. Baker and E. Rothstein, in *Handbuch der Katalyse*, G. M. Schwab, Ed., Vol. II, Springer, Vienna, 1940, p. 45.

3. W. P. Jencks, *Catalysis in Chemistry and Enzymology*, McGraw-Hill Book Co., New York, 1969.

4. J. O. Edwards and R. G. Pearson, *J. Am. Chem. Soc.*, **84,** 16 (1962).

5. R. G. Pearson, in *Mechanisms of Inorganic Reactions*, R. F. Gould, Ed., *Advances in Chemistry Series*, No. 40, ACS Publications, Washington, 1971.

6. R. G. Pearson, *J. Am. Chem. Soc.*, **85,** 3533 (1963); *Science*, **151,** 172 (1966).

7. R. G. Pearson and J. Songstad, *J. Am. Chem. Soc.*, **89,** 1827 (1967).

8. R. G. Pearson, *J. Chem. Ed.*, **45,** 581, 643 (1968).

9. M. L. Bender, *Mechanisms of Homogeneous Catalysis from Protons to Proteins*, Wiley-Interscience, New York, 1971.

10. M. L. Bender and L. J. Brubacher, *Chemistry and Enzyme Action*, McGraw-Hill Book Co., New York, 1973.

11. D. G. Oakenfull, T. Riley, and V. Gold, *Chem. Commun.*, **1966,** 385.

12. W. Turnquest and M. L. Bender, *J. Am. Chem. Soc.*, **79,** 1652 (1957).

13. M. L. Bender, *Chem. Rev.*, **60,** 53 (1960).

14. D. L. Leussing and N. V. Raghavan, *J. Am. Chem. Soc.*, **102,** 5635 (1980).

15. E. H. Cordes and W. P. Jencks, *J. Am. Chem. Soc.*, **84,** 826 (1962).

16. R. Breslow, *J. Am. Chem. Soc.*, **79,** 1762 (1957).

17. R. Breslow, *J. Am. Chem. Soc.*, **80,** 3719 (1958).

18. V. Franzen and L. Fikentscher, *Liebig's Ann. Chem.*, **613,** 1 (1958).

19. J. F. Kirsch and W. P. Jencks, *J. Am. Chem. Soc.*, **86,** 845 (1964).

20. P. B. D. De La Mare and J. H. Ridd, *Aromatic Substitution*, Academic Press, Inc., New York, 1959.

21. D. E. Metzler, M. Ikawa, and E. E. Snell, *J. Am. Chem. Soc.*, **76,** 648 (1954).

22. E. E. Snell, A. E. Braunstein, E. S. Severin, and Y. M. Torchinsky, *Pyridoxal Catalysis, Enzymes and Model Systems*, Interscience Publishers, New York, 1968.

8 | Coenzymes

8.1. INTRODUCTION

As proteins, enzymes contain many acidic, basic, and nucleophilic functionalities as indicated in previous chapters and as shown in Table 8.1. However, proteins do not contain many specialized chemical functionalities such as those that participate in oxidation–reduction reactions, aldol condensation reactions, the transamination reaction, the condensation of amino acids, the methylation of amines, transacylation, and phosphorylation. The systems that, in combination with the proteins, facilitate these reactions are coenzymes, also shown in Table 8.1.

There are a multitude of enzymatic reactions that proceed in conjunction with coenzymes. Coenzymes effectively catalyze reactions only when they combine with appropriate proteins (apoenzymes), although they can be moderately active even in the absence of apoenzymes.

The most important equation in coenzyme chemistry is thus Eq. (8.1) where the coenzyme is usually an organic moiety of some kind, the apoenzyme is usually a protein, and the holoenzyme is the whole enzyme.

$$\text{coenzyme} + \text{apoenzyme} = \text{holoenzyme} \qquad (8.1)$$

Enzymes containing coenzymes can be subdivided into enzymes containing nonredox coenzymes and those containing redox coenzymes. As do most of the chapters in this book, this one emphasizes mechanism, the mechanism by which these coenzymes function.[1,2]

Table 8.1.   Catalytic Constituents of Enzymatic Active Sites

Coenzymatic Catalysts or Reactants

Oxidation–Reduction Systems
  Nicotinamide adenine dinucleotide (NAD)
  Nicotinamide adenine dinucleotide phosphate (NADP)
  Flavin nucleotides such as flavin mononucleotide (FMN) and flavin adenine
    dinucleotide (FAD)
  Metal porphyrin complexes such as those found in the cytochromes, cobamide,
    peroxidase, and catalase
  Ascorbic acid
  Lipoic acid (thioctic acid)
  Coenzyme Q (ubiquinone)
  Metal ions such as $Cu^{2\oplus}$
Nonoxidation–Reduction Systems
  Thiamine pyrophospate
  Pyridoxal phosphate
  Folic acid (pteroyl-L-glutamic acid)
  Biotin
  Glutathione
  S-Adenosylmethionine
  Coenzyme A
  Adenosine monophosphate, diphosphate, and triphosphate (AMP, ADP, ATP)
  Uridine phosphate
  Various metal ions, mainly $Zn^{2\oplus}$, $Mn^{2\oplus}$, $Mg^{2\oplus}$, $Cu^{2\oplus}$, $Co^{2\oplus}$
  4'-Phosphopantetheine

Constituents of the Protein

Nucleophiles
  Carboxylate ion of aspartate or glutamate
  Alcoholic hydroxyl group of serine or threonine
  SH group of cysteine
General Acids (Bases)
  Carboxylic acids (carboxylate ion) of aspartate or glutamate
  Phenolic hydroxyl group (phenoxide ion) of tyrosine
  Ammonium ion (amine) of lysine
  Imidazolium ion (imidazole) of histidine
  Guanidinium ion (guanidine) of arginine
Other
  Indole ring of tryptophan
  SCH_3 group of methionine
  Peptide or amide group

Source: M. L. Bender, *Encyclopedia of Polymer Science and Technology*, Vol. 6, John Wiley &
Sons, Inc., New York, 1967, p. 1.

8.2. REDOX COENZYMES

Many enzymes contain redox cofactors including: nicotinamide coenzymes such as NADH (nicotinamide adenine dinucleotide) and NADP (nicotinamide adenine dinucleotide phosphate); flavin nucleotides such as FMN (flavin mononucleotide) and FAD (flavin adenine dinucleotide); metallic porphyrins; metallic porphyrins containing an organometallic bond such as the cobamide (vitamin B_{12}) coenzymes; ascorbic acid; lipoic acid (thioctic acid); coenzyme Q (ubiquinone); various metal ions that can change valence; and ferridoxin. Most of these coenzymes behave as extensions of the principles enunciated in the preceding chapters.

8.2.1. NAD$^{\oplus}$ Coenzymes

In conjunction with the appropriate apoenzymes, NAD$^{\oplus}$ (or NADH) catalyzes a variety of redox reactions such as: the oxidation (reduction) of alcohols (ketones), α-hydroxy-carboxylic acids (α-ketoacids), α-amino acids (α-mino acids), and the dihydroflavins (flavins); and the oxidation of 3-hydroxyacyl-CoA. NAD$^{\oplus}$ coenzyme was once referred to as DPN (diphosphopyridine nucleotide), and earlier as CoI (coenzyme I). This coenzyme, a pellagra preventative, is nutritionally beneficial and is often included in foods.

The structure of NAD$^{\oplus}$, the oxidized form, consists of a nicotinamide moiety, an adenine moiety, two ribose groups, and two phosphate groups, as shown in **8.1**.

NAD$^{\oplus}$

8.1

The oxidation of ethanol by NAD$^{\oplus}$ (with the enzyme alcohol dehydrogenase) gives acetaldehyde and NADH, when two electrons and a hyrogen atom are transferred from ethanol to NAD$^{\oplus}$. In Eq. (8.2), the R group represents the remainder of the NAD$^{\oplus}$ molecule shown in **8.1**. Note the direct involvement of the NAD$^{\oplus}$ in the reaction. It is used up stoichiometrically and there is a *direct hydrogen transfer*.[3] The enormous effort expended on these enzymes derives not only from their importance, but also from the fact that the reactions they catalyze can be easily followed. There is a large change in absorbancy at 340 nm, occurring during the redox reactions involving NAD$^{\oplus}$ cofactors. This change in absorbancy results from the reduction of the pyridine ring on the left to a dihydropyridine ring on the right of Eq. (8.2).

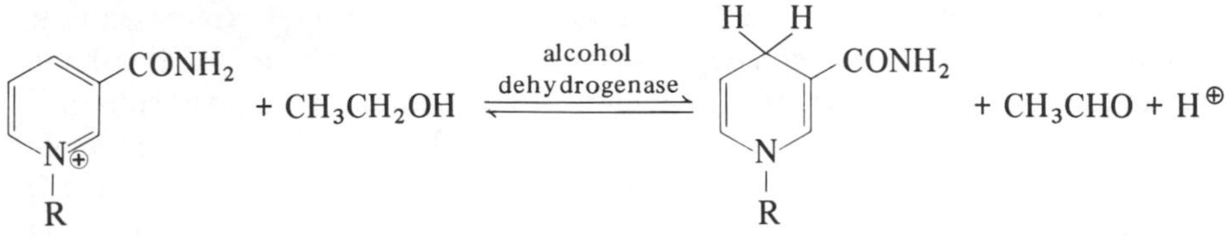

$$\text{(pyridinium-CONH}_2) + CH_3CH_2OH \xrightleftharpoons[\text{dehydrogenase}]{\text{alcohol}} \text{(dihydropyridine-CONH}_2) + CH_3CHO + H^\oplus$$

(8.2)

The reaction of Eq. (8.2) is followed by the reoxidation of NADH, during the reduction of 1,3-diphosphoroglyceric acid which is catalyzed by the enzyme 3-phosphoglyceraldehyde dehydrogenase [Eq. (8.3)]. This enzyme contains a flavin coenzyme.

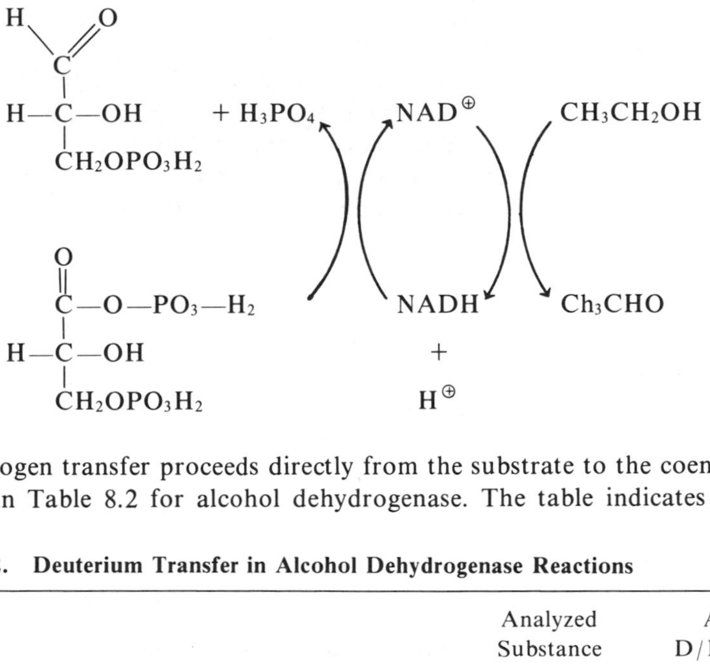

$$\begin{array}{c} H\diagdown \text{C} \diagup O \\ | \\ H-C-OH \\ | \\ CH_2OPO_3H_2 \end{array} + H_3PO_4 \quad \begin{array}{c} NAD^\oplus \end{array} \quad CH_3CH_2OH$$

$$\begin{array}{c} O \\ \| \\ C-O-PO_3-H_2 \\ | \\ H-C-OH \\ | \\ CH_2OPO_3H_2 \end{array} \qquad NADH \qquad Ch_3CHO$$

$$+ \\ H^\oplus$$

(8.3)

Hydrogen transfer proceeds directly from the substrate to the coenzyme as shown in Table 8.2 for alcohol dehydrogenase. The table indicates that no

Table 8.2.   Deuterium Transfer in Alcohol Dehydrogenase Reactions

	Analyzed Substance	Atoms D/Molecule
$CH_3CH_2OH + NAD^\oplus \xrightarrow{D_2O}$ $CH_3CHO + NADH + H^\oplus$	NADH	0.02
$CH_3CD_2OH + NAD^\oplus \xrightarrow{H_2O}$ $CH_3CDO + NADD + H^\oplus$	NADD	1.01; 0.99
	CH_3CDO	1.00

Source: B. Vennesland and F. H. Westheimer, in The Mechanism of Enzyme Action, W. D. McElroy and B. Glass, Eds., Johns Hopkins Press, Baltimore, 1954, p. 357.

hydrogen atoms come from the medium; rather, a hydrogen atom is directly transferred from the substrate to NAD$^\oplus$. Furthermore, "turnover" with excess CH_3CD_2OH introduces a single, specific deuterium atom into NADD. This result indicates that the enzymatic reaction is stereospecific for NAD$^\oplus$ (or NADH), which contains a prochiral carbon atom.

8.2.2. Flavin Coenzymes

There are two principal flavin coenzymes (called flavin or riboflavin): flavin mononucleotide (FMN) and flavin adenine dinucleotide (FAD). The structures of these materials are shown in **8.2** and **8.3**. Flavoenzymes may be separated into three groups based on their function:

1. Biological dehydrogenation systems.
2. Systems for the activation of molecular oxygen and transfer of one or two oxygen atoms from $^3O_2$ to the substrate.
3. Systems responsible for electron transport.

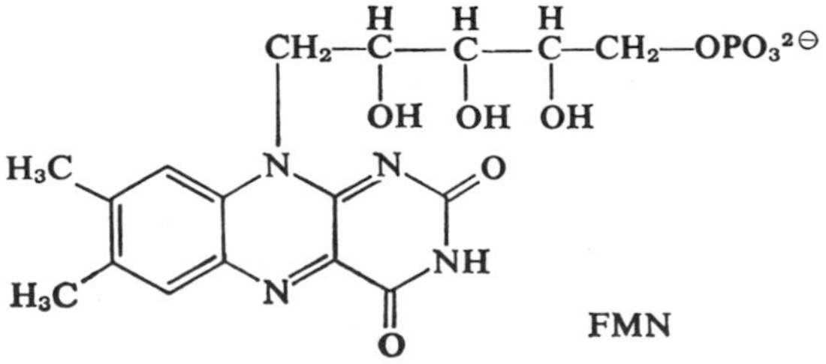

2

Riboflavin phosphate (FMN)

8.2

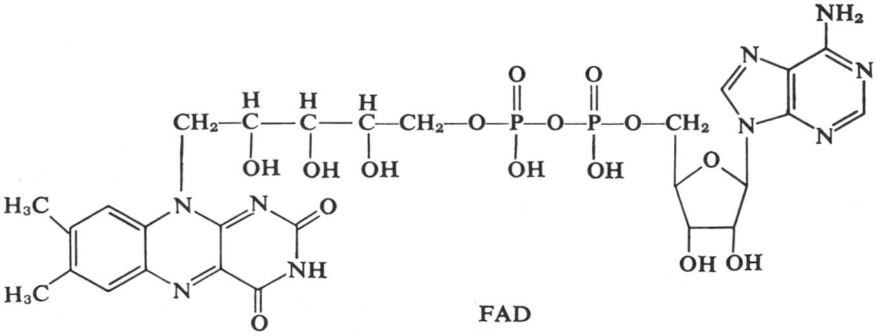

3

Flavin adenine dinucleotide (FAD)

8.3

Several models that mimic the action of redox enzymes requiring NAD$^\oplus$ have investigated. The zinc complex of phenanthroline-2-carboxyaldehyde was treated with *N*-propyldihydronicotinamide in acetonitrile as solvent; nonenzymatic oxidation–reduction occurs with direct hydrogen transfer at a moderate rate at room temperature. The chelation of the zinc ion enormously increases the polarity of the carbonyl group so as to promote the transfer of an incipient hydride ion from the dihydronicotinamide to the aldehyde carbon atom. The reaction fails in water, presumably because water hydrates (and thus deactivates) the zinc ion.

(8.4)

Several other models use an aldehyde containing a highly electronegative carbonyl group compound with hydrogen bonding; the latter served almost as well as a metal ion to further enhance the reactivity of the carbonyl. The effect of hydrogen bonding in accelerating such reactions had been noted previously.

More recently, ethylbenzoyl formate was reduced with *N*-benzyldihydronicotinamide in acetonitrile as solvent, using magnesium perchlorate as catalyst; the reaction proceeded in 86% yield in 17 hours at room temperature. Although this rate falls far short of the corresponding enzymatic ones, the example is especially noteworthy because the needed metal ion was not chelated to the substrate as in the first example, but was free in solution. Similar reductions with chiral dihydronicotinamides gave partially resolved products. The metal ion was chelated to the reducing agent rather than to the substrate. In all of these examples, the rates are low and much remains to be discovered before the action of the enzymes can be duplicated.[4]

In spite of the extensive investigation of flavoenzymes and flavin model systems, very few of the reaction mechanisms are well understood.[5] However, let us point out some of the mechanistic features of several of these reactions that are fairly well established. The oxidation of an alcohol α to a carbonyl as, for example, in the conversion of lactic acid to pyruvic acid by lactic acid oxidase (which contains a flavin coenzyme) is a two-step process:

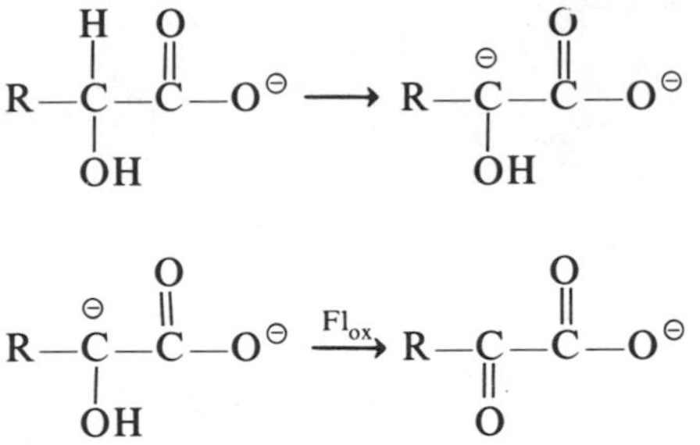

$$R-\underset{\underset{OH}{|}}{\overset{\overset{H}{|}}{C}}-\overset{\overset{O}{\|}}{C}-O^{\ominus} \longrightarrow R-\underset{\underset{OH}{|}}{\overset{\ominus}{C}}-\overset{\overset{O}{\|}}{C}-O^{\ominus}$$

(8.5)

$$R-\underset{\underset{OH}{|}}{\overset{\ominus}{C}}-\overset{\overset{O}{\|}}{C}-O^{\ominus} \xrightarrow{Fl_{ox}} R-\underset{\underset{O}{\|}}{\overset{}{C}}-\overset{\overset{O}{\|}}{C}-O^{\ominus}$$

The lactate is first converted to the corresponding anion by a general base of the enzyme followed by oxidation to pyruvic acid. The mechanism of the oxidation is unclear. Although one electron transfer has been suggested, the oxidations can also occur by the transfer of a hydride equivalent not involving radicals. Although a number of uncertainties exist as to the oxidation of alcohols α to carbonyls, the mechanism of thiol oxidation seems quite clear (Eq. 8.6):

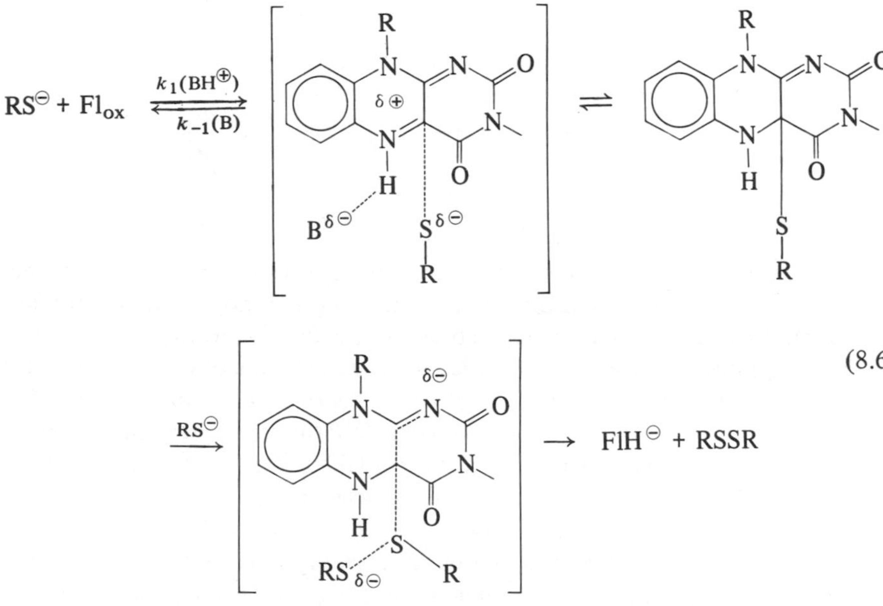

$$RS^{\ominus} + Fl_{ox} \underset{k_{-1}(B)}{\overset{k_1(BH^{\oplus})}{\rightleftarrows}}$$

(8.6)

The reaction is general acid catalyzed and spectral evidence for the 4a (bridgehead)-thiol adduct has been demonstrated for the enzyme lipoamide dehydrogenase.

Again, as with the oxidation of alcohol, the mechanism of the flavin activation and delivery of oxygen still remains somewhat uncertain. However, it is known that the kinetics of the N^5-alkyl-4a-(hydroperoxy)-oxidation of substrates conform to Eq. (8.7)

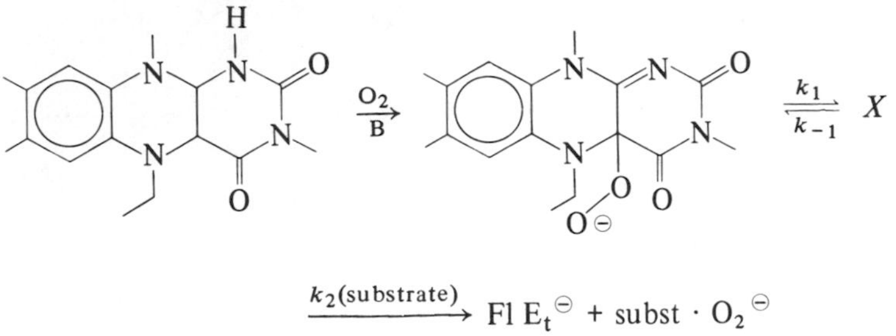

(8.7)

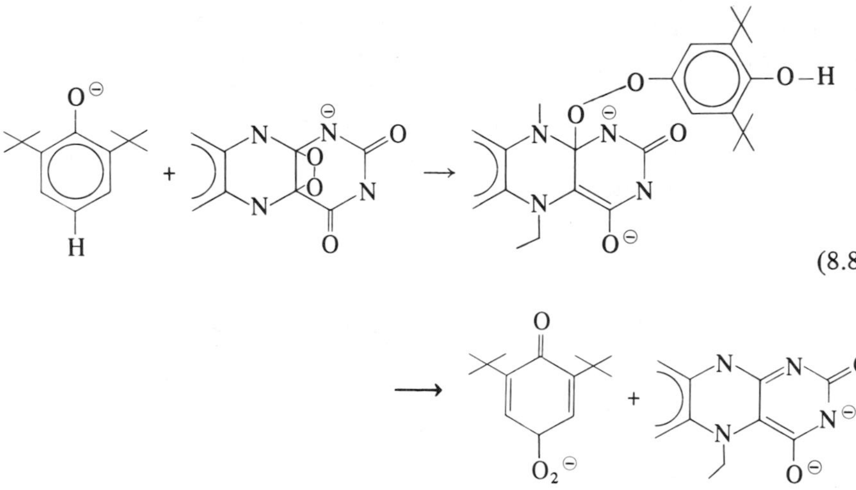

$$\xrightarrow{k_2(\text{substrate})} \text{Fl E}_t{}^{\ominus} + \text{subst} \cdot \text{O}_2{}^{\ominus}$$

(8.8)

Equation (8.8) represents current thinking about the structure of the intermediate X and how it transfers oxygen.[5] There is now evidence for an enediol mechanism for the enzyme, glyoxalase I, from model studies.[6]

The acyloin condensation of aldehydes, catalyzed by thiazolium ion bound to cetyltrimethylammonium bromide (CTAB) micelles, can be diverted by the addition of flavin to the oxidation reaction to afford the corresponding carboxylic acids. It was found, however, that when the aldehyde concentration is elevated or the aldehyde is relatively reactive, intermolecular flavin (3-methyltetra-O-acetylriboflavin, MeFl) cannot trap the intermediates (active aldehydes) formed from thiazolium ion and aldehydes completely, leading to a competition between the conventional acyloin condensation and the flavin oxidation. Intramolecular catalysis (Chapter 10) was applied to this system in order to suppress the acyloin condensation relative to the flavin oxidation. The first utilized quasi-intramolecular flavin oxidation in which hydrophobic 10-dodecylisoalloxazine (10-DodFl) and N-hexadecylthiazolium bromide (HxdT) are bound to a CTAB micelle aggregate. The second is a flavin–thiazolium biscoenzyme (Fl-T) oxidation in which the intermediates on the thiazolium moiety were oxidized efficiently by the intramolecular flavin. When 4-chlorobenzaldehyde (100 mM) was employed as substrate, the trapping efficiency (=flavin oxidation product/

sum of acyloin condensation products) for MeFl was 1.6. The trapping efficiency for the quasi-intramolecular flavin oxidation was improved up to 15–33-fold, due to the enhanced local concentration of 10-DodFl in the micelle phase; efficiency for the biscoenzyme system was further enhanced (>115-fold). A kinetic examination has established that the reaction was zero order in MeFl for the intermolecular flavin oxidation of 4-chlorobenzaldehyde, whereas it becomes first order in MeFl for the oxidation of more reactive pyridine-4-carboxaldehyde (pyCHO). This indicates that the rate-limiting step changes depending on the reactivity of aldehyde: the deprotonation from the thiazolium-aldehyde adduct is rate limiting in the oxidation of 4-chlorobenzaldehyde, whereas the oxidation of the deprotonated active aldehyde by MeFl becomes rate limiting in the oxidation of pyCHO. On the other hand, quasi-intramolecular flavin oxidation of pyCHO was zero order in 10-DodFl at low pyCHO concentrations (<10 mM) and was approximated by a first-order equation at high pyCHO concentrations (>50 mM). In the biscoenzyme oxidation of pyCHO, the zero-order decrease was always observed for up to 60% reaction, indicating the high efficiency of intramolecular flavin as a trapping agent. The present system is a relevant model for pyruvate oxidase, which requires FAD and thiamine pyrophosphate as cofactors and catalyzes the conversion of pyruvic acid to acetic acid.[7]

Evidence is presented in support of free-radical mechanisms for the oxidation of ionizable carbon acids by oxidized flavin. Activation of molecular oxygen by reduced flavin is shown to occur through the formation of a 4a-hydroperoxy-flavin that may, dependent upon conditions and substrate, transfer one or two oxygen atoms.[8]

New oxidation–reduction enzymes have been prepared by the covalent modification of hydrolytic enzymes with coenzyme analogs. Among the semisynthetic enzymes that have been generated, the most effective one that has been studied to date is the covalent flavin–papain complex produced by the reaction of the sulfhydryl group of Cys-25 in the papain active site with 7-bromoacetyl-10-methylisoalloxazine. The kinetics of the oxidation of dihydro-nicotinamides observed using this flavopapain as the catalyst indicated that saturation occurs at low substrate concentrations and that the rate accelerations exceed an order of magnitude relative to models. Furthermore, examination of the stereochemistry of the oxidation of labeled NADH derivatives revealed that the flavopapain shows a marked preference for removing hydrogen from the 4A-position. This has demonstrated the feasibility of tampering very significantly with an enzyme active site without destruction of the enzyme as a catalytic species.[9]

8.2.3.   Cobamide and Cobalamin Coenzymes (Vitamins B₁₂)

Cobamide and cobalamin are rather large cobalt complexes. The structure of cobamide and cobalamin are given in structures **8.4** and **8.5**, respectively. They consist of a cobalt atom in the center fixed to a corrin ring that provides four

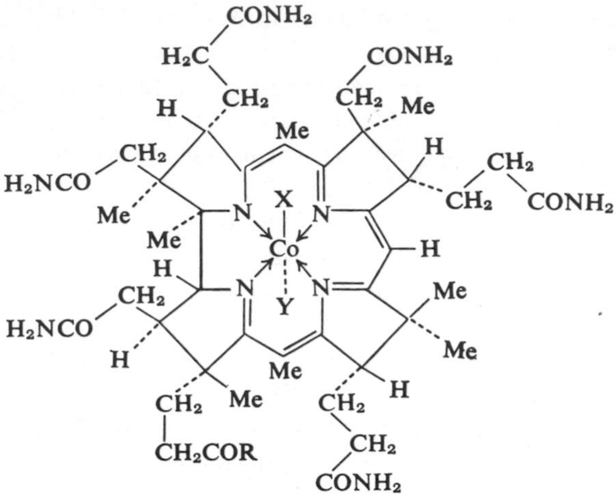

8.4 Cobamide

$$R = NHCH_2CH(OH)CH_3$$

8.5 Cobalamin

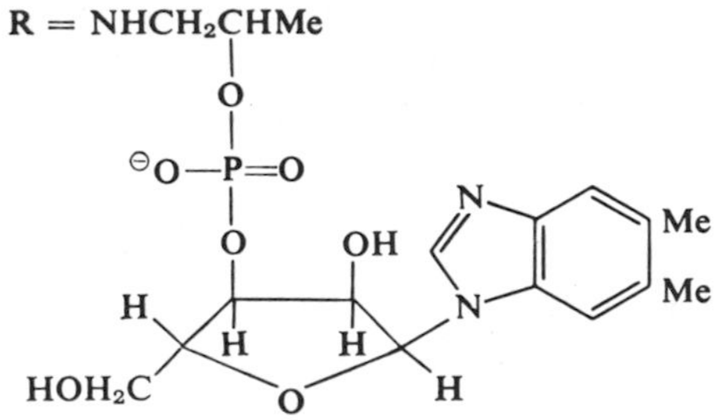

$$R = NHCH_2CHMe$$

The structural formulae of cobalamins and cobamides

8.5   8.4

coordinating nitrogen atoms. The cobalt also has one (or two) axial ligands, X and Y. If one of the axial ligands, Y, is the benzimidazole of the naturally occurring nucleotide, the series of complexes is called cobalamins (**8.5**) while if the base is removed by hydrolysis and Y = H$_2$O, the series is referred to as the cobamides (**8.4**).

One of the more interesting reactions involving B$_{12}$ is the conversion of 1,2-propanediol to propionaldehyde.[10] This is a classical organic reaction that is usually carried out with sulfuric acid, but with a B$_{12}$ enzyme, it can be carried out under mild conditions. The reaction is essentially an intramolecular redox reaction, and it is known that B$_{12}$ is essential since when it is removed by charcoal, the enzyme is inactive, but when B$_{12}$ is readded, the enzyme is once again active. When dl-1,2-propanediol-1-t is converted to propionaldehyde in the presence of the enzyme dialdehydrase containing cobalamin coenzyme, tritium is transferred to the coenzyme. The coenzyme is tritiated exclusively at the C-5' position of the adenosyl moiety. Based on these results, Eq. (8.9) seems to be a reasonable mechanism.

$$SH + \boxed{\overset{\overset{\displaystyle CH_2-R}{|}}{Co}} \longrightarrow \boxed{\overset{\overset{\displaystyle S}{|}}{Co}} + CH_3R \longrightarrow \boxed{\overset{\overset{\displaystyle P}{|}}{Co}} + CH_3R \longrightarrow PH + \boxed{\overset{\overset{\displaystyle CH_2R}{|}}{Co}} \qquad (8.9)$$

Bisdehydrocorrin (BDHC) and corrinoid complexes are comparable in the electronic nature of nuclear cobalt. Consequently, the double bonds at the periphery of the A and D rings of BDHC are not in conjugation with the interior double bonds. The steric effect is more pronounced for Co(I)(BDHC) than for vitamin B_{12s} when the reaction with alkyl halides is carried out, and is attributed to the 1,3-diaxial-type interaction between angular methyl groups placed at the C(1) and C(19) position of BDHC and an approaching alkyl ligand. The photolysis of the methylated and ethylated Co(III)(BDHC) complexes results in the normal homolytic Co–C cleavage under anaerobic conditions. On the other hand, the Co–C bond in the isopropyl derivatives undergoes heterolytic cleavage to yield the isopropyl anion and Co(III)(BDHC). The Co(BDHC) complex can be used as a catalyst for selective hydrogenation of primary alkyl halides using sodium hydroborate as the stoichiometric reducing agent.[11] SH, PH, and CH_3R represent substrate, product, and 5-deoxyadenosine, respectively. In this reaction, hydrogen is abstracted from C-1 of dl-1,2-propanediol substrate and transferred to the coenzyme, where it becomes equivalent with at least one but probably both hydrogens of the C-5' position. This results in a reduced form of the coenzyme and a molecule derived through the oxidation of the substrate. In a subsequent step, the hydrated form of propionaldehyde is formed by a transfer of hydrogen from the reduced coenzyme to the intermediate derived from the substrate.

Another important reaction of vitamin B_{12} (with enzyme) is the isomerization of methylmalonylCoA to succinylCoA.

$$\begin{array}{ccc}
\overset{\displaystyle O}{\overset{\|}{C}}-S-CoA & & \overset{\displaystyle O}{\overset{\|}{C}}-S-CoA \\
| & \rightleftharpoons & | \\
CH-CH_3 & & CH_2 \\
| & & | \\
COOH & & CH_2 \\
& & | \\
& & COOH
\end{array} \qquad (8.10)$$

It is equivalent to an (organic) intramolecular 1,2 shift of a CoA bound carboxyl group.[12]

8.2.4. Ascorbic Acid (Vitamin C)

Ascorbic acid is a strong reducing reagent and is converted to dehydroascorbic acid when oxidized [Eqs. (8.11) and (8.12)]. For reaction (8.11), the mechanism in

Eq. (8.12) is proposed,[7] where the metal ion serves as a conjugating link between the oxidant and reductant, allowing facile electron transfer (see Chapter 9).

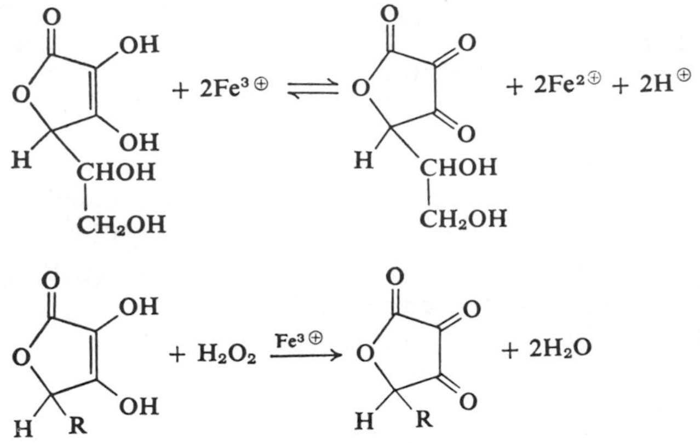

$$(8.11)$$

Ascorbic acid   Dehydroascorbic acid

8.6a   8.6b

Ascorbic acid   8.6a

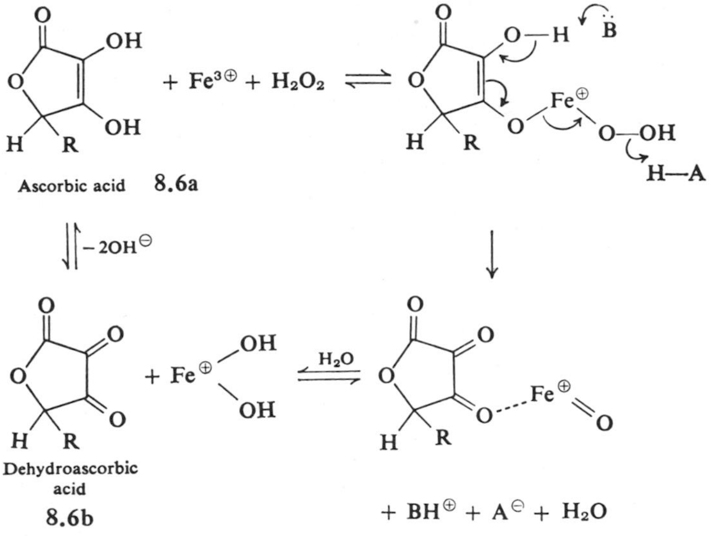

$$(8.12)$$

Dehydroascorbic
acid

8.6b $+ BH^{\oplus} + A^{\ominus} + H_2O$

8.2.5. Ferridoxin

Ferridoxin is a nonheme iron containing coenzyme. It is involved in such diverse functions as pyruvic acid metabolism and nitrogen fixation. Ferridoxins have iron–sulfur moieties. They are attached to low molecular weight proteins and have strongly negative redox potentials. They have been obtained from many green plants as well as from some photosynthetic and nonphotosynthetic bacteria. Ferridoxins are reported to contain four to seven iron atoms per molecule, depending on the species (seven in *Clostridia*) probably linked to an equal number of cysteine residues in the protein. They also contain an equal number of acid-labile sulfur atoms (either inorganic sulfide or very labile organic residues giving rise to H_2S on acidification). A model of the active site is shown in **8.7**. Simple inorganic models of the iron–sulfur system are now well characterized and resemble the coenzyme both structurally and electronically.[13]

Ferridoxin
8.7

Two reaction systems, I and II, that result in the assembly of the biologically relevant $[Fe_4S_4(SPh)_4]^{\ominus 2}$ cluster from simple reactants have been examined by spectrophotometry and ^1H NMR spectroscopy in order to define reaction sequences and identify intermediates. Systems I and II were based on the reactant mole ratios $PhS^{\ominus}$: $FeCl_3$: $S = 3.5:1:1$ and $\geq 5:1:1$, respectively, and were conducted in acetonitrile and methanol solutions. In system I, the first identifiable intermediate is the adamantanelike species $[Fe_4(SPh)_{10}]^{\ominus 2}$, which reacts with sulfur in an all-or-nothing process in both solvents to afford I. No other intermediates were identified in this system. In system II, the first recognizable species formed is tetrahedral $[Fe(SPh)_4]^{\ominus 2}$. Reaction of this cluster with sulfur in acetonitrile yields the binuclear cluster $[Fe_2S_2(SPh)_4]^{\ominus 2}$. Thus system II provides the first demonstrated instance of elaboration of a tetranuclear cluster through mononuclear and binuclear intermediates. In methanol, the overall reactions affording I from initial reactants are quantitative when assayed spectrophotometrically.[14]

8.3. NONREDOX COENZYMES

There are a multitude of nonredox coenzyme systems, including thiamine pyrophosphate (vitamin B_1), pyridoxal phosphate (vitamin B_6), folic acid (pteroylglutamic acid), biotin, S-adenosylmethionine, vitamins A, D, E, and K, Coenzyme A, metal ions such as $Zn^{\oplus 2}$, $Mg^{\oplus 2}$, $Cu^{\oplus 2}$, and $Co^{\oplus 2}$ (most of which do not undergo easy electron transfer), the "acyl-carrier protein," adenosine monophosphate, diphosphate, and triphosphate, glutathione, and others. Most of these coenzymes will be covered here.

8.3.1. Thiamine Pyrophosphate (Vitamin B₁)

Thiamine pyrophosphate, a coenzyme of universal occurrence in living systems, was originally detected as a nutritional factor required for the prevention of polyneuritis in birds and beriberi in man. Its structure is given in formula **8.8**. The crystalline vitamin (B₁) was first obtained in 1925 and, as can be seen from the structure, incorporates two heterocyclic rings, a pyrimidine ring, and a thiazolium ring.[15] This vitamin participates in a variety of enzymatic reactions, including nonoxidative decarboxylation of α-keto acids; oxidative decarboxylation of α-keto acids; and, the formation of α-ketols (acyloins). For details of its mechanism of action, see Chapter 7.

Thiamine pyrophosphate
8.8

The compound methyl 2-hydroxy-2-(2-thiamin)ethylphosphonate, phosphalactylthiamin, a phosphonate analogue of 2-(1-carboxyl-1-hydroxyethyl) thiamin, is the initial intermediate in the thiamin-catalyzed decarboxylation of pyruvic acid. Crystal-structure analysis of phosphalactylthiamin reveals that the thiamin portion of the molecule assumes the S conformation that is characteristic of other C(2)-substituted thiamins. However, in contrast to previously studied derivatives, the conformation of the phosphalactyl substituent is unique in that its hydroxyl is in close contact with the bridging methylene instead of the thiazolium ring sulfur and the bond to the phosphonate is oriented perpendicular to the ring plane. This structural feature is also consistent with the NMR spectrum. The structure suggests that the principles of least motion and maximum orbital overlap can be applied to thiamin catalysis of the decarboxylation of pyruvate since the observed structure conforms to theoretical expectations for 2-(α-lactyl)thiamin diphosphate.[16]

8.3.2. Pyridoxal Phosphate (Vitamin B₆)

Pyridoxal phosphate **8.9** (vitamin B₆) (PyCHO) is involved in many transformations of amino acids, including racemization, decarboxylation, transamination, β-substitution, elimination, and condensation reactions. The vitamin also exists as the alcohol (pyridoxine or pyridoxol), and the corresponding amine(pyridoxamine) (PyCH₂NH₂). Both the aldehyde and phenolic groups of pyridoxal phosphate are essential for catalysis, whereas the phosphate group is only

$$\begin{array}{c} \text{CHO} \\ \text{H}_2\text{O}_3\text{POCH}_2 \underset{\text{N}}{\overset{\text{OH}}{\bigcirc}} \text{CH}_3 \end{array}$$

Pyridoxal phosphate
8.9

involved in electrostatic binding of the coenzyme to the enzyme. The phenolic group exists in the anionic form near neutrality.

Pyridoxal itself has catalytic functions similar to pyridoxal phosphate enzymes although the rate of a reaction catalyzed by pyridoxal is a millionfold slower than that catalyzed by the combination of pyridoxal phosphate and the appropriate enzyme.[17] The mechanism of transamination by pyridoxal, proceeding by isomerization between two Schiff bases, is shown in Eq. (8.13).[18] A more detailed mechanism of the first reaction of Eq. (8.12) (the formation of pyridoxamine) is shown in Eq. (8.13). Clearly, the phenolic group or its anion functions as an intramolecular general acid or base catalyst. The positive charge at the

$$\text{PyCHO} + \text{R}-\underset{\underset{\text{NH}_2}{|}}{\overset{\overset{\text{H}}{|}}{\text{C}}}-\text{CO}_2\text{H} \rightleftharpoons \left[\text{PyCH}=\text{N}-\text{C} \overset{\nearrow^{\text{R}}}{\underset{\searrow_{\text{CO}_2\text{H}}}{\overset{\nwarrow^{\text{H}}}{}}} \right] \rightleftharpoons$$

$$\left[\text{PyCH}_2-\text{N}=\overset{\overset{\text{R}}{|}}{\text{C}}-\text{CO}_2\text{H} \right] \rightleftharpoons \overset{\overset{\text{R}}{|}}{\underset{\underset{\text{O}}{||}}{\text{C}}}-\text{CO}_2\text{H} + \text{PyCH}_2\text{NH}_2$$

(8.13)

$$\text{PyCH}_2\text{NH}_2 + \text{R}'-\overset{\overset{\text{O}}{||}}{\text{C}}-\text{CO}_2\text{H} \rightleftharpoons \text{PyCH}_2\text{N}=\overset{\overset{\text{R}'}{|}}{\text{C}}-\text{CO}_2\text{H} \rightleftharpoons$$

$$\text{PyCH}=\text{N}-\underset{\underset{\text{H}}{|}}{\overset{\overset{\text{R}'}{|}}{\text{C}}}-\text{CO}_2\text{H} \rightleftharpoons \text{PyCHO} + \text{R}'-\underset{\underset{\text{NH}_2}{|}}{\overset{\overset{\text{H}}{|}}{\text{C}}}-\text{CO}_3\text{H}$$

$$\text{R}-\underset{\underset{\text{NH}_2}{|}}{\overset{\overset{\text{H}}{|}}{\text{C}}}-\text{CO}_2\text{H} + \text{R}'-\overset{\overset{\text{O}}{||}}{\text{C}}-\text{CO}_2\text{H} \rightleftharpoons \text{R}'-\underset{\underset{\text{NH}_2}{|}}{\overset{\overset{\text{H}}{|}}{\text{C}}}-\text{CO}_2\text{H} + \text{R}-\overset{\overset{\text{O}}{||}}{\text{C}}-\text{CO}_2\text{H} \quad \text{Net}$$

nitrogen atom of pyridoxal enhances the electrophilicity of the carbonyl carbon atom, facilitating the formation of the Schiff base. Furthermore, delocalization of positive charge in the pyridine ring in all the intermediates decreases the activation energy of reaction. These three factors are important for the catalysis by pyridoxal and are probably the origin of the effective catalysis of pyridoxal phosphate.

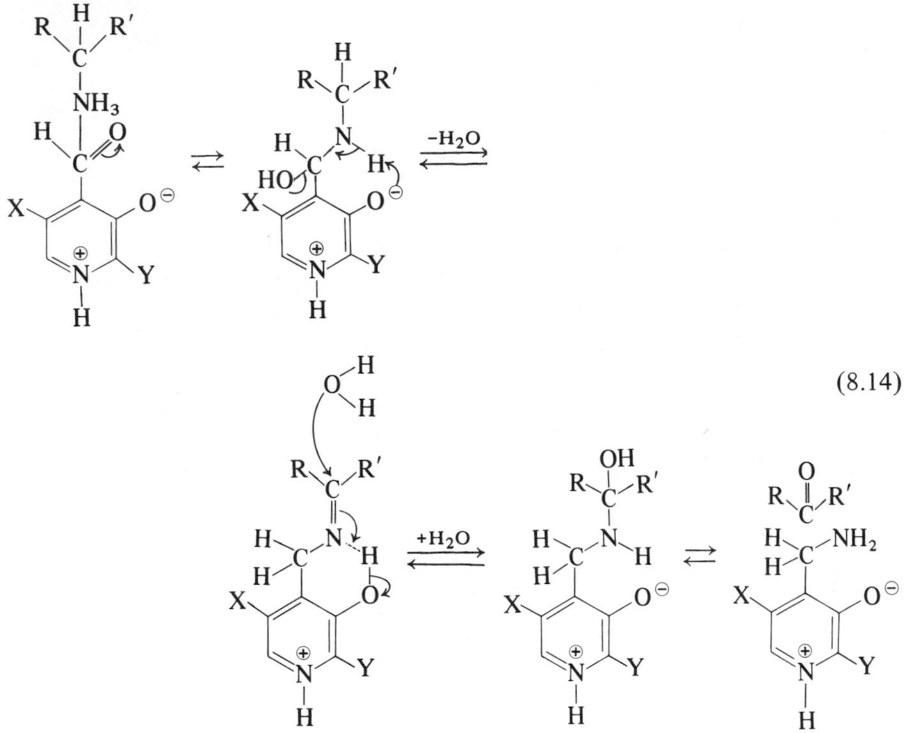

(8.14)

$$R = O\overset{O}{\overset{\|}{C}}CH_2CH_2- \quad R' = -\overset{O}{\overset{\|}{C}}-O^\ominus \quad X = HOCH_2- \quad Y = -CH_3$$

The transamination reactions are catalyzed by metal ions[18] (aluminum and ferric ions) or by imidazole[19] in model systems. The metal chelate complex is shown in **8.10**. The function of the metal ion is to labilize

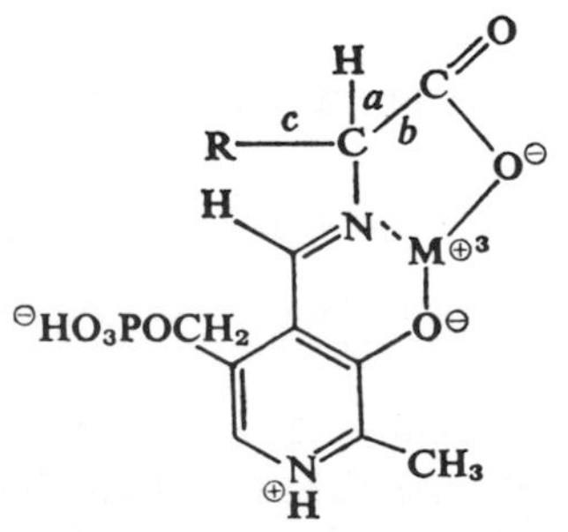

Metal chelate of pyridoxal phosphate and amino acid

8.10

the π electron system. Also shown in **8.10** is that the cleavage of different bonds in the chelate lead to different reactions: *a* leads to transamination *b* leads to decarboxylation, and *c* leads to condensation. All these reactions, with the exception of decarboxylation, have been observed in model systems.

The imidazole catalysis of the transamination reaction, shown in part in Fig. 8.1 at various pH's, shows dependence of the catalytic rate constant on the product of imidazole and imidazolium ion concentrations.

Plots such as Fig. 8.1 (or 11.3) at various pH's show constancy of the binding constant calculated in terms of a complex of aldimine (or ketimine) with one molecule of imidazole and one molecule of imidazolium ion. The reactant α-aminophenylacetic acid also complexes with imidazole as shown by solubility studies. The formation constants of this complex are quite similar to the constants determined kinetically for the imine complexes if calculated on the basis of the formation of a complex composed of one imidazole molecule, one imidazolium ion, and one amino acid molecule in the form of the zwitterion. On the basis of these experiments, a multiple catalysis (Chapter 11) of the prototropic shift by imidazole and imidazolium ion has been suggested, as shown in Eq. (8.15).[19]

$$(8.15)$$

The mechanistic difference between the pyridoxal (coenzyme) reaction [Eqs. (8.12) and (8.13] and the enzymatic reactions is that a Schiff base is first formed between pyridoxal phosphate and an amino group of a lysine residue of the enzyme in enzymatic reactions; this is followed by nucleophilic attack of the amino group of the amino acid on the Schiff base, resulting in the formation of the second Schiff base through transimination [Eq. (8.16)].

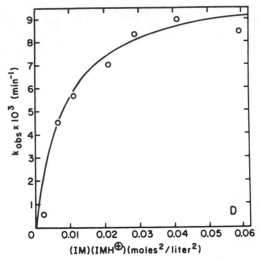

Fig. 8.1. Dependence of the catalytic rate constant of transamination rate constant on the concentrations of imidazole and imidazolium ions. From T. C. Bruice and R. M. Topping, *J. Am. Chem. Soc.*, **85**, 1488 (1963). Copyright © 1963 by the American Chemical Society. Reproduced by permission of the copyright owner.

(8.16)

The contribution of each of the several catalytic functions of pyridoxal phosphate is governed by the particular enzyme. For example, Eq. (8.17) shows that pyridoxal phosphate is involved in enzymatic decarboxylation, transamination, racemization, and amino acid synthesis.

Transamination using an achiral pyridoxamine and an α-ketoacid produces a D,L-amino acid. However, transamination using a chiral pyridoxamine analog, (R) or (S)-15-aminomethyl-14-hydroxy-5,5-dimethyl-2,8-dithio[9](2,5)pyridinophane, with various α-ketoacids in the presence of $Zn^{\oplus 2}$ leads to the asymmetric induction, of either L or D α-amino acids. This provides an excellent model system for enzymatic transamination as shown in Fig. 8.2. The most successful reaction involved the chiral pyridoxamine with phenylpyruvic acid to give R-phenylalanine in 23–61% enantiomeric purity.[20]

It was postulated long ago that the enzymes containing PLP (pyridoxal phosphate) in amino transferases (transaminases) orient the amino acid–PLP compound so that the α hydrogen is nearly orthogonal to the plane of the conjugated pyridine ring. The enzymes accept only a single enantiometer of an amino acid to form an α-keto acid. They then generate the analogous enantiomer of a new amino acid from a different α-keto acid. In the enzymes examined so far, the intermediate imine is always protonated from the *si* face at C-4′ of the cofactor so that the *pro*-S hydrogen (H_8) of PMP (pyridoxamine phosphate) is added or removed. A significant amount of the α hydrogen of L-alanine is incorporated at C-4′ by alanine aminotransferase, thereby indicating that a

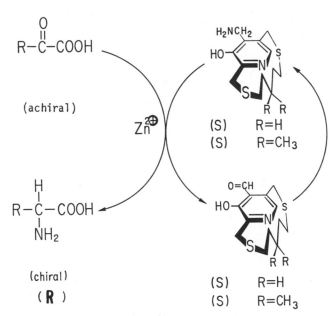

Fig. 8.2. Asymmetric induction of a chiral amino acid by transimination of an achiral α-keto acid with a chiral pyridoxamine derivative. (From Y. Tachibana, M. Ando, and H. Kuzuhara, *Chem. Lett.*, 1765 (1982).

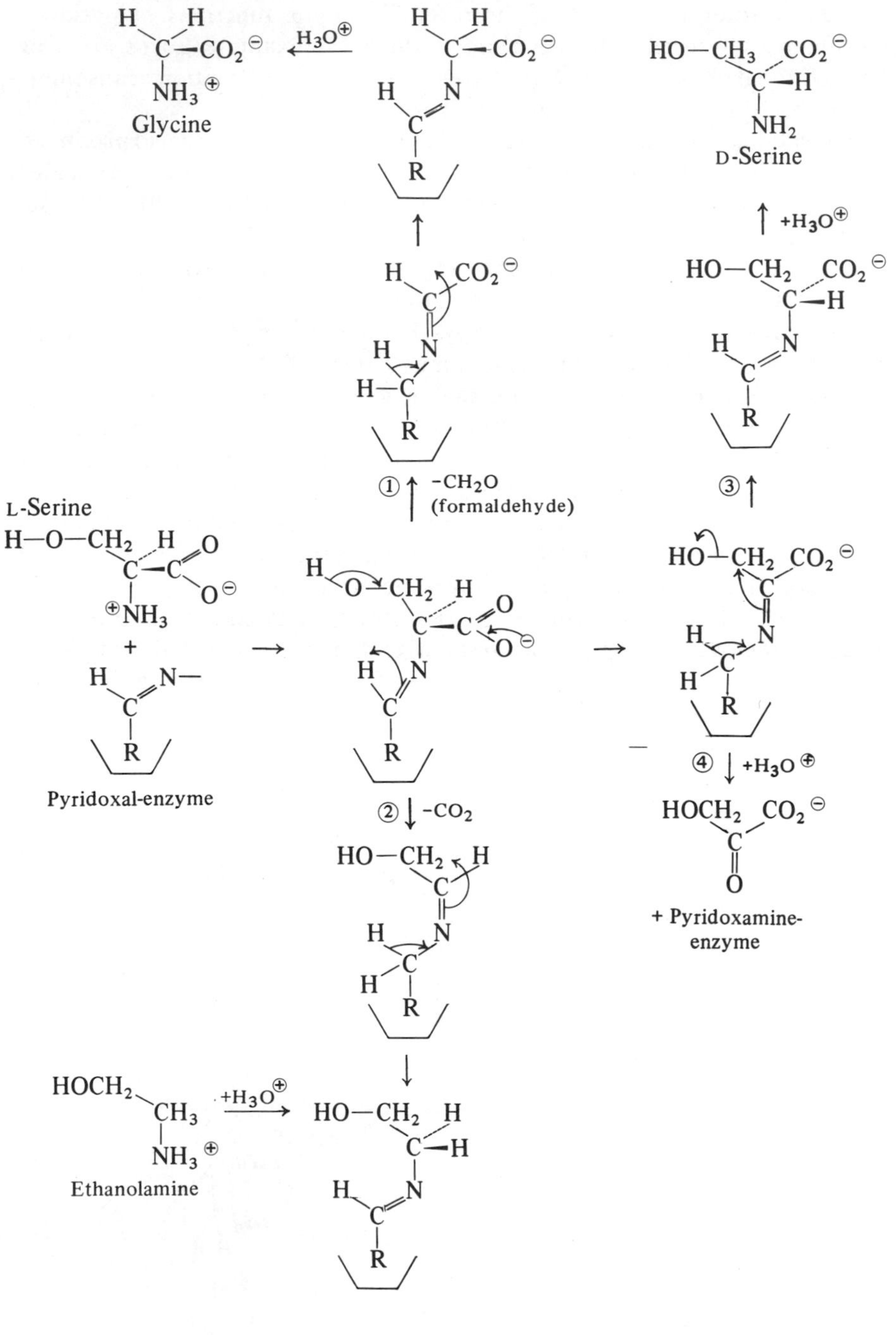

H₃O⁺ → Glycine

L-Serine / Pyridoxal-enzyme

① ↑ -CH₂O (formaldehyde)

② ↓ -CO₂

Ethanolamine +H₃O⁺

D-Serine

+H₃O⁺

③ ↑

④ ↓ +H₃O⁺

+ Pyridoxamine-enzyme

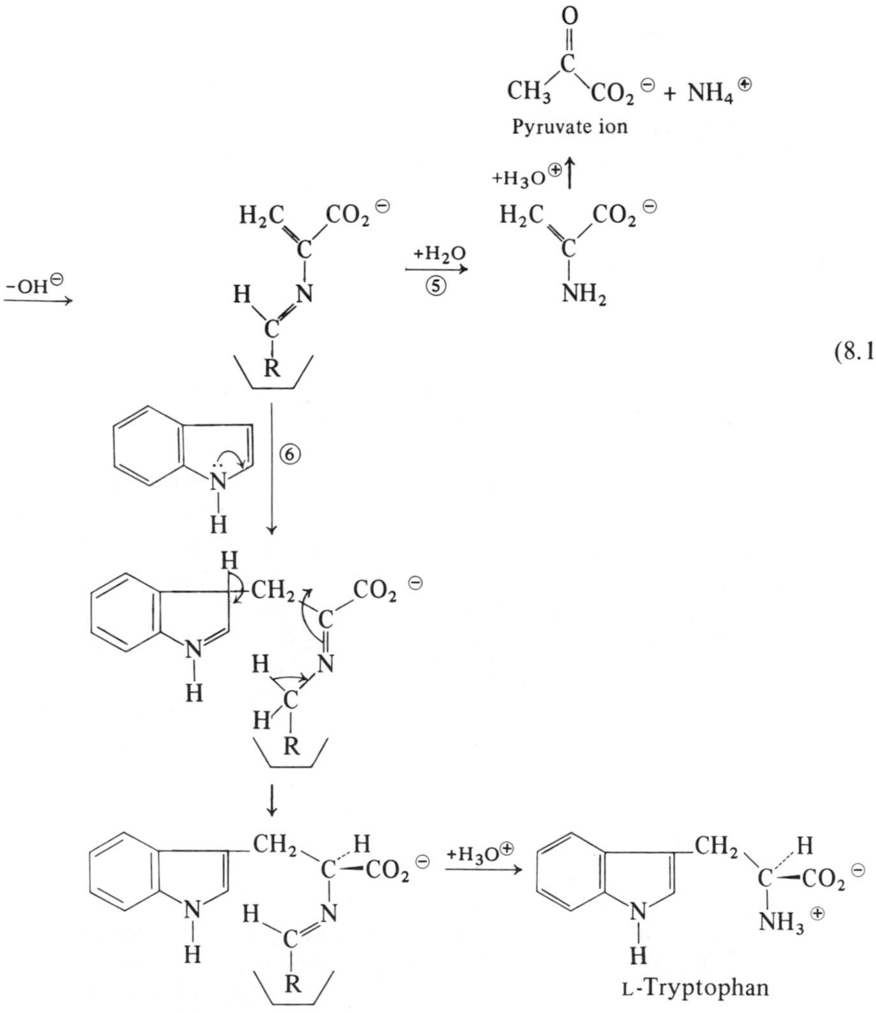

$$CH_3 \overset{\displaystyle O}{\underset{}{\text{C}}}\; CO_2^{\ominus} + NH_4^{\oplus}$$

Pyruvate ion

$+H_3O^{\oplus}\uparrow$

$$H_2C \;\; CO_2^{\ominus}$$
$$NH_2$$

$\xrightarrow{\;-OH^{\ominus}\;}$

$$H_2C \;\; CO_2^{\ominus}$$
$$H-N$$
$$C$$
$$R$$

$\xrightarrow[\text{⑤}]{+H_2O}$

⑥

L-Tryptophan

$\xrightarrow{+H_3O^{\oplus}}$

$$CH_2 \;\; H$$
$$C-CO_2^{\ominus}$$
$$NH_3^{\oplus}$$

(8.17)

179

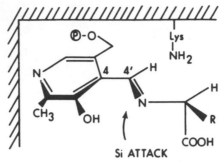

Fig. 8.3. Preferential protonation of an imine intermediate in the pyridoxal phosphate-catalyzed coenzyme reactions.

single basic group on the enzyme accomplishes an intramolecular suprafacial 1,3-prototropic shift. Transamination of the enzyme adduct by L-alanine frees an ε-amino group of lysine near the active site, but apparently this group is not essential to retain the stereochemistry of protonation in closely related aspartate aminotransferase for which the results suggest involvement of an active-site histidine.

If an enzyme orients the amino acid–PLP compound such that the carboxyl group is perpendicular to the conjugated system, stereoelectronic effects favor decarboxylation. Considerable evidence has accumulated that α-decarboxylases control the conformation about the α-carbon–nitrogen bond by binding a distal group in the fully extended side chain of the amino acid moiety. The intermediate imine normally protonates at the α carbon to ultimately release an amine. Although the protonation could, in theory, occur from either side of the planar imine, the decarboxylations of L-tyrosine, L-lysine, L-glutamate, and L-histidine by their respective decarboxylases have been shown to proceed in a retentive mode. Sometimes a small amount of protonation occurs at C-4′ of the cofactor to ultimately produce an inactive pyridoxamine phosphate-enzyme compound. In the case of glutamate decarboxylase, this protonation at C-4′ occurs from the *si* face just as in the transaminases. It thus appears that all reactions occur on a single side of the coenzyme substrate compound, the other being inaccessible (see Fig. 8.3).[21]

8.3.3. Tetrahydrofolic Acid

Tetrahydrofolic acid (FH_4) is composed of a reduced pteridine, *p*-aminobenzoic acid, and L-glutamic acid. Its structure is shown in **8.11**. FH_4 coenzymes are involved in the transfer of one-carbon fragments at the oxidation levels of formate, formaldehyde, and methanol. At the oxidation level of formaldehyde, the adducts may exist in four structural forms: N^{10}-formyl FH_4, N^5-formyl FH_4, N^5-formimino FH_4, or N^5,N^{10}-methylene FH_4.

Thus the amino acid glycine is converted to the amino acid serine by **8.10** and **8.11**. In this reaction, both components are activated: Pyridoxal phosphate (**8.10**) activates the glycine, and tetrafolic acid (**8.11**) activates the formaldehyde.

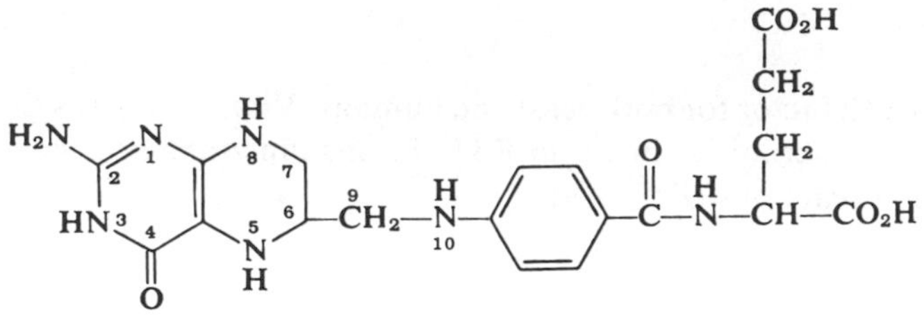

Tetrahydrofolic acid (FH₄)

8.11

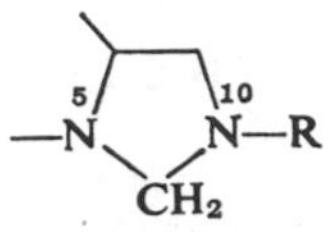

N^5, N^{10}-Methylene FH₄

8.12

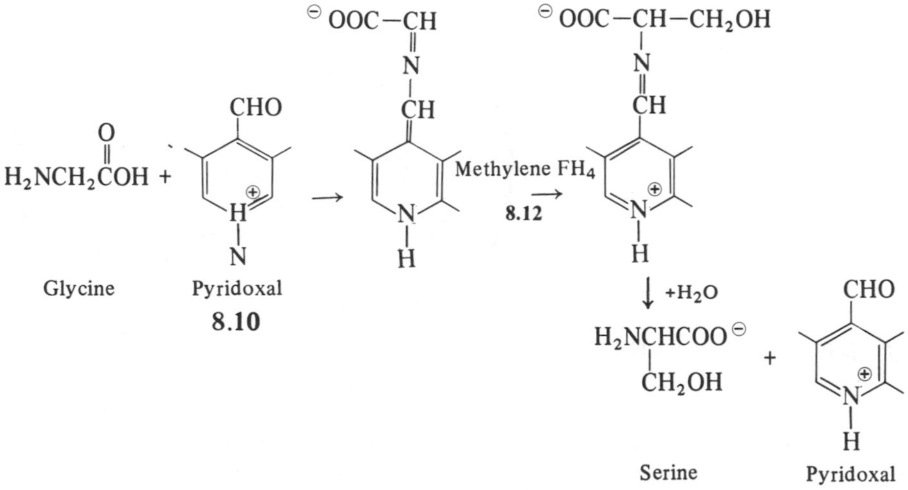

(8.18)

Suitably substituted imidazoline and imidazolidine derivatives function as carbon transfer reagents via mechanisms analogous to the mode of action of N^5, N^{10}-methenyl- and N^5, N^{10}-methylene–tetrahydrofolate coenzymes, respectively. While in biological processes, the role of the tetrahydrofolate coenzymes is limited to the transfer of one-carbon fragments at various oxidation levels, the models provide an opportunity of transferring complex carbon fragments incorporating a variety of functional groups. Based on the latter concept, a number of folic acid models have been prepared.[22]

Dihydrobiopterin, a folic acid derivative, is the coenzyme for the hydroxylation of phenylalanine.

8.3.4. Biotin

Biotin, a growth factor for both yeast and humans (Vitamin H) was first isolated by Kogl; the structure is shown in **8.13**. Biotin functions as a coenzyme for carboxylation reactions (Eq. 8.19).

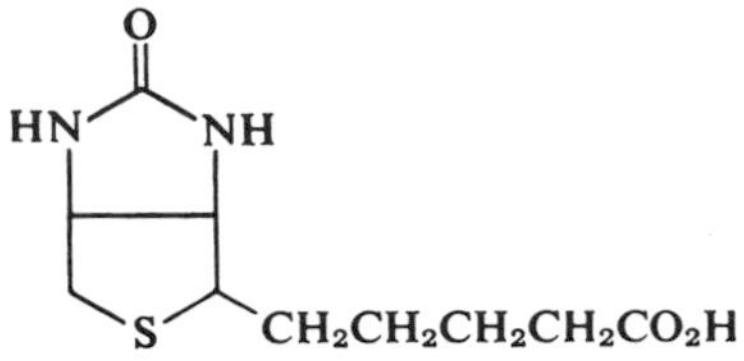

biotin

(8.13)

Biotin is bound to an enzyme via an amide linkage involving its carboxyl group and a lysine residue of the enzyme. Biotin-catalyzed carboxylation proceeds as shown by Eq. (8.19).

(a) E-biotin + ATP + $HCO_3^{\ominus}$ $\rightleftharpoons$

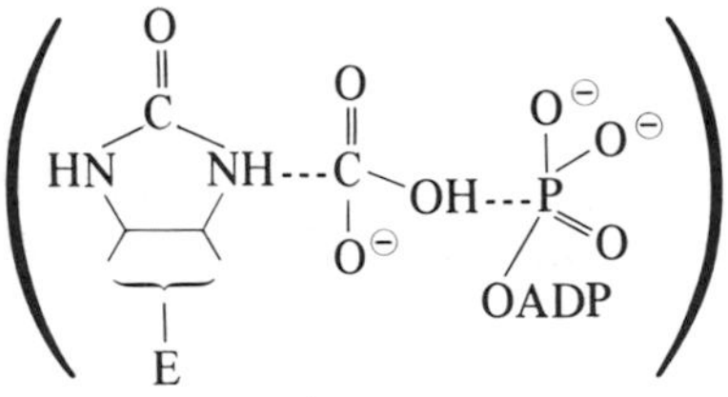

$$E\text{-Biotin-}CO_2^{\ominus} + ADP + P_i$$ (8.19)

(b) E-biotin-$CO_2^{\ominus}$ + RH $\rightleftharpoons$ E-biotin + R—$CO_2^{\ominus}$

RH is usually an acyl group in acyl-coenzyme A. The utilization of this cofactor may provide the enzyme with certain catalytic advantages as compared with a path involving the direct reaction of an acyl coenzyme A with ATP-activated carbon dioxide or bicarbonate.

Recently three irreversibly acidified model compounds of N'-carboxybiotin were prepared to assess the importance of prior protonation of the biotin ring system on the CO_2 transfer potential of the N'-carboxy group. Two were considered model compounds of N'-carboxybiotin in which protonation has occurred at the urido carbonyl oxygen atom. Conversely, a compound was synthesized to evaluate the CO_2-transfer potential of the N'-carboxy group, if protonation occurred at the N'-nitrogen atom. The reactivity of each substrate with nucleophiles has been evaluated. Of these three compounds, only one led to efficient transfer to the carbomethoxy group upon treatment with nitrogen-containing nucleophiles (morpholine, cyclohexylamine, and diisopropylamine). With smaller nucleophiles (i.e., water, methanol) reaction was centered at the ring C-2 position. Correspondingly, treatment of another compound with nucleophiles (i.e., alcohols, amines) led to products that can be explained in terms of two competing reactions. One pathway involves initial attack of the nucleophile at the C-2 position of the imidazolinium cation (an $A_{AC}2$ process) to give a tetrahedral intermediate which then undergoes bond cleavage in either of two directions. The competing pathway observed was an irreversible S_N2 displacement reaction (an $A_{AL}2$ process) at the methylene position of the O-alkyl side chain. Factors are presented that account for the overall product distribution obtained from these reactions. Finally, the products obtained from the treatment of another related compound with nucleophiles (i.e., alcohols, amines) could be accounted for solely by reactions which occurred at the C-2 position of the ring. (an $A_{AC}2$ process). The corresponding S_N2 pathway is not a viable route for this substrate.[23]

8.3.5. S-Adenosylmethionine

Methionine has been known to be a biochemical "methyl donor" for a number of years and the sulfonium salt S-adenosylmethionine (**8.14**) has been demonstrated to be the compound corresponding to "active" methionine which performs the transfer of a methyl group to various nucleophiles. A typical reaction is [Eq. (8.20)] the synthesis of creatine from guanidylacetic acid (an enzymatic nucleophilic displacement reaction):

S-Adenosylmethionine
8.14

$$(8.20)$$

This reaction in organic chemical terms is the reaction of a nucleophile (amine) with a sulfonium salt yielding a sulfide and substituted amine. There are many similar reactions of S-adenosylmethionine with other nucleophiles such as nicotinamide, imidazole, histamine, and catechols of the epinephrine type.

The steric course of an enzyme utilizing S-adenosylmethionine carrying a methyl group made chiral by labeling with ^{1}H, ^{2}H, and ^{3}H in an asymmetrical arrangement has been investigated.[21] Incubation of the two diastereomers of this substrate with the acceptors epinephrine or protocatechuic acid and the enzyme catechol O-methyltransferase gave the corresponding methylated catechols with inversion of configuration proving that a direct S_N2 reaction occurred without the intervention of a methylated enzyme intermediate.[24]

8.3.6.   Coenzyme A

Coenzyme A is involved in acyl group transfer. It was identified as a heat-stable cofactor required for certain acetylations and was implicated as a cofactor for the incorporation of acetate into acetoacetate and citrate. Its structure is shown in **8.15**. This molecule contains a multiplicity of functional groups, but the most significant group from the point of view of mechanism is the sulfhydryl group.

Coenzyme A
8.15

It is this sulfhydryl group that is involved in the formation of acetyl coenzyme A. In other words, acyl derivatives of coenzyme A are thiol esters. In fact, coenzyme A is often abbreviated as CoASH. One may ask the question, What is special about thiol esters as opposed to oxygen esters? In alkaline hydrolysis, the two show the same rate constants. In acid hydrolysis, thiol esters react faster than corresponding oxygen esters. In addition, thiol esters have been shown to be more reactive in condensations involving carbanions. Acyl derivatives of CoASH are usually formed in ATP-dependent reactions and undergo: (1) a number of reactions that involve attack of a nucleophile on the acyl carbon atom, resulting in transfer of the acyl moiety to the nucleophile with release of CoASH; (2) additions to species having a double bond in the acyl moiety; (3) condensation at the α-carbon atom of the thiol ester; and (4) reactions involving acyl group interchange. The free energy of a thiol ester (hydrolysis) is larger than that of the corresponding oxygen ester, which may be one reason for the preference for thiol esters.

8.3.7. Purine Phosphates

The two purine phosphates of interest are guanosine and adenosine phosphates. We concentrate on the latter. The structure of adenosine triphosphate (ATP) is shown in **8.16**. There are, of course, the corresponding monophosphate and diphosphate. The number of individual enzymes known to require ATP as the coenzyme, substrate, and energy source is very large.

Adenosine triphosphate (ATP)
8.16

ATP is involved in many biological oxidations and, therefore, there is a question of whether it should be considered here or in the section on redox systems. The fundamental role of the hydrolysis of ATP is as the driving force for biochemical processes which is presumably derived from the ultimate hydrolysis of the pyrophosphate linkages of the molecule.

ATP and ADP are in the "high energy" states. That is, hydrolysis of ATP to ADP and that of ADP to AMP, respectively, are accompanied by large decreases of free energy (-7.3 and -6.5 kcal/mole, respectively, at pH 7.0), which can be used as the driving force for other energetically unfavorable but biologically essential reactions. Hydrolyses of ATP and ADP are catalyzed by divalent metal ions such as $Mg^{\oplus 2}$ and $Zn^{\oplus 2}$; this is due to formation of a chelate such as **8.17** (see Chapter 9).

Metal complex of ATP
8.17

All ATP-dependent biosynthetic processes involve the formation of a covalent bond between two substrate molecules coupled with the cleavage of one of the pyrophosphate links of ATP. A number of ATP-dependent reactions involve carboxyl group activation. ATP can be cleaved either to ADP and phosphate ion or to AMP and pyrophosphate ion. An enzyme-bound acyl adenylate (formula 8.18) is formed in the latter reactions. Acetyl CoA is presumably synthesized as shown in Eq. (8.21).

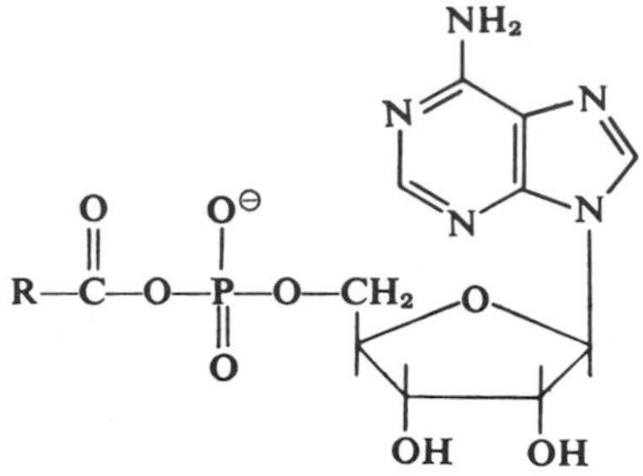

acyl adenylate
8.18

$$CH_3-CO_2H + ATP \underset{}{\overset{Mg^{\oplus 2}}{\rightleftharpoons}} \left[CH_3 - \overset{O}{\overset{\|}{C}} - O - AMP \right] + PP_i$$

$$\left[CH_3 - \overset{O}{\overset{\|}{C}} - O - AMP \right] + HSCoA \rightleftharpoons CH_3 - \overset{O}{\overset{\|}{C}} - SCoA + AMP \qquad (8.21)$$

8.3.8. Pyrimidine Phosphates

There are several pyrimidine phosphates: uridine, cytosine, and inosine phosphates. Uridine nucleotides have been shown to be important in carbohydrate metabolism. The general structure for UDP-sugars is shown in 8.19.[25] The role of UDP-sugar coenzymes is twofold. Such coenzymes are involved in a variety of glycosyl transfer reactions. Also the sugar itself can be transformed either by oxidation or by oxidation and decarboxylation.

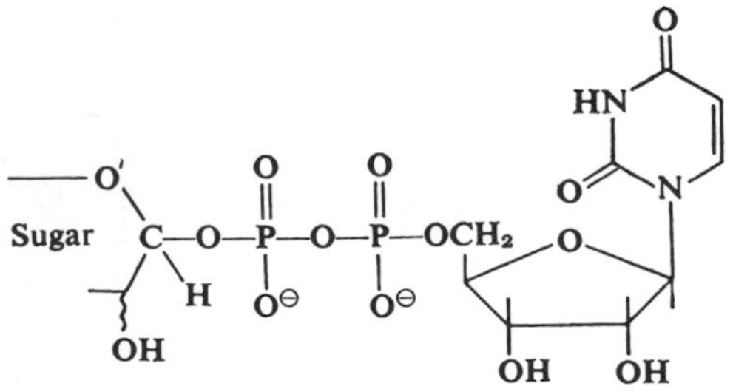

UDP-sugar
8.19

8.3.9.   Glutathione

The structure of glutathione was established as the tripeptide L-glutamyl-L-cysteinyl-L-glycine as shown in **8.20**. Although a large number of biochemical functions have been ascribed to glutathione, the number of instances in which glutathione is known to function as a coenzyme is limited. Of these probably the most widely studied is the glyoxalase reaction.

$$
\underset{H}{\overset{CO_2H}{H_2N-\overset{|}{\underset{|}{C}}-CH_2-CH_2-\overset{O}{\overset{||}{C}}-\overset{H}{\overset{|}{N}}-\overset{CH_2-SH}{\overset{|}{CH}}\text{———}\overset{O}{\overset{||}{C}}-\overset{H}{\overset{|}{N}}-CH_2-CO_2H}}
$$

Glutathione
8.20

$$
CH_3-\overset{O}{\overset{||}{C}}-\overset{O}{\overset{\nearrow}{\underset{\diagdown}{C}}}-H \xrightarrow{\text{glutathione}} CH_3-\overset{OH}{\underset{H}{\overset{|}{\underset{|}{C}}}}-CO_2H \qquad (8.22)
$$

An elegant model system has been developed for the glyoxalase reaction [Eq. (8.23)].[26]

$$(8.23)$$

Phenylglyoxal is smoothly converted to mandelic acid in aqueous solution at room temperature in the presence of N,N-dimethyl-β-mercaptoethylamine. Neither thiols alone nor amines alone are catalysts for this reaction, although a mixture of the two is somewhat effective. The free thiol group is required for catalysis, since the *S*-methyl derivative of the catalyst is inactive. Likewise, the trimethylammonium derivative of this catalyst is devoid of catalytic activity, indicating the necessity of a basic site on the catalyst molecule. The model reaction, as with glyoxalase I, proceeds with internal hydrogen transfer. On the basis of this information, the mechanism in Eq. (8.23) appears reasonable for the model system.

8.3.10. Vitamins A, D, E, and K

In addition to the coenzymes discussed earlier, Vitamins A, D, E, and K should be considered. The structures of these vitamins are shown in **8.21**. These coenzymes are involved in electron transport (redox) systems.

The term vitamin D originally designated the antirachitic principle in preparations of the "fat-soluble A" factor. However, before the actual isolation (now called vitamin D_3 or cholecalciferol) from fish liver oils, it was shown that an antirachitic compound (vitamin D, calciferol, or ergocalciferol) could be produced in the laboratory by the irradiation of the plant sterol ergosterol. The term vitamin D_1 has been discarded since the material to which it was first applied has been found to be a mixture of calciferol and other sterols.

Vitamin A plays an essential role in various biochemical and physiological processes. As a pharmacological agent, the vitamin or preferably synthetic derivatives known as "retinoids" are useful in the treatment of certain dermatitides. There has also been considerable interest in the apparent ability of these compounds to interfere with carcinogenesis.

In animals, the signs of deficiency of vitamin E include structural and functional abnormalities of many organs and organ systems. Attending these morphological alterations are biochemical defects that appear to involve fatty acid metabolism and numerous other enzyme systems. Notable is the fact that many signs and symptoms of vitamin E deficiency in animals superficially resemble disease states in humans; however, there is little unequivocal evidence that vitamin E is of nutritional significance in man or is of any value in therapy.

Vitamin K is a dietary principle essential for the normal biosynthesis of several factors required for clotting of blood. It was observed that chickens fed on inadequate diets developed a deficiency disease in which the outstanding symptom was spontaneous bleeding, apparently due to a low content of prothrombin in the blood. In subsequent studies, it was reported that, although the condition was not cured by any of the known vitamins, it could be rapidly alleviated by feeding an unidentified fat-soluble substance. This substance was given the name "vitamin K" (*Koagulation* vitamin).

In higher animals, vitamin D deficiency causes abnormalities in calcium and phosphate metabolism and results in structural changes in bones and teeth; the syndrome characteristic of a severe deficiency in children is called rickets, in adults, osteomalacia. The ingestion of excessive amounts of vitamin D also produces toxic symptoms. Initially there is a rise in the blood calcium level followed by metastatic calcification of various internal organs and, ultimately, by decalcification of skeletal structures.

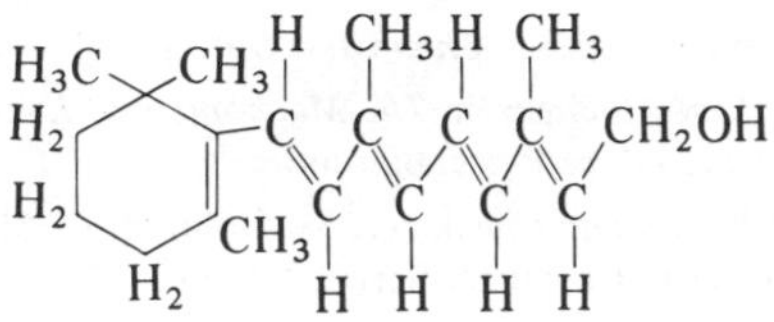

Vitamin A₁

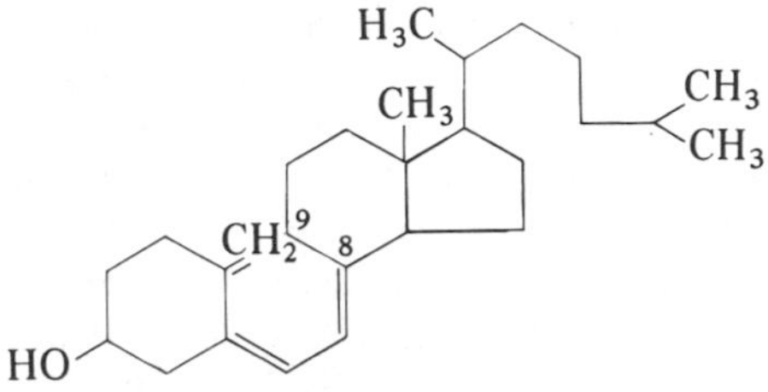

Vitamin D₃

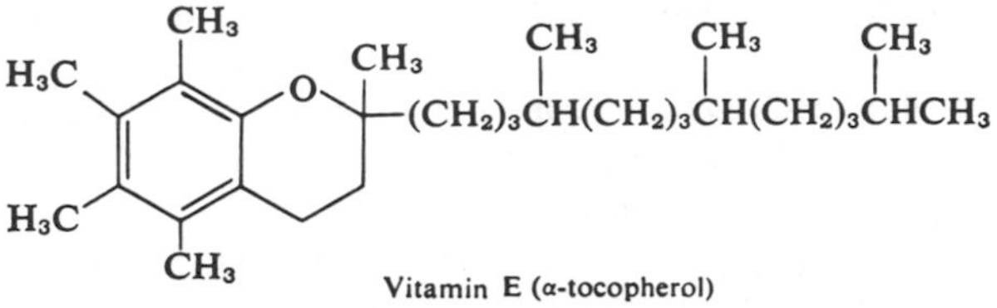

Vitamin E (α-tocopherol)

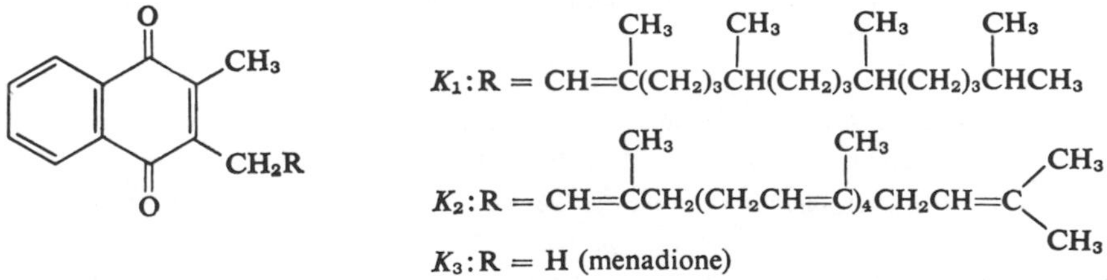

$$K_1:R = CH=C(CH_2)_3\overset{CH_3}{\underset{|}{C}}H(CH_2)_3\overset{CH_3}{\underset{|}{C}}H(CH_2)_3\overset{CH_3}{\underset{|}{C}}HCH_3$$

$$K_2:R = CH=\overset{CH_3}{\underset{|}{C}}CH_2(CH_2CH=C)_4CH_2CH=\overset{CH_3}{\underset{\diagdown}{C}}\overset{CH_3}{\diagup}$$

$K_3:R = H$ (menadione)

Vitamins A, D, E, and K

8.21

The D vitamins stimulate the absorption of $Ca^{\oplus 2}$ from the intestinal tract, but do not appear to exert a direct effect on the absorption of phosphate; the lowered accumulation of bone salts on avitaminotic animals is chiefly a result of an impaired ability to absorb calcium. In addition, vitamin D appears to function in the internal tissues. For example, the amount of citrate present in the bones and internal organs (kidney, heart) of vitamin D-deficient rats rises rapidly when the vitamin is given. Although it is generally agreed that the D vitamins play an important role in the process of growth and especially in the formation and maintenance of bones, the biochemical functions of this group of vitamins remain obscure.

REFERENCES

1. M. L. Bender, *Mechanism of Homogeneous Catalysis from Protons to Proteins*, Wiley-Interscience, New York, 1971, Chapter 17.

2. H. Dugas and C. Penney, *Bioorganic Chemistry*, Springer-Verlag, New York, 1981.

3. B. Vennesland and F. H. Westheimer, in *The Mechanism of Enzyme Action*, W. D. McElroy and B. Glass, Eds., John Hopkins Press, Baltimore, 1954, p. 357.

4. F. H. Westheimer, in *Biomimetic Chemistry*, D. Dolphin, C. McKenna, Y. Murakami, and I. Tabushi, Eds., Advances in Chemistry Series, No. 191, Washington, D.C., 1980, p. 31.

5. T. C. Bruice, *Acc. Chem. Res.*, **13**, 256 (1980).

6. S. Shinkai, T. Yamashita, Y. Kusano, and O. Manabe, *J. Am. Chem. Soc.*, **103**, 2070 (1981).

7. S. Shimkai, T. Yamashita, Y. Kusano, and O. Manabe, *J. Am. Chem. Soc.*, **104**, 563 (1982).

8. T. C. Bruice, in *Biomimetic Chemistry*, D. Dolphin, C. McKenna, Y. Murakami, and I. Tabushi, Eds., Advances in Chemistry Series, No. 191, Washington, D.C., 1980, p. 89.

9. E. T. Kaiser, in *Biomimetic Chemistry*, D. Dolphin, C. McKenna, Y. Murakami, and I. Tabushi, Eds., Advances in Chemistry Series, No. 191, Washington, D.C., 1980, p. 35.

10. A. O. Hill, J. M. Pratt and R. J. P. Williams, *Chem. Brit.*, **156** (1969).

11. Y. Murakami, in *Biomimetic Chemistry*, D. Dolphin, C. McKenna, Y. Murakami, and I. Tabishi, Eds., Advances in Chemistry Series, No. 191, Washington, D.C., 1980, p. 179.

12. T. C. Stadtman, *Science*, **172**, 859 (1971).

13. Y. Yang, K. H. Johnson, R. H. Holm, and J. G. Norman, Jr., *J. Am. Chem. Soc.*, **97**, 6596 (1975).

14. K. S. Hagen, J. G. Reynolds, and R. H. Holm, *J. Am. Chem. Soc.*, **103**, 4054 (1981).

15. R. Breslow, *J. Am. Chem. Soc.*, **80**, 3719 (1958).

16. A. Turano, et al., *J. Am. Chem. Soc.*, **104**, 3089 (1982).

17. E. E. Snell, A. E. Braunstein, E. S. Severin, and Y. M. Torchinsky, *Pyridoxal Catalysis: Enzymes and Model Systems*, Interscience Publishers, New York, 1968.

18. D. E. Metzler and E. E. Snell, *J. Am. Chem. Soc.*, **74**, 979 (1952).

19. R. M. Topping and T. C. Bruice, *J. Am. Chem. Soc.*, **85**, 1488 (1963).

20. Y. Tachibana, M. Ando, and H. Kuzuhara, *Chem. Lett.*, 1765 (1982).

21. J. C. Vederas and H. G. Floss, *Acc. Chem. Res.*, **13**, 455 (1980).

22. U. K. Pandit and H. Bieräugel, *Abstracts of the Twenty-eighth Congress of IUPAC* (1981).

23. M. J. Cravey and H. Kohn, *J. Am. Chem. Soc.*, **102**, 3928 (1980).

24. M. D. Tsai, H. G. Floss, P. A. Crooks, and J. Coward, *J. Biol. Chem.*, **255**, 9124 (1980).

25. R. Caputto, L. F. Leloir, C. E. Cardini, and A. C. Paladini, *J. Biol. Chem.*, **184**, 333 (1950).

26. V. Franzen, *Chem. Ber.*, **90**, 623 (1957).

9 | Metal Ion Catalysis

9.1. INTRODUCTION

Metal ions can produce substantial accelerations of reactions in homogeneous solution. Ultimately, catalysis occurs because the metal ion has the ability to coordinate with the substrate or reactant (see Fig. 9.1) and thereby stabilize the transition state. However, the mode of action of metal ions in the stabilization of transition states is varied.

Although distinctions among the various modes of metal-ion catalysis are somewhat arbitrary and overlapping, the discussion can be divided into two main parts:

1. Superacid catalysis, where the metal ion acts as if it were a proton of magnified charge (and can act even at neutrality where the proton concentration is 10^{-7} M).
2. Catalysis involving redox reactions, where the metal ion serves as a carrier of electrons.

9.2. SUPERACID CATALYSIS

Metal ions catalyze many organic reactions by means of a mechanism akin to acid or electrophilic catalysis. Since a hydronium ion contains a formal charge of plus one while a metal ion may be a multivalent cation, catalysis by metal ions is

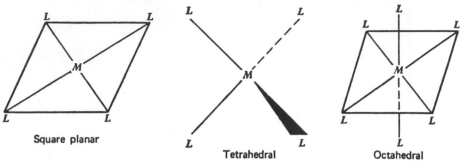

Square planar   Tetrahedral   Octahedral

Fig. 9.1. Some examples of metal (M)–ligand (L) coordination. From M. L. Bender, *Mechanisms of Homogeneous Catalysis from Protons to Proteins*, Wiley-Interscience, New York, 1971, Chapter 8.

analogous to that seen with hydronium ions and has been dubbed "superacid" catalysis.[1] This type of metal-ion catalysis occurs in many nucleophilic reactions of organic compounds.

In hydronium ion catalysis, the high rate of proton transfer with respect to other covalent changes is one of the factors ensuring efficient catalysis. It is therefore of interest to ascertain the speed of transfer of metal ions from one ligand to another. Since much of our discussion centers around reactions in aqueous solution, we will consider the rate constants for water substitution in the inner coordination sphere of metal ions. Figure 9.2 shows rate constants of water

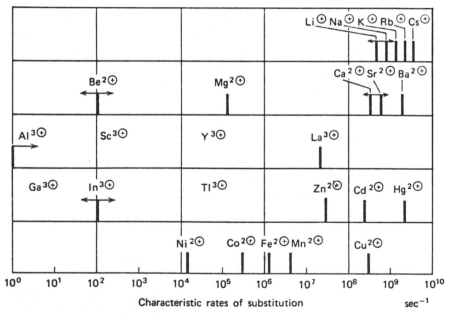

Fig. 9.2. Characteristic rate constants for H_2O substitution in the inner coordination sphere of metal ions (abscissa in sec^{-1}). From M. Eigen, *Pure Appl. Chem.*, **6**, 105 (1963). Reproduced by permission of the International Union of Pure and Applied Chemistry and Butterworths Scientific Publications.

substitution for some alkali metal, alkaline earth metal, transition metal, and other metal ions. Clearly, most rate constants are larger than $10^4 \sec^{-1}$, indicating that, in general, ligand substitution is rapid compared to most covalent changes.

9.2.1. Cleavage of Carbon–Carbon and Carbon–Hydrogen Bonds

Polyvalent metal ions catalyze the decarboxylation of oxaloacetic acid and many other β-keto acids containing a second carboxylic acid group adjacent or nearly adjacent to the ketonic function, including dihydroxyfumaric acid, acetone-dicarboxylic acid, oxalosuccinic acid, and dihydroxytartaric acid. On the other hand, neither monocarboxylic acids nor β-keto acids without a second carboxylic acid group are decarboxylated by metal ions. Many metal ions were tested as catalysts in the decarboxylation of oxaloacetic acid: aluminum, ferric, ferrous, and cupric ions were the most efficient; sodium, potassium, and silver ions were inactive. Furthermore, the monoester of dimethyloxaloacetic acid, in which the carboxylic acid group adjacent to the ketone is esterified, is not subject to metal ion catalysis, while the dianion, in which the carboxylic acid group adjacent to the ketone is not esterified, is. This evidence indicates that the metal ion is coordinated with the carboxylate ion gamma to the one that is lost. Cupric ion, which usually shows square planar coordination (Fig. 9.1), gives excellent catalysis. Hence the metal ion is probably not coordinated with the carbonyl oxygen atom and with both of the carboxylate ion groups simultaneously. On this basis, the coordination of the metal ion is postulated to involve the γ-carboxylate ion and the ketonic oxygen atom, as in Eq. (9.1).

$$(9.1)$$

Since aluminum ion is an excellent catalyst for the decarboxylations listed and does not undergo valency change easily, it appears that a redox reaction is not involved in the catalysis. The absence of catalysis by such highly charged ions as $Co(NH_3)_6^{\oplus 3}$ indicates that the cationic acceleration described here is due to an interaction of a specific short-range character. During the decarboxylation, an electron pair initially associated with the carboxylate ion group is transferred to the rest of the molecule. A metal ion complexed with the carbonyl group should assist this transfer because of its position and positive charge. Thus mechanism (9.1) assigns an electrophilic function to the metal ion catalysis.[2]

Many attempts have been made in this and other metal ion–catalyzed reactions to correlate catalytic rate constants of different metal ions with association constants of the metal ions of either the reactant or product. Two such attempts in metal ion–catalyzed decarboxylation show a linear free-energy correlation between the rate constants and the product association constants, but not with the reactant association constants (Fig. 9.3). This result indicates that the transition state of the decarboxylation reactions resembles the product, which has considerable enolate ion character, rather than the reactant.[3]

Negative ions such as citrate and acetate ions diminish the catalytic activity of cupric ion, the amount of diminution being much greater for citrate ion than for acetate ion. In the decarboxylation of acetonedicarboxylic acid, the neutral species $Cu^{\oplus2}A^{\ominus2}$ and the monoanionic species $CuOAc^{\oplus}A^{\ominus2}$ are catalytically active, but the dianionic species $Cu(ortho-Ac)_2A^{\ominus2}$ is completely inactive. These facts are in agreement with the hypothesis that any ligand reducing the *effective* charge of the metal ion–substrate complex reduces the catalytic effectiveness of the metal ion. Conversely, a complex-forming agent that does not destroy the charge on the cupric ion enhances the catalytic activity. Thus pyridine, which readily forms complexes with cupric ion, but has no charge, promotes the cupric ion–catalyzed decarboxylation of dimethyloxaloacetic acid. Likewise, *ortho*-phenanthroline ehances the catalytic activity of manganous ion sixteenfold.

Metal ions do not catalyze the decarboxylation of β-ketomonocarboxylic

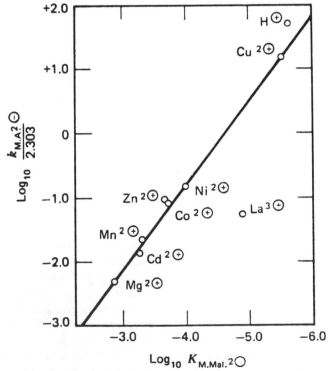

Fig. 9.3. Linear free-energy relationship in the decarboxylation of acetone–dicarboxylic acid by various metal ions. The log of the rate constant is plotted against the association constant of the same ions with malonate ion. From J. E. Prue, *J. Chem. Soc.* **1952**, 2337.

acids because a more favorable free energy of metal ion complexation occurs in the ground state than in the transition state. Reversal of this argument suggests that metal ions can catalyze carboxylation of substances containing active hydrogen atoms since the metal ion complex should be more favorable in the transition than in the ground state in this reaction. For example, magnesium methyl carbonate may be used to carboxylate both ketones [Eq. (9.2)] and aliphatic nitro compounds, an elegant synthetic method as well as an interesting catalytic phenomenon.[4] In addition, the magnesium complexes of the enolate ions of β-keto acids may be further alkylated [Eq. (9.2)] to form ketones.

$$R-\overset{\overset{O}{\|}}{C}CH_2R' + (CH_3O\overset{\overset{O}{\|}}{C}O^\ominus)_2Mg^{\oplus 2} \longrightarrow R-\underset{\underset{\underset{R'}{|}}{C}}{C}\overset{\overset{Mg^{2\oplus}}{\overset{.\,.\,.\,.}{\diagup\;\diagdown}}}{\underset{\diagdown\;\diagup}{\overset{\ominus O\qquad O^\ominus}{|\qquad\quad|}}}C=O + 2CH_3OH + CO_2 \qquad (9.2)$$

$$\downarrow \begin{array}{l} 1)\ R'X \\ 2)\ H_3O^\oplus, -CO_2 \end{array}$$

$$R-\overset{\overset{O}{\|}}{C}-\underset{\underset{R''}{|}}{C}HR'$$

In addition to cleavage of carbon–carbon bonds, the cleavage of carbon–hydrogen bonds is facilitated by metal ions in systems where the metal ion can coordinate with the substrate in the proper position. Many multivalent cations (in order of efficiency: cupric, nickelous, lanthanum, zinc, manganous, cadmium, magnesium, and calcium ions) are catalysts for the bromination of ethyl acetoacetate and 2-carboethoxycyclopentanone. Likewise, zinc ion catalyzes the iodination of pyruvate and o-carboxyacetophenone. The rate-determining step in these ketone halogenations is the formation of an enol through a proton transfer to a general base. The metal ion can catalyze the reaction in a manner similar to that for decarboxylation by stabilizing the anionic charge generated in the cleavage of the carbon–hydrogen bond. The relative catalytic efficiency of the series of metal ions listed parallels their order of stability in complexes with salicylaldehyde, which is consistent with the enolate ion description of the metal ion–catalyzed decarboxylation.

In an enzymatic process involving metal ion catalysis, the enzyme (1) provides specificity toward the substrate and (2) imparts specific complexation of the metal ion in a manner enhancing its activity.

Multivalent metal ions catalyze aldol condensation reactions. For example, in basic solution, cupric diglycinate condenses with formaldehyde, producing cupric complexes of the amino acid serine, or with acetaldehyde, producing complexes of the amino acids threonine and allothreonine.[4] The catalytic activity of the metal ion in these reactions stems from the facilitation of carbon–hydrogen bond cleavage, yielding enolate ion necessary for condensation. The mechanism is shown in Eq. (9.3).

$$(9.3)$$

9.2.2.   Addition to Carbon–Oxygen and Carbon–Nitrogen Double Bonds

The hydrolysis of many amino acid esters is subject to catalysis by metal ions.[5] Structurally, all these esters contain a functional group that can serve as ligand for the metal ion. For instance, metal ions such as cobaltous, cupric, manganous, calcium, and magnesium ions effectively catalyze the hydrolysis of α-amino acid esters. In glycine buffer at pH 7.3, for example, glycine methyl ester and phenylalanine ethyl ester undergo a facile hydrolysis catalyzed by cupric ion. Under these conditions, the reactions follow first-order kinetics in the substrate, and it is thus possible to compare the rate constants of hydroiysis of DL-phenylalanine ethyl ester catalyzed by hydronium, hydroxide, and cupric ions at pH 7.3 and 25°C: $H_3O^{\oplus}$, $1.46 \times 10^{-11} \, \mathrm{sec}^{-1}$; $OH^{\ominus}$, $5.8 \times 10^{-9} \, \mathrm{sec}^{-1}$; and $Cu^{\oplus 2}$ (0.0775 M), $2.67 \times 10^{-3} \, \mathrm{sec}^{-1}$.[6] Although the last constant is a complex constant not directly comparable to the first two, it clearly shows facile catalysis by cupric ion. As with all amines, these α-amino acids form coordination compounds with heavy metal ions.[7] When the concentration of metal ion is varied, the rate constant of the hydrolysis varies, reaching a maximal value as the [metal ion]:[ester] ratio approaches unity, indicating that the most active species is a complex of *one* metal ion and *one* ester molecule. On the basis of this evidence, a 1:1 complex between the metal ion and the α-amino acid ester may be postulated in which the metal ion chelates with the α-amino group and the carbonyl oxygen atom of the ester. Thus the mechanism of these catalyses is expressed either by Eq. (9.4) or by (9.5).

$$
\begin{array}{c}
\underset{\substack{\text{H}_2\text{C}\text{---}\text{C}=\text{O}\\ \text{H}_2\text{N}\quad\text{O}^{\ominus}\\ \diagdown\text{Cu}^{\oplus 2}\\ \text{H}_2\text{N}\quad\text{O}\\ |\quad\ \ ||\\ \text{RCH}\text{---}\text{C}\text{--OCH}_3}}{}
\xrightarrow{\text{OH}^{\oplus}}
\underset{\substack{\text{H}_2\text{C}\text{---}\text{C}=\text{O}\\ \text{H}_2\text{N}\quad\text{O}^{\ominus}\\ \diagdown\text{Cu}^{\oplus 2}\\ \text{H}_2\text{N}\quad\text{O}\\ |\quad\ \ |\\ \text{RCH}\text{---}\text{C}\text{--OCH}_3\\ |\\ \text{OH}}}{}
\xrightarrow[-\text{CH}_3\text{OH}]{+\text{H}^{\oplus}}
\text{hydrolysis products}
\end{array}
\tag{9.4}
$$

$$
\begin{array}{c}
\underset{\substack{\text{H}_2\text{C}\text{---}\text{C}=\text{O}\\ \text{H}_2\text{N}\quad\text{O}^{\ominus}\\ \diagdown\text{Cu}^{\oplus 2}\text{CH}_3\\ \text{H}_2\text{N}\quad\text{O}\\ |\\ \text{RCH}\text{---}\text{C}=\text{O}}}{}
\xrightarrow{\text{OH}^{\oplus}}
\underset{\substack{\text{H}_2\text{C}\text{---}\text{C}=\text{O}\\ \text{H}_2\text{N}\quad\text{O}^{\ominus}\\ \diagdown\text{Cu}^{\oplus 2}\text{CH}_3\\ \text{H}_2\text{N}\quad\text{O}\\ |\\ \text{RCH}\text{---}\text{C}\text{--O}^{\ominus}\\ |\\ \text{OH}}}{}
\xrightarrow[-\text{CH}_3\text{OH}]{+\text{H}^{\oplus}}
\text{hydrolysis products}
\end{array}
\tag{9.5}
$$

As evidenced by the preceding relative rates of D, L-phenylalanine ethyl ester hydrolysis, the effect of copper ions is beyond a simple electrostatic one and corresponds to superacid catalysis. However, with histidine, cysteine, and aspartic acid esters, the rate of cupric ion–catalyzed hydrolysis is only a hundredfold greater than that of the neutral substrate. In these systems, the metal ion can chelate at two sites *other* than the ester bond; hence a catalytic effect is seen that corresponds only to an electrostatic effect. Clearly, then, when one of the two sites with which the metal ion can complex is the reactive ester linkage itself, superacid catalysis occurs. It must be pointed out that the cupric ion is regenerated in all of these reactions although the complete reactions are not indicated.

Metal ions, including transition metal and rare earth ions, catalyze the hydrolysis of a variety of amides. In these compounds the metal ion can complex with one or more amine or carboxylate ion ligands in addition to the amide group. Thus the structural prerequisites for metal ion catalysis of amide hydrolysis parallel those for ester hydrolysis. However, metal ion catalysis in amide hydrolysis is not nearly as striking as in ester hydrolysis. For example, hydrolysis of glycinamide in the presence of 0.02 M cupric ion is only twentyfold faster than the spontaneous hydrolysis. This result is at first surprising since most of the infrared evidence for the interaction of metal ions with carboxylic acid derivatives shows stronger interaction with amides than esters, presumably because of the greater basicity of the former. This indicates that stronger complexing of the metal ion with the *ground* state than with the *transition* state must necessarily lead to a slower rather than to a faster reaction.

Thiol esters are particularly susceptible to cleavage by heavy metal ions such as mercuric, lead, and silver ions. Since these cations will cleave simple esters containing no secondary ligand groups, chelation does not appear to be essential in these reactions as it was in the hydrolysis of oxygen esters and amides.

Presumably the coordination of the sulfur atom with the heavy metal ion is the principal driving force of this reaction.

Reactions of carbonyl compounds and imines are also catalyzed by metal ions. The hydrolyses of some imines are catalyzed by divalent metal ions such as

(9.6)

cupric or nickelous ions. Spectrophotometric evidence indicates the formation of a metal ion–substrate complex with subsequent facile cleavage of the complex. The hydrolyses of imines derived from salicylaldehyde are, however, retarded by metal ions, presumably because the reactant complex is stabler than the transition state (product) complex.

In some reactions, however, the hydroxide ion instead of the carbonyl group of the substrate coordinates to the metal ion followed by an intracomplex nucleophilic reaction [path B of Eq. (9.6)]. Thus the role of metal ions here is to enhance the availability of hydroxide ion for the reaction rather than to polarize the carbonyl group [path A of Eq. (9.6)].[7] For example, in the hydrolysis of glycinamide by a Co(III) complex, a stable complex, path B involving the coordination of hydroxide ion to the metal ion is slightly more efficient than path A involving the coordination of carbonyl group of the substrate.[4] It is also possible that some of the preceding reactions which were attributed to the polarization of the carbonyl groups by metal ion are due to coordination of hydroxide ion to metal ion, since it is impossible to distinguish these two mechanisms kinetically.

The mechanism involving the coordination of hydroxide ion to a metal ion [path B in Eq. (9.6)] was borne out in the $Zn^{\oplus 2}$-catalyzed hydrolysis **9.1**, a model for the anhydride intermediate in carboxypeptidase A catalysis, **9.2**. The two nitrogen atoms of **9.1** coordinate the $Zn^{\oplus 2}$, which produces a 10^3-fold acceleration of the hydrolysis of the anhydride moiety. The pH dependence indicated that only the attack of hydroxide ion, not that of water, was enhanced by $Zn^{\oplus 2}$. This fact would not be expected if $Zn^{\oplus 2}$ were functioning as a Lewis acid facilitating the attack on the coordinated anhydride by an external nucleophile, since in a more reactive coordinated anhydride the preference for a better nucleophile (hydroxide ion) should be decreased not increased. Thus the mechanism shown in **9.2** was proposed.[8]

9.1   carboxypeptidase A   9.2

Zinc ion catalyzes the reduction of an aldehyde, 1,10-phenanthroline-2-carboxaldehyde, by N-propyl-1,4-dihydronicotinamide, an NADH analog, in acetonitrile [Eq. (9.9)]. Here the aldehyde is coordinated to or, at least, is located in the proximity of a metal ion and its carbonyl group is activated by polarization. The aldehyde-Zn$^{\oplus 2}$ complex first forms an intermediate with the NADH analog, followed by direct hydrogen transfer as it occurs in the enzymatic reaction. The catalytic efficiency of the zinc ion is significant, since no reaction could be detected in its absence. This model system may mimic the enzyme alcohol dehydrogenase, which has zinc ion(s) at the inactive site of the enzyme.

(9.7)

Zinc ion complexed with Tris in the pH range 7.5–10 is a very effective catalyst for hydrolysis and aminolysis (by Tris) of benzylpenicillin. Both Cu$^{\oplus 2}$ and Ni$^{\oplus 2}$ ions are nonreactive in this system. The rate of loss of penicillin from solution was found to be a linear function of zinc ion at concentrations up to 10^{-5} M, while the dependence on Tris concentration is maximal at about 0.02–0.03 M at each pH studied. At high penicillin concentrations, saturation kinetics are observed. Product assays show the major product to be a penicilloic acid at low Tris concentrations and N-(penicilloyl)-Tris at high Tris concentrations. The proposed mechanism suggests that the reaction is mediated by a ternary complex in which the metal ion acts to bring the reactants (penicillin and Tris) into close proximity and to lower the pK_a of a Tris hydroxyl group creating a strong nucleophile (Fig. 9.4).[9] This mechanism also explains the results of the product assays and the lack of reactivity of the other metal ions tested. This reaction may be related to the mechanism for a known zinc-dependent enzyme β-lactamase.

Examination of the schematic representation of the ternary complex shows that the expected product would be a Tris ester of penicilloic acid. These esters are known to hydrolyze rather readily and are also subject to aminolysis. Thus it would be expected that this intermediate would partition to the acid and amide

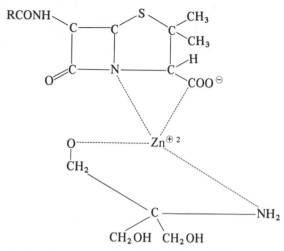

Fig. 9.4. Schematic representation of zinc–Tris–penicillin ternary complex.

with the relative proportion of the latter increasing with increasing Tris concentration but independent of the rate of penicillin loss. The results of the product assays are totally consistent with the ternary complex pathway.

The lack of reactivity of Cu^{+2} and Ni^{+2} ions in this system at pH 8 and metal ion concentrations up to 10^{-3} M seems to be the result of chelation of the metal ion by Tris, essentially preventing interaction with penicillin. The formation constants for the metal ion–Tris chelates are shown in Table 9.1.

Metal ions catalyze the hydrolysis of phosphoric and phosphonic acid halides, phosphate esters,[10] and various phosphoric acid anhydrides including acyl phosphates, pyrophosphates, and triphosphates. The hydrolyses of diisopropyl phosphorfluoridate (DFP) and isopropyl methylphosphonofluoridate (Sarin) are susceptible to catalysis by many metal salts and chelates such as MoO_4^{-2}, Ce(III), Mn(II), Cu(II), WO_4^{-2}, and CrO_4^{-2}. In the hydrolysis of Sarin, catalysis by cerous, cupric, and manganous ions in the form of bifunctional species containing a nucleophilic center (hydroxide ion) and an electrophilic center (metal ion) is particularly effective. Although the hydroxymetallic ions are considerably weaker bases than hydroxide ion itself, they are catalytically more active by a factor of 10.

Table 9.1.   Formation Constants for Tris Complexes

Metal Ion	Log K_1	Log K_2
Cu^{+2}	3.98	3.49
Ni^{+2}	2.80	2.10
Zn^{+2}	2.26	(1.4)

Source: Reference 9.

The solvolysis of tetrabenzyl pyrophosphate catalyzed by the general base lutidine is additionally catalyzed by calcium ion [Eq. (9.8)]. In the presence of 0.02 M calcium ion and 0.2 M lutidine (2,6-dimethylpyridine), the rate of solvolysis of tetrabenzyl pyrophosphate (with cleavage of the P–O–P bond) is increased by a factor close to a millionfold over the uncatalyzed reaction, each catalyst making approximately a thousandfold contribution.[11] The divalent cation can chelate two oxygen atoms of the pyrophosphate ester, making the phosphorus atom more susceptible to nucleophilic attack.

$$(PhCH_2O)_2\overset{\overset{O}{\|}}{P}\overset{\overset{O}{\|}}{O}P(OCH_2Ph)_2 + Ca^{\oplus 2} \longrightarrow (PhCH_2O)_2P\underset{O}{\overset{\overset{Ca^{\oplus 2}}{\underset{O \quad O}{}}}{}}P(OCH_2Ph)_2 \qquad (9.8)$$

$$\downarrow \text{lutidine}$$

$$\text{hydrolysis product}$$

Nucleophilic reactions of acetyl phosphate are catalyzed by cations such as magnesium, calcium, cobalt, manganese, nickel, zinc, and lithium. Calcium ion catalyzes a neutral as well as a basic hydrolysis. The chelate shown in Eq. (9.10) is postulated as an intermediate in the metal ion–catalyzed hydrolysis. In reactions proceeding with phosphorus–oxygen fission [Eq. (9.9)] the metal ion may facilitate metaphosphate ion formation by stabilizing the leaving group in the same manner as a proton.

$$CH_3\overset{O}{\underset{O}{C}}\underset{O^{\ominus}}{\overset{Mg^{2\oplus}}{\underset{O \quad O}{P{=}O}}} \longrightarrow CH_3{-}\overset{O}{C}\underset{O}{} \;+\; \underset{O \quad O}{\overset{Mg^{2\oplus}}{P}}\overset{O^{\ominus}}{} \xrightarrow[\text{fast}]{H_2O} H_2PO_4^{\ominus} \qquad (9.9)$$

Enzymatic phosphorylation reactions in which adenosine triphosphate (ATP) participates require magnesium ion. Although the metal ion may act as a chelating agent between nucleotide and enzyme in these reactions, several arguments suggest that part of the metal ion function is associated with superacid catalysis of the kind described here. Certainly in nonenzymatic hydrolysis of ATP, as well as simpler polyphosphates and triphosphates, the facilitation of hydrolysis by various divalent metal ions such as calcium, magnesium, manganous, cupric, and cadmium ions must be attributed to superacid catalysis. For instance, the rate of hydrolysis of ATP is accelerated tenfold by calcium ion at pH 9 and sixtyfold by cupric ion at pH 5. Divalent metal ions also catalyze the nonenzymatic transphosphorylation of ATP and γ-phenylpropyl triphosphate with inorganic phosphate and carboxylate ions. In the uncatalyzed hydrolysis of γ-phenylpropyl diphosphate, the monoprotonated species hydrolyzes approximately 2000 times faster than the fully ionized species. Consequently, a metaphosphate ion intermediate has been invoked here.[12] On this basis, the mechanism of metal ion catalysis of hydrolysis of the terminal phosphate bond is postulated as in Eq. (9.10), which is similar to that for the hydrolysis of acetyl phosphate. Again metaphosphate ion formation is facilitated by the superacid properties of the metal ion chelate. The identical rate constants of γ-phenylpropyl triphosphate and ATP in metal ion–catalyzed reactions with inorganic phosphate indicate that the adenine moiety is not involved (kinetically important) in complexing with the metal ion. NMR measurements show that magnesium, calcium, and zinc ions bind predominantly to the β- and γ-phosphates of ATP and that cupric ion binds predominantly to the α- and β-phosphates, whereas manganous ion binds to all three. Therefore, mechanism (9.10) is an arbitrary designation of the position of the complexing of the metal ion. It is also arbitrary with respect to the assumption of a 1:1 complex.

$$(9.10)$$

9.2.3.   Factors in Superacid Catalysis

In reactions where the metal ion is consumed stoichiometrically, the term "metal ion promotion" as opposed to "metal ion catalysis" is a more accurate expression.

The most important characteristic in a metal ion promoted or catalyzed reaction is its positive charge.[13] In many reactions, variation in catalytic effectiveness can be directly correlated with variation in the magnitude of cationic charge. Since this charge consists of the effective charge on the metal ion complex, and not on the metal ion alone, the electrostatic nature of the ligands attached to the metal ion is as important as the inherent charge on the metal ion. In several of the reactions cited earlier, catalysis by a polyvalent metal ion is reduced to zero by complexation of the metal ion with anionic ligands. Furthermore, charge density may be more important than net charge. The force between two charges or dipoles depends on the inverse square of the distance between them. Therefore, for maximal effect, the metal ion should be directly associated with the substrate molecule to be catalyzed. More specifically, it must be in some way electronically linked to the reactive bond of the substrate to be broken. Thus stereospecific coordination of the metal ion is of utmost importance. With transition metal ions, the electrostatic effect of the ion is also affected by the shielding of the ligand from the nuclear charge of the metal ion by its d electrons and its ligand field.

All organic reactions are influenced by electronic changes within a molecule. The principal function of the metal ion catalyst is to effect such changes. The simplest manifestation of electronic distortion by a metal ion is seen in the fact that the acidity of a water molecule coordinated with a cupric ion is 10^7 times greater than the acidity of a free water molecule. The acidity of an organic substrate complexed with cupric ion might therefore also be substantially greater than that of the organic substrate alone. This hypothesis means that a considerable shift of electron density can be brought about by coordination of a metal ion with an organic substrate; therefore, reactions dependent on electronic movement should be effectively facilitated by a metal ion.

Many reactions require the presence of an "electron sink" in the molecule to accommodate electron density shifts in the course of the reaction. The introduction of a suitably positioned metal ion in a substrate undergoing such a reaction can serve this purpose. Other reactions require the neutralization of negative charge to reduce electrostatic repulsion during reaction. A metal ion will also serve this purpose. Still other reactions require the polarization of a particular bond to effect reaction. Again, introduction of metal ion in a specific position in the substrate molecule will accelerate reaction in this way. Finally, the stabilization of leaving groups will often facilitate reaction. When the leaving groups are halide, phosphate, mercaptide, or other anions, metal ions can facilitate reaction.

Either a proton or a metal ion can introduce a positive charge into a substrate molecule, producing the aforementioned electronic changes. A metal ion, however, is superior to a proton on several grounds:

1. A metal ion can introduce a multiple positive charge into an organic molecule, whereas a proton can introduce only a single positive charge.
2. A metal ion can operate in neutral solution, whereas a proton cannot.
3. A metal ion can coordinate several donor atoms, whereas a proton can coordinate only one.

In listing the ways in which metal ions may promote organic reactions, the requirement that the metal ion be suitably positioned within the substrate molecule was emphasized. Specific complexation or chelation of the metal ion with the substrate appears to be an absolute requirement of metal ion catalysis of nucleophilic reactions. In many reactions, chelation appears to be the rule; this requirement means that the substrate must contain either one or two donor atoms, in addition to the reaction center, with which the metal ion must coordinate. Many attempts have been made to correlate the effectiveness of catalysis by a series of metal ions with the relative formation constants of the complexes. Successful analyses require a complex closely approximating the transition state of the reaction rather than the ground state. This result indicates that the metal ion complex must stabilize the transition state of the reaction in order to assist the reaction effectively and that metal ion complex formation in the ground state depresses the reaction rate.

Since coordination compounds of metal ions may involve a large number of ligands, it is possible to form not only a metal ion complex with the substrate, but also a complex with both the substrate and the nucleophilic agent simultaneously. The metal ion can thus serve as a template for both components of a bimolecular reaction and assist reaction by making the entropy of activation more positive.

A metal ion complex containing hydroxide ion as a ligand serves as a carrier of hydroxide ion in neutral solution, just as the metal ion itself can be considered to approximate a superproton in neutral solution. Thus a metal ion complex containing hydroxide ion (or another nucleophile) can be considered to be a bifunctional catalyst, the metal ion serving as a general acid and the hydroxide ion serving as general base or nucleophile.

9.3. CATALYSIS VIA REDOX REACTIONS

The role of metal ions in superacid catalysis is that of an electrophile or general acid and depends on the magnitude of the cationic charge as well as complex-forming ability. Except for these factors, there is no distinction between transition, inner transition, and other metal ions. In catalysis of oxidation–reduction reactions, however, only transition metal ions play a role, since the key feature in these reactions is the ability of the metal ion to exist in solution in more than one oxidation state.[14] The ability to complex plays a secondary, but still necessary, role.

9.3.1. Metal Ion Redox Reactions

Even if oxidizing and reducing agents have the proper redox potentials (or standard free energies) for reaction with each other, the reaction may still be slow because of a large free energy of activation. This is particularly true of organic reducing agents. In such a case, a metallic ion of variable valence may greatly accelerate the rate by providing an easier reaction path.

Cupric and silver ions, as well as other cations exhibiting two or more stable oxidation states serve to "transport" electrons in redox reactions through a chain mechanism in which the catalytic metal ion is successively oxidized and reduced. For example, reaction (9.11) is catalyzed by Cu(II) since the rate of the reaction shows a dependence both on the vanadium(III) and copper(II) concentrations but not the iron(III) concentration.

$$V(III) + Fe(III) \longrightarrow V(IV) + Fe(II) \qquad (9.11)$$

The mechanism of this catalyst is given by

$$V(III) + Cu(II) \longrightarrow V(IV) + Cu(I)$$

$$Cu(I) + Fe(III) \xrightarrow{\text{fast}} Fe(II) + Cu(II) \qquad (9.12)$$

Since the electron transfer reaction

$$Tl(I) + 2Ce(IV) \longrightarrow Tl(III) + 2Ce(III) \qquad (9.13)$$

involves either a three-body collision or an unavailable intermediate valency state [either thallium(II) of cerium(II)], reaction will proceed slowly. However, in the presence of manganous ion, both of these difficulties are avoided by the introduction of an alternate catalytic reaction path (Chapter 1).

$$Ce(IV) + Mn(II) \longrightarrow Ce(III) + Mn(III)$$

$$Mn(III) + Ce(IV) \longrightarrow Ce(III) + Mn(IV)$$

$$Mn(IV) + Tl(I) \longrightarrow Mn(II) + Tl(III) \qquad (9.14)$$

9.3.2. Simple Redox Catalyses

Coordination often greatly increases the ease with which an electron transfer can occur. An example is the catalytic effect of Mn(III) on the reaction between chlorine and oxalic acid. The key step is the internal oxidation–reduction of $MnC_2O_4^{\oplus}$, followed by reaction with chlorine.

$$H_2C_2O_4 + Mn^{3\oplus} \rightleftharpoons MnC_2O_4^{\oplus} + 2H^{\oplus}$$
$$MnC_2O_4^{\oplus} \longrightarrow Mn^{\oplus 2} + CO_2 + CO_2^{\ominus}$$
$$CO_2^{\ominus} + Cl_2 \longrightarrow CO_2 + Cl\cdot + Cl^{\ominus}$$
$$Cl\cdot + Mn^{\oplus 2} \longrightarrow Cl^{\ominus} + Mn^{\oplus 3}$$
$$\overline{\rule{0pt}{1.2em}\hspace{1em}}$$
$$H_2C_2O_4 + Cl_2 \longrightarrow 2H^{\oplus} + 2Cl^{\ominus} + 2CO_2 \qquad (9.15)$$

The copper(0)–copper(I)–copper(II) system has long been a favorite catalytic system for reactions that may be accelerated by electron transfer, because of favorable redox properties of this sytem. Both the Sandmeyer and the Meerwein reaction (the substitution of an aromatic diazonium salt with molecules containing activated double bonds such as acrylonitrile) are catalyzed by copper(I) salts. These reactions have been postulated to proceed via the phenyl free radical as shown by Eq. (9.16).

$$\text{(9.16)}$$

The decomposition of hydrogen peroxide is accelerated by many metal ions. This iron-catalyzed reaction may be described by

$$\text{Fe(II)} + H_2O_2 \longrightarrow \text{Fe(III)}(OH^{\ominus}) + HO\cdot$$

$$\text{Fe(III)} + H_2O_2 \longrightarrow \text{Fe(II)} + H^{\oplus} + HO_2$$

$$\text{Fe(III)} + HO_2 \longrightarrow \text{Fe(II)} + H^{\oplus} + O_2 \qquad \text{(9.17)}$$

A somewhat more complicated system, involving hydrogen peroxide (or molecular oxygen) in the presence of ferrous ions, ascorbic acid, and ethylenediaminetetraacetic acid is capable of hydroxylating aromatic compounds. It bears some resemblance to enzymatic hydroxylations of aromatic compounds by oxygen or hydrogen peroxide and has been the subject of many investigations.

The autooxidation of organic compounds, that is, the reaction of oxidizable materials with molecular oxygen is catalyzed by transition metal ions and inhibited by easily oxidized materials such as phenols, aromatic amines, and secondary alcohols. Metal ions of variable valence are effective in these reactions. Thus copper, cobalt, iron, and manganese salts are good catalysts, whereas aluminum, magnesium, zinc, and lead salts are inactive or very poor catalysts. The oxidation of hydrocarbons, probably of most practical interest, is carried out commonly with metal naphthenates, resinates, or stearates. The mechanism of the reaction for the cobalt case is given by Eq. (9.18).

$$\text{Co(II)} + \text{ROOH} \longrightarrow \text{Co(III)} + \text{OH}^{\ominus} + \text{RO}\cdot$$
$$\text{Co(III)} + \text{ROOH} \longrightarrow \text{Co(II)} + \text{H}^{\oplus} + \text{ROO}\cdot$$
$$\left.\begin{array}{r}\text{ROO}\cdot + \text{RH} \longrightarrow \text{ROOH} + \text{R}\cdot \\ \text{R}\cdot + \text{O}_2 \longrightarrow \text{ROO}\cdot \end{array}\right\} \text{chain propagation}$$
$$\text{RO}\cdot + \text{RH} \longrightarrow \text{ROH} + \text{R}\cdot$$
$$2\text{ROO}\cdot \longrightarrow \text{inactive products} \qquad\qquad (9.18)$$

Cobalt (II), copper (II), nickel (II), and iron (II) salts are the preferred catalysts. The start of the reaction coincides with the oxidation of Co (II) to Co (III).

In Eq. (9.18), the first two initiating steps also regenerate the catalyst. The next two steps are the main chain-carrying steps leading to the major initial product, hydroperoxide. Further oxidation and decomposition of the hydroperoxide produce ketone and carboxylic acid.

9.3.3.   Bridging Between One Substrate and Another by Metal Ion

In reactions discussed so far, metal ions have been reversibly oxidized and reduced. Metal ions, however, can perform another catalytic function (in redox reactions): as bridging and electron transfer agents between one substrate and another.[15] Ferric ion catalyzes the oxidation of ascorbic acid by hydrogen peroxide (Chapter 8). This process can be understood in terms of the following:

$$(9.19)$$

In this mechanism, ferric ion forms a mixed complex with both the enediol and hydrogen peroxide. This mixed complex can then undergo both an acid–base reaction, perhaps involving water molecules of the solution, and an electron transfer reaction in which the ferric ion serves as a bridging group between ascorbic acid and hydrogen peroxide so that ascorbic acid becomes oxidized and hydrogen peroxide reduced (Chapter 8).

The mechanism of aromatic hydroxylation by aqueous hydrogen peroxide may be probed most easily when ferric ion and catechol are used as catalysts. Although ferric ion cannot be replaced by Cr (III), Co (III), Zn (II), Mn (II), Al (III), or Mg (II), cupric ion may act as an inefficient catalyst for the reaction.

Athough 1,2-dihydroxy- or 1,4-dihydroxyaromatic compounds are catalysts for the reaction, monohydroxy- or 1,3-dihydroxyaromatic compounds are not. Kinetic results are not consistent with a free-radical chain mechanism, but suggest that the hydroxylating agent is a complex of ferric ion, the enediol, and hydrogen peroxide similar to Eq. (9.19). The isomer distribution of the phenols formed in the reaction and the relative reactivity of the aromatic substrates indicate that the oxidizing agent is very nonselective and confirm that the hydroxyl radical is not the hydroxylating species. The actual oxidizing agent is complexed iron oxide, formed by the elimination of a molecule of water from the intermediate containing ferric ion, hydrogen peroxide, and the enediol catalyst. The catalytic nature of the reaction is easily seen in the following:

(9.20)

The only reactants consumed in the cycle are hydrogen peroxide and the aromatic compound; the products are substituted phenols and water. The oxidation of saturated and unsaturated aliphatic hydrocarbons by ascorbic acid, ferric or ferrous ion, and oxygen can be described in a similar fashion.

9.4. ENZYME MODELS INVOLVING METAL IONS

Several systems involving metal ions have already been shown to serve as enzyme models (see Section 9.2.2). However, it is probably appropriate at this point to underscore the importance of metals in model systems with several additional examples. A very interesting carbonic anhydrase mimic is found in the tris-(4,5-diisopropylimidazole-2-yl)phosphine Zn(II) system. Tris(4,5-diisopropylimidazol-2-yl)phosphine appeared from molecular models to be a reasonable mimic of the active site in carbonic anhydrase. Several physicochemical studies with tris-(4,5-diisopropylimidazole-2-yl)phosphine Zn(II) complex such as spectral studies and catalysis of the $CO_2 \rightleftarrows HCO_3^-$ interconversion indicate that it can be considered a reasonable but deficient model of the enzyme carbonic anhydrase.[16] It was further shown that tris-(4,5-dimethyl-2-imidazolyl)methylphosphine oxide cobalt (II) mimics the spectroscopic properties of Co(II) carbonic anhydrase and also catalyzes the interconversion of CO_2 and HCO_3^-.[17]

A model study of the enzyme urease established that O-coordination of a urea derivative to Ni(II) promotes nucleophilic attack of solvent on the otherwise unreactive carbonyl group of the urea. The rate of the nonenzymatic degradation of urea in aqueous media is independent of pH between pH 2 and 12, falling below pH 2 and rising above pH 12. It was demonstrated at pH values of 7, 13, and 14 that the reaction is an *elimination*, yielding as the sole products ammonia and cyanic acid (9.21). The data support an invariant mechanism of degradation over the entire pH range [Eq. (9.21)], the falloff below pH 2 being reasonably ascribed to the protonation of urea and the dependence on [$^\ominus OH$] above pH 12 being ascribed to specific base catalysis of the elimination reaction [Eqs. (9.22a and b)]. The latter chemistry is adequately supported by a variety of models.

$$H-N=C=O + NH_3 + H_2O \qquad (9.21)$$

$$H_2N-\overset{\overset{\displaystyle O}{\|}}{C}-NH_2 + {}^\ominus OH \rightleftharpoons H-\overset{\ominus}{N}-\overset{\overset{\displaystyle O}{\|}}{C}-NH_2 + H_2O \qquad (9.22a)$$

$$H-\overset{\ominus}{N}-\overset{\overset{\displaystyle O}{\|}}{C}-NH_2 + H-OH \longrightarrow H-N=C=O + NH_3 + {}^\ominus OH \qquad (9.22b)$$

The enzyme urease catalyses the hydrolysis of urea to form carbamate ion [Eq. (9.23)]. At pH 7.0 and 38°C, the urease-catalyzed hydrolysis of urea must be at least 10^{14} times as fast as the spontaneous hydrolysis of urea, which has never

$$H_2N-\overset{\overset{\displaystyle O}{\|}}{C}-NH_2 + H_2O \longrightarrow H_2N-COO^\ominus + \overset{\oplus}{N}H_4 \qquad (9.23)$$

been observed. On the balance of available evidence from model studies, and especially from consideration of the structure of molecules which are and are not substrates for urease, it was recently postulated that all substrates for urease (thus far, urea, N-hydroxyurea, N-methylurea, semicarbazide, formamide, and acetamide) are activated toward nucleophilic attack on carbon by virtue of O-coordination to an active-site Ni(II) ion as in **9.3** in which the metalloenzyme acts as a superacid. The detailed mechanism arising from this postulate was without precedent in the chemistry of ureas or amides. Hydroxamic acids reversibly inhibit urease from all sources that have been tested. Wherever the dependence of urease on metals was determined, nickel has been found to be involved. We may confidently expect, therefore, that all hydrolytic ureases will be nickel metalloenzymes. With the preceding information in hand, a model system containing Ni(II) and urea, which successfully mimics elements of the enzyme-catalyzed reactions is described.

$$
\left.\begin{array}{c} U \\ R \\ E \\ A \\ S \\ E \end{array}\right| \!\!\! \rightsquigarrow Ni^{\oplus 2}\!-\!O\!=\!C\!\!\begin{array}{c} \diagup NH_2 \\ \diagdown R \end{array}
$$

9.3

$(R = NH_2, NHOH, NHCH_3, NHNH_2, H, CH_3)$

In aqueous ethanol in the presence of excess $NiCl_2$, N-(2-pyridylmethyl) urea, undergoes ethanolysis to give ethyl N-(2-pyridylmethyl)carbamate and hydrolysis to give (2-pyridylmethyl) amine in accordance with Eq. (9.24).

$$R-NH-\overset{\overset{\displaystyle O}{\|}}{C}-NH_2 + EtOH \longrightarrow R-NH-\overset{\overset{\displaystyle O}{\|}}{C}-OEt + NH_3 \quad (9.24a)$$

$$R-NH-\overset{\overset{\displaystyle O}{\|}}{C}-NH_2 + H_2O \longrightarrow R-NH_2 + H_2CO_3 + NH_3 \quad (9.24b)$$

The disappearance of N-(2-pyridylmethyl)urea and the appearance of products from measurements on (2-pyridylmethyl)amine were both first order, with virtually identical rate constants. Values of k_{obs} (Table 9.2) were essentially independent of the initial concentration of N-(2-pyridylmethyl)urea but increased linearly with the concentration of water.

Equations (9.24a) and (b) describe parallel pseudo-first-order reactions for which $k_{obs} = k'_{EtOH} + k'_{H_2O} + k'_{H_2O} = k_{obs}[(2\text{-pyridylmethyl})amine]_\infty/[N\text{-(2-pyridylmethyl)urea}]_0$. Values of k'_{EtOH} have a slight dependence on $[H_2O]$ (Table 9.2), and linear extrapolation to zero $[H_2O]$ gives a limiting value, $k'^{lim}_{EtOH} = 1.60 \times 10^{-4}$ sec^{-1} at 80.25°C. A calculated second-order rate constant for

Table 9.2. Ethanolysis and Hydrolysis of N-(2-Pyridylmethyl)urea in the Presence of 0.39 M NiCl$_2$[a]

Temperature (°C)	$[1]_0$, (M)	$[H_2O]$, (M)	$10^5 k_{obs}$[b] (sec^{-1})	$\dfrac{[3]_\infty}{[1]_0}$[c]	$10^5 k'_{EtOH}$ (sec^{-1})	$10^5 k'_{H_2O}$ (sec^{-1})
80.25	0.025	0.78	17.7	0.064	16.6	1.1
80.25	0.050	0.78	17.1	0.067	15.9	1.1
80.25	0.100	0.78	17.8	0.078	16.4	1.4
80.25	0.050	1.67	20.3	0.100	18.3	2.0
80.25	0.050	3.00	23.1	0.185	18.8	4.3
80.25	0.050	5.22	27.7	0.269	20.2	7.5
70.04	0.050	0.78	5.10	0.086	4.66	0.44
60.02	0.050	0.78	1.50	0.089	1.36	0.134
50.00	0.085	0.78	0.409	0.091	0.372	0.037

Source: Reference 18.
[a]NiCl$_2 \cdot 6H_2O$ was partially dehydrated at 82°C in vacuo. The concentration of NiCl$_2$ was determined specrophotometrically at 395 nm in 1 M HCl. Distilled ethanol was dried over 4A molecular sieves.
[b]$\pm 4\%$ (2 standard errors); from loss of **1**.
[c]**1** = N-(2-pyridylmethyl)urea; **3** = (2-pyridylmethyl)amine.

reaction with ethanol may be defined as $k_{EtOH} = k'^{lim}_{EtOH}/[EtOH]$ and has the value $9.4 \times 10^{-6}\ M^{-1}\ sec^{-1}$ based on the molarity of pure ethanol (17.0 M at 25°C). Values of k'_{H_2O} are directly proportional to $[H_2O]$, leading to a calculated second-order rate constant k_{H_2O} of $1.45 \times 10^{-5}\ M^{-1}\ sec^{-1}$ at 80.25°C.

In ethanol, nickel chloride displays the three electronic absorption peaks in the visible and near infrared regions that are characteristic of octahedral coordination (Fig. 9.5). The spectrum is slightly but significantly altered by addition of N-(2-pyridylmethyl)urea or N-(2-pyridylmethyl)carbamate in equimolar amounts, establishing that an octahedral complex of Ni(II) is formed by each. The spectral differences between nickel(II) and the nickel(II)-N-(2-pyridylmethyl)urea complex near 400 nm are unchanged between 0.16 and 0.019 M. This indicates that the dissociation constant of the complex in ethanol must be less than $\sim 0.01\ M$. It follows that N-(2-pyridylmethyl)urea exists almost completely as a Ni(II) complex under the conditions of the kinetic studies.

$$(9.25)$$

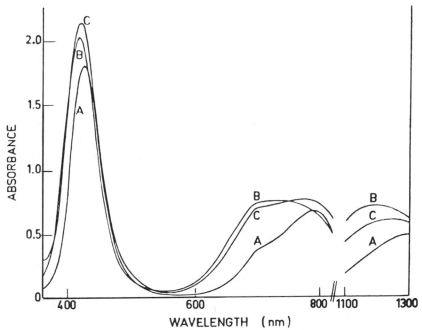

Application of transition-state theory to k'_{EtOH} leads to $\Delta H^{\ddagger} = 27.5 \pm 0.3$ kcal mol^{-1} and $\Delta S^{\ddagger} = 1.45 \pm 0.06$ cal mol^{-1} K^{-1}. For k'_{H_2O}, $\Delta H^{\ddagger} = 25.2 \pm 1.5$ kcal mol^{-1}. The similarity in $\Delta H^{\ddagger}$ for k'_{EtOH} and k'_{H_2O} strongly implies that the two reactions have the same mechanism. This is supported by similar values of the corresponding second-order rate constants ($k_{EtOH}/k_{H_2O} = 0.65$ at 80.25° C).

Spontaneous degradation of N-(2-pyridylmethyl)urea in anhydrous ethanol at 80.25° C in the presence of 1 mM 8-hydroxyquinoline is undetectable in 1200 h, which indicates a rate constant less than 2.3×10^{-9} sec^{-1}. The rate enhancement [Eq. (9.24a)] due to coordination of N-(2-pyridylmethyl)urea to Ni(II) is therefore greater than 7×10^4. Salts that promoted the ethanolysis of N-(2-

Fig. 9.5.   Absorption spectra of NiCl$_2$ in ethanol which contains ≤ 0.04 M H$_2$O: (A) NiCl$_2$ only; (B) NiCl$_2$ + equimolar N-(2-pyridylmethyl)urea; (C) NiCl$_2$ + equimolar ethyl N-(2-pyridylmethyl)-carbamate. λ_{max} (nm), ϵ_{max} (M^{-1} cm^{-1}): (A) 426, 10.9; 787, 4.1; 1310, 3.0; (B) 418.5, 12.2; 730, 4.6; 1190, 4.4; (C) 421, 13.1; 772, 4.65; 1250, 3.7.

pyridylmethyl)urea were $NiCl_2$, $CoCl_2$, and $MnCl_2$ in order of decreasing efficiency.

A reasonable mechanism for the $NiCl_2$-promoted ethanolysis of N-(2-pyridylmethyl)urea is shown in (9.25). In essence, the $Ni^{\oplus 2}$ ion acts as a superacid, just as the enzyme does, to promote nucleophilic attack of ethanol on the carbonyl group of the urea to form a tetrahedral intermediate. After a prototropic shift, ammonia is ejected to form the Ni(II) complex of the product.[18]

The kinetic activity of a [20]paracyclophane (PCP) connected covalently to 1,4-dihydronicotinamide (HNA) and 2-pyridinecarboxylic acid (Py) moieties, for the reduction of hexachloroacetone, HNA—PCP—Py, has been investigated as an alcohol dehydrogenase model. The reduction ability of PCP—HNA was significantly lowered as it underwent complex formation with zinc. On the other hand, HNA—PCP—Py showed an apparent rate maximum in a relatively lower concentration range of $ZnCl_2$. The kinetic behavior was analyzed on the basis of the formation of two kinds of zinc complexes of HNA—PCP—Py: The 1:1 complex

$$
\begin{array}{c}
\quad\quad Py \text{-\,-\,-} \\
PCP \quad\quad\quad Zn^{II} \\
\quad\quad HNA \text{-\,-\,}
\end{array}
$$

in which both Py and HNA moieties are simultaneously coordinated to the same zinc ion, showed a decreased reactivity relative to metal-free HNA—PCP—Py; while the 2:1 complex (HNA—PCP—Py—Zn^{II}—Py—PCP—HNA), in which HNA is free from metal coordination, exercised a much enhanced activity, seven times as reactive as metal-free HNA—PCP—Py.

The role of zinc ion in the reduction of a carbonyl compound with alcohol dehydrogenase is to polarize the carbonyl group through metal coordination and consequently facilitate the hydride transfer (in a formal sense) from the 1,4-dihydronicotinamide moiety of the coenzyme NADH to the carbonyl carbon atom. Hexachloroacetone, an electron-deficient carbonyl compound, may have the least tendency to coordinate to a metal ion. On the basis of these considerations, the ultimate goal of the present study has been to clarify whether or not the zinc ion accelerates the reduction of hexachloroacetone, a *nonspecific* substrate, as regards the metal-coordination tendency, by a reaction mechanism schematically shown in **9.4**. The acceleration effect provided by the zinc ion, involved in the 2:1 complex (HNA—PCP—Py—Zn^{II}—Py—PCP—HNA), is the first successful example along this line. As for the overall rate effect provided by the zinc ion on the present reaction, the major complex formed with HNA—PCP—Py

$$
\begin{array}{c}
\quad\quad Py \\
PCP \quad\quad\quad Zn^{II} \\
\quad\quad HNA
\end{array}
$$

precluded the reactivity of HNA—PCP—Py itself. In order to improve this situation, a ligand for coordination with the zinc ion must have significantly larger metal-binding ability than the dihydronicotinamide moiety. Furthermore, dihydronicotinamide and ligand moieties need to be constrained in a certain limited space so that the cooperation of the former and the zinc ion bound to the latter toward a substrate molecule is geometrically possible under the condition that the zinc binding with the ligand should not result ultimately in metal bridging between the dihydronicotinamide and ligand moieties.[19]

9.4

REFERENCES

1. F. H. Westheimer, *Spec. Publ. Chem. Soc.*, **8**, 1 (1957).
2. R. Steinberger and F. H. Westheimer, *J. Am. Chem. Soc.*, **73**, 429 (1951).
3. J. E. Prue, *J. Chem. Soc.*, 2337 (1952).
4. M. Stiles and H. L. Finkbeiner, *J. Am. Chem. Soc.*, **81**, 505 (1959).
5. M. L. Bender, *Adv. Chem. Series*, No. 37, Washington, 1963; No. 70, Washington, 1968.
6. M. L. Bender and B. W. Turnquest, *J. Am. Chem. Soc.*, **81**, 505 (1959).
7. D. A. Buckingham, D. M. Foster, and A. M. Sargeson, *J. Am. Chem. Soc.*, **92**, 6151 (1970).
8. R. Breslow, D. E. McClure, R. S. Brown, and J. Eisenbach, *J. Am. Chem. Soc.*, **97**, 194 (1975).
9. M. A. Schwartz, *Bioorg. Chem.*, **11**, 14 (1982).
10. B. Anderson et al., *J. Am. Chem. Soc.*, **99**, 2652 (1977).
11. G. O. Dudek and F. H. Westheimer, *J. Am. Chem. Soc.*, **81**, 2641 (1959).
12. D. Miller and F. H. Westheimer, *J. Am. Chem. Soc.*, **88**, 1514 (1966).
13. M. L. Bender, *Mechanisms of Homogeneous Catalysis from Protons to Proteins*, Wiley-Interscience, New York, 1971, p. 232.
14. M. F. Ansell and B. C. L. Weedon, *Chem. Britain*, **3**, 306 (1967).
15. G. A. Hamilton, *Adv. Enzymol.*, **32**, 55 (1969).
16. R. S. Brown, N. J. Curtis and J. Huguet, *J. Am. Chem. Soc.*, **103**, 6953 (1981).
17. R. S. Brown, D. Salmon, N. J. Cirtos, and S. Kusuma, *J. Am. Chem. Soc.*, **104**, 3188 (1982).
18. R. L. Blakeley, A. Treston, R. K. Andrews and B. Zerner, *J. Am. Chem. Soc.*, **104**, 612 (1982).
19. Y. Murakami, Y. Aoyama, and Jun-ichi Kikuchi, *Bull. Chem. Soc. Jpn.*, **55**, 2898 (1982).

10 | Intramolecular Catalysis and Intramolecular Reactions

Intramolecular reactions generally occur more readily than corresponding intermolecular reactions. Although classical neighboring group participation was first recognized in nucleophilic substitution, it has since been seen to cover the entire gamut of catalysis, from general acid–base catalysis to nucleophilic-electrophilic catalysis.[1] The elucidation of the mechanism of intramolecular catalysis frequently involves the differentiation of general base from nucleophilic catalysis, general base catalysis from a combination of general acid–hydroxide ion catalysis, and general acid catalysis from general base–hydronium ion catalysis. In other words, all the ambiguities of intermolecular systems are also found in intramolecular systems. Deciding on the catalytic mechanisms is generally done in the same way as described for intermolecular catalyses discussed in the previous chapters.

In recent years, considerable advances in the elucidation of intramolecular catalysis have been made. This has been stimulated by the hypothesis that intramolecular reactions can serve as models for the intracomplex catalysis exhibited by enzymes. Intramolecular catalysis has been referred to as the single most important contributor to enzymatic catalysis.[2]

An intramolecular catalyst must have a high effective local concentration, and if its sterochemistry is correct, its reaction should occur readily. A corollary to this argument is that the ratio of the effective concentration of the intramolecular general catalyst to the concentration of external hydronium or hydroxide ion must be large, and thus the intramolecular catalyst should effectively compete with external catalysts.

It is critical to point out that many of the examples we will consider in this chapter do not represent true catalyses in the sense that the catalyst is regenerated, but rather serve to exemplify the importance of intramolecular reactions in enzyme catalysis.

The systems described show rather astonishingly large rate accelerations relative to their intermolecular analogs. However, the "accelerated" reaction occurrs only once, unlike a real catalytic system, where the catalysts can in principle operate an infinite number of times. Again, the main purpose of this chapter is to point out the advantage of having all the components of a reaction attached to some sort of common backbone.

10.1. INTRAMOLECULAR GENERAL ACID OR BASE ACCELERATION

10.1.1. Acceleration by the Carboxylate Ion or the Carboxylic Acid Group or Both Groups

The enolization of *ortho*-isobutyrylbenzoic acid involves four kinetic terms corresponding to the four terms in Eq. (10.1). The principal kinetic term in the region pH 2.5–9 is the third term, involving carboxylate ion. Because this term is at least 50 times more important than the carboxylic acid term, the reaction is likely to proceed via intramolecular general base acceleration by carboxylate ion [Eq. (10.2)].

$$k_{obs} = k_H[H^\oplus]\,[RCOOH] + k_{OH}[OH^\ominus]\,[RCOO^\ominus] + k_{RCO_2^\ominus}\left[\begin{array}{c}\text{(structure)}\end{array}\right] +$$

$$k_{RCO_2H}\left[\begin{array}{c}\text{(structure)}\end{array}\right] \tag{10.1}$$

$$\tag{10.2}$$

The mechanism of Eq. (10.3), which depicts external hydroxide ion acceleration facilitated by the internal carboxylic acid group, is kinetically indistinguishable from Eq. (10.2), but is ruled out on grounds that:

1. Such acceleration is not seen in intermolecular reactions.
2. The rate constant of such a reaction would be too close to diffusion control for proton transfer.
3. The rate constant is higher for the isopropyl derivative described here than for the corresponding methyl derivative, whereas in intermolecular reactions, the reverse reactivity occurs.

The only reasonable explanation of this inversion of reactivity is through mechanism (10.2), which postulates that stereochemically freezing the rotation of the molecule increases the reaction rate of the isopropyl over that of the methyl compound.[3]

$$\ominus HO \longrightarrow H \qquad\qquad (10.3)$$

The classical example of intramolecular rate acceleration is the hydrolysis of aspirin. The pH-rate constant profile of this reaction consists of an acid-promoted reaction, a base-promoted reaction, and a reaction independent of pH near neutrality. Of course, the latter reaction, usually called a water reaction, can be, and is, a spontaneous reaction of the substrate. Closer inspection of Fig. 10.1 (Ref. 3) indicates that the pH-independent region is dependent on the ionization of a basic group of pK 4, such as the carboxylate ion.

Because the hydrolysis of aspirin is fifty times faster than that of its *para*-isomer, the *ortho*-carboxylate group must be involved in this hydrolysis directly.[4] It can can do so either as a general base or as a nucleophile. Recent evidence indicates a general base mechanism on the basis of $H_2 {}^{18}O$ studies [Eq. (10.4)].

$$\longrightarrow \quad + CH_3CO_2H \qquad\qquad (10.4)$$

General acid acceleration by an internal carboxylic acid group is also seen in the hydrolysis of acetals.[5] This is characteristic of internal catalysis since these

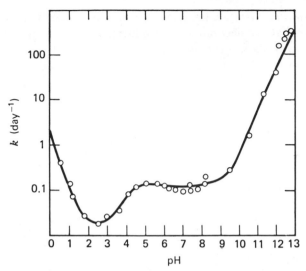

Fig. 10.1.   The hydrolysis of aspirin at 25°C. From L. J. Edwards, *Trans. Faraday Soc.*, **46**, 723 (1950).

reactions are specific acid-promoted when the catalyst is external. Thus compound **10.1**, which has an adjacent carboxyl group, is hydrolyzed much faster than **10.2** or **10.3**, which do not have a carboxyl group.[5]

OCH$_2$OMe   OCH$_2$OMe   MeOCH$_2$O

CO$_2$H   CO$_2$Me   CO$_2$H

10.1   **10.2**   **10.3**

650   1   2.2

(relative rates)   (10.5)

This large acceleration by an adjacent carboxylic acid group in **10.1** is attributable to its general acid promoted acceleration as shown in Eq. (10.6).

$$
\text{(scheme)} \quad \longrightarrow \quad + \; CH_2{=}\overset{\oplus}{O}Me \tag{10.6}
$$

$$
\Big\downarrow H_2O
$$

$$
CH_2O + MeOH
$$

Equations (10.7) and (10.8) involve electrophilic–nucleophilic mechanisms in which the *ortho*-carboxylate ion reacts with the protonated substrate, forming intermediates that can decompose to products. The postulated intermediate in Eq. (10.7) is, however, stable under the reaction conditions, while the postulated intermediate in Eq. (10.8) leads to the products at a rate faster than the overall reaction.

$$\text{(10.7)}$$

$$\text{(10.8)}$$

At this point, it is important to consider the relationship between the acceleration observed in the preceding systems and catalysis. Although the hydrolysis of aspirin is substantially faster than the hydrolysis of phenyl acetate, the former reaction, as presented, is not a catalyzed reaction. Each aspirin molecule is only hydrolyzed once, and there is no regeneration of catalyst. So what is the point of this exercise?

Consider the imaginary system

$$R-\overset{\text{O}}{\underset{\|}{C}}-OR' + H_2O \rightleftharpoons RCO_2H + R'OH \tag{10.9}$$

$$\text{(10.10)}$$

In (10.9), we see a simple ester hydrolysis; in (10.10), a catalyzed hydrolysis. The catalyzed hydrolysis depends on the presence of salicylic acid, a catalyst that is ultimately regenerated. What is most important is that the intermediate salicylate ester is hydrolyzed very fast to the acid. The reason for this rapid hydrolysis is of course associated with the intramolecular reaction described earlier. We will focus on these intramolecular reactions in this chapter. In an enzyme active site, the salicylic acid moiety could be regarded as being attached to the protein backbone. An enzyme may attach even more groups together. Reactions susceptible to such catalysis include hydration and dehydration of olefins and reactions of carboxylic acid derivatives.

Probably the first multiple intramolecular catalysis to be found was the hydrolysis of succinylaspirin (**10.4**).[6] Whereas aspirin (**10.5**) and the monomethyl ester (**10.6**) are hydrolyzed in the pH range of 3 to 9 at rates proportional to the ionized carboxylate ion, **10.4** is hydrolyzed at a rate proportional to the concentration of a species containing both carboxylate ion and a free carboxylic acid, assuming pK's of 3.6 and 4.5 for the salicylic and succinic carboxylic acid groups, respectively (see Fig. 10.2). Two kinetically indistinguishable pathways are possible, each involving attack by carboxylate ion with electrophilic assistance from an unionized carboxylic acid group. If the reaction proceeds as shown in Eq. (10.11), the compound **10.4** is calculated to hydrolyze 24,000 times as fast as **10.5**, which lacks only the second unionized carboxylic acid group. If the reaction proceeds through Eq. (10.12), however, compound **10.6** is calculated

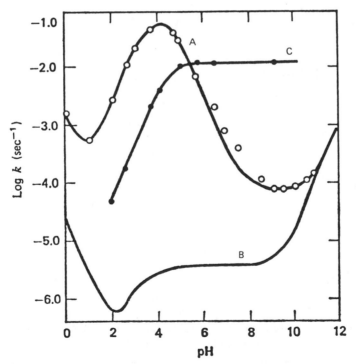

Fig. 10.2.   Hydrolysis of aspirin and related compounds at 25° C. A, **14.0**; B, **10.5**; C. **10.6**. From H. Morawetz and I. Oreskes, *J. Am. Chem. Soc.,* **80,** 2591 (1958). Copyright © 1958 by the American Chemical Society. Reprinted by permission of the copyright owner.

to hydrolyze 66 times as fast as **10.5**. The lower ratio must be accepted as long as definitive evidence to distinguish these possibilities is absent.

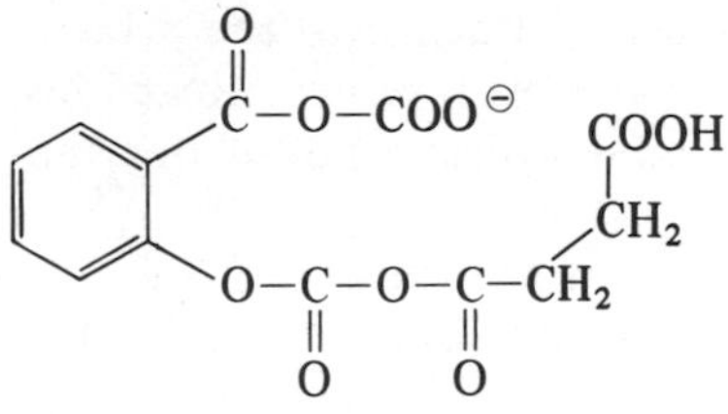

10.4

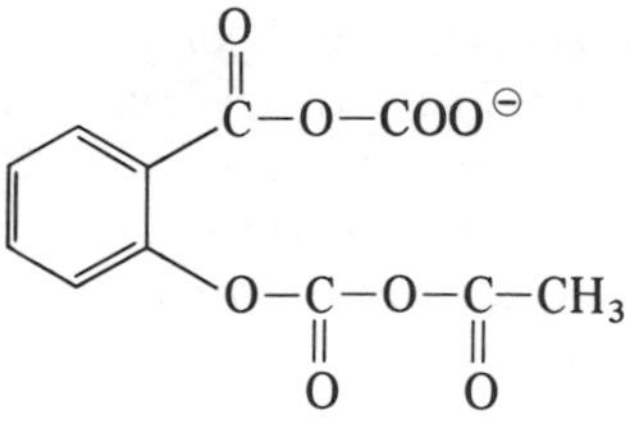

10.5

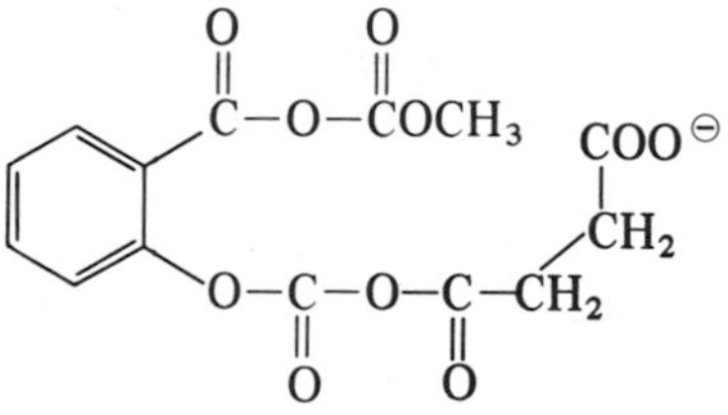

10.6

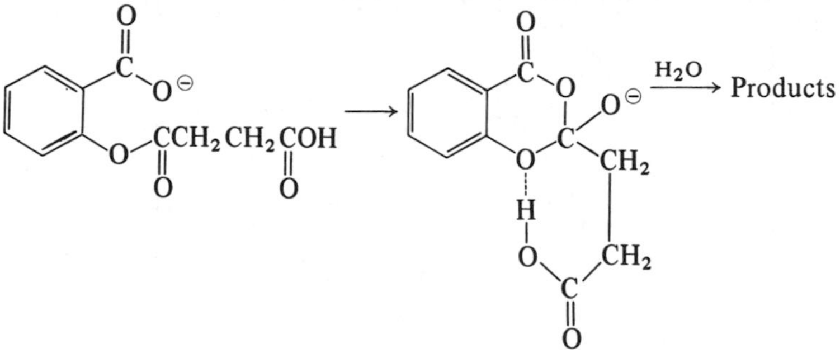

(10.11)

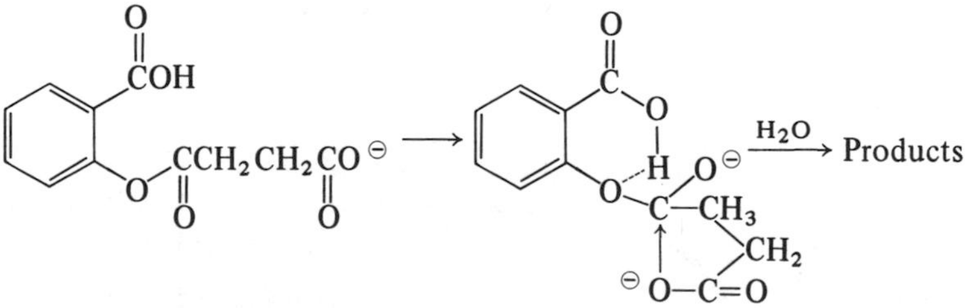

(10.12)

The pH rate constant profiles for the hydrolysis of the phthalamic acid derivatives **10.7–10.10** show a similar rate difference. Compound **10.7** hydrolyzes with the largest rate constant and exhibits a bell-shaped pH rate constant profile with a maximum at a pH corresponding to the maximum concentration of the singly ionized species. Compounds **10.8** and **10.10** of this series show sigmoid curves, similar to the behavior of phthalamic acid itself (Section 10.2.1). Compound **10.9** is unreactive under the conditions investigated while compound **10.8** possesses reactivity similar to phthalamic acid.[6,7] Since the reaction proceeds through an anhydride intermediate, **10.7a** rather than **10.7b** is the reactive species. On this basis, the rate enhancement of the multiple catalysis over the single catalysis can be calculated as the rate constant ratio **10.7/10.8** = 80. Correction for inductive effect differences reduces the ratio to 40, which is due to general acid catalysis by the adjacent carboxyl group. The magnitude of this effect is similar to that found in the ester hydrolysis discussed earlier.[11]

The hydration of fumaric acid to malic acid may be brought about by external hydronium ion or hydroxide ion catalysis, but hydration of the monoanion of fumaric acid needs no assistance from external catalysts. Neither the diacid nor the dianion of fumaric acid shows this uncatalyzed reaction, nor does crotonic acid, a similar molecule containing only one carboxylic acid group. Both a carboxylic acid group and an ionized carboxylate ion must participate together for hydration as in the hydration of fumaric acid monoanion. Two possible mechanisms for this multiple intramolecular catalysis are apparent: either a general acid–base catalysis or a general acid–nucleophilic catalysis [Eqs. (10.13) and (10.14), respectively]. Distinction between these mechanisms was made on the basis of experiments on the reverse reaction. Using optically active malate in the reverse reaction allows one to make both polarimetric and spectrophotometric measurements of the rate constant. Equation (10.13) requires that the two rate constants be identical, but this is not necessary for Eq. (10.14), since the intermediate present in this reaction may be partitioned in both directions. The observation that the polarimetric rate constant is larger than the spectrophotometric rate constant is incompatible with Eq. (10.13), but consistent with Eq. (10.14). Therefore, this intramolecular multiple catalysis is brought about by the combination of general acid and nucleophilic catalysts.[8]

$$\tag{10.13}$$

$$\tag{10.14}$$

The molecules catechol salicylate and salicyl salicylate both contain two phenolic hydroxyl groups, each of which is capable of acting individually as an intramolecular catalyst. When these two catalysts are placed in the same molecule, they might be expected to act in concert. Such is not the case, however. Two reasons may be suggested for these failures: the conformation of these two hindered molecules does not allow simultaneous interaction of both catalytic functions and/or the loss of a good leaving group does not require an acid catalyst.

However, simultaneous interaction of two catalysts can occur in the hydrolysis of methyl 2,6-dihydroxybenzoate. The hydrolysis of this ester shows a bell-shaped pH-rate constant profile depending on two groups of pK 8.2 and 11.6 (Fig. 10.3). The rate constant at the maximum of the bell is approximately equal to the rate constant of methyl salicylate in its completely ionized form.

10.1.2.   Acceleration by the Phosphoric Acid Group

Monoanions of phosphate monoesters hydrolyze with phosphorus–oxygen fission at rates much larger than those exhibited by either the *dianion* or the *diacid*. The pH-rate constant profile for the hydrolysis of a typical compound,

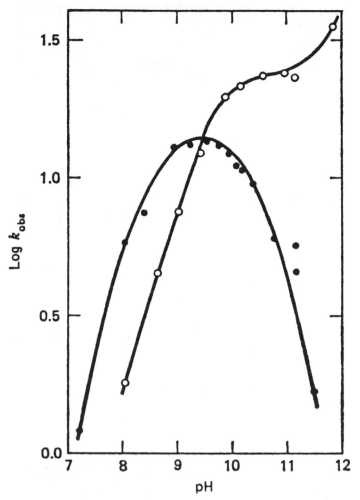

Fig. 10.3. pH-rate profile for the kinetics of hydrolysis of methyl 2, 6-dihydroxybenzoate (●) and methyl salicylate (○) in 1.15% acetonitrile–water at 60° C. From F. L. Killian and M. L. Bender, *Tetrahedron Lett.*, **16**, 1255 (1969).

methyl hydrogen phosphate, is shown in Fig. 10.4. This profile can be analyzed in terms of the sole reactivity of the *monoanion*, with a small contribution from a hydronium ion–catalyzed reaction of the neutral diacid in the very low pH region. The hydrolysis proceeds via intramolecular general acid acceleration by the hydroxyl group facilitating the cleavage of the P–O bond through one of the transition states shown in Eq. (10.15). The first three transition states of Eq. (10.15) portray the rate-determining formation of monomeric *metaphosphate ion* $(PO_3^{\ominus})^9$ that would presumably add water in a fast step to give ortho-phosphate. The last two transition states of Eq. (10.15) indicate the direct formation of *orthophosphate ion*.

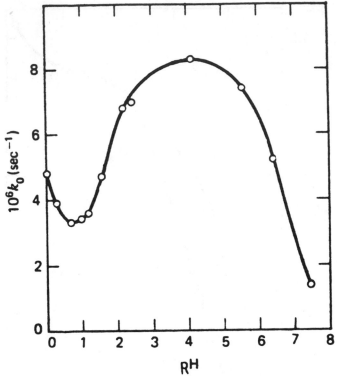

Fig. 10.4. The hydrolysis of methyl hydrogen phosphate at 100°C. From C. A. Bunton, D. R. Llewellyn, K. G. Oldham, and C. A. Vernon, *J. Chem. Soc.*, 3574 (1958).

The hydrolysis of acetyl phosphate monoanion is also enhanced by the internal hydroxyl group [as shown in Eq. (10.16)] with the intermediacy of a *metaphosphate ion* which has complete trigonal symmetry involving one-third of a negative charge on each oxygen atom and 1.67 bonds between each phosphorus atom and oxygen atom.

10.1.3. Acceleration by the Alcohol and Alkoxide Groups

Hydroxyl groups are common substituents in many organic molecules and have been shown to serve efficiently as intramolecular catalysts.

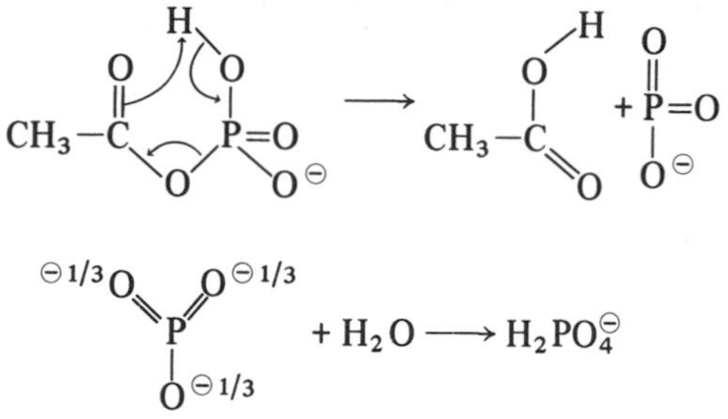

$$(10.16)$$

In the hydrolysis of salicylate esters and amides, the *ortho*-hydroxyl group is an important participant. The complex pH-rate constant profile for hydrolysis of *p*-nitrophenyl-5-nitrosalicylate, shown in Fig. 10.5, can be interpreted in terms of three reactions:

1. Reaction of the neutral substrate with water in the left-hand region.
2. Reaction of the anionic substrate with water in the central region, including the rising portion and the central plateau.
3. Reaction of the anion of the substrate with hydroxide ion in the right-hand region dependent on the concentration of hydroxide ion.

Of the three pH regions in the salicylate hydrolyses, the middle one is the most interesting. The reaction in this region proceeds via general base catalysis by *ortho*-phenoxide ion [Eq. (10.17)] since addition of nucleophiles containing no proton (e.g., azide ion) has little kinetic effect.[10]

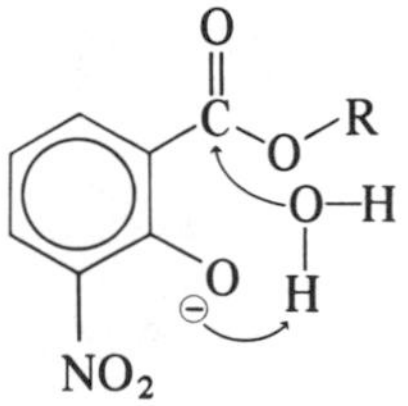

$$(10.17)$$

Intramolecular general acid catalysis by the hydroxyl group has been documented in the hydrolysis of biochemically important phosphates. The dianion of glucose 6-phosphate is approximately five times more reactive in hydrolysis than the monoanion. This observation, which is contrary to the discussion in Section 10.1.1 indicates that some special structural feature of glucose-6-phosphate must be involved. The 1-hydroxyl group of glucose is relatively acidic, with a pK of 10.8 at 100°C. Since β-glucose 6-phosphate can assume a conformation in which the hydrogen atom of the 1-hydroxyl group can

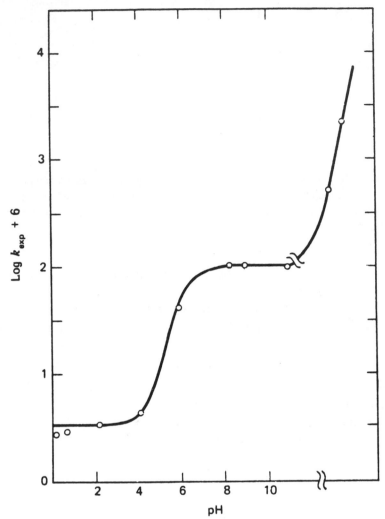

Fig. 10.5.   The hydrolysis of p-nitrophenyl 5-nitrosalicylate, in 34.4% dioxane–water at 25°C. From M. L. Bender, F. J. Kezdy, and B. Zerner, *J. Am. Chem. Soc.,* **85,** 3017 (1963). Copyright © 1963 by the American Chemical Society. Reprinted by permission of the copyright owner.

interact with the phosphate group, a plausible mechanism to explain this result involves proton transfer from the 1-hydroxyl group to the 6-oxygen atom, assisting cleavage of the phosphorus–oxygen bond. The presence of two negative charges on the phosphate group provides great driving force for this acceleration.[11]

An aliphatic hydroxyl group located in proximity to an ester bond may facilitate its alkaline hydrolysis. Thus the alkaline hydrolyses of **10.11** and **10.12**, which have hydroxyl groups adjacent to the ester function, are much faster than

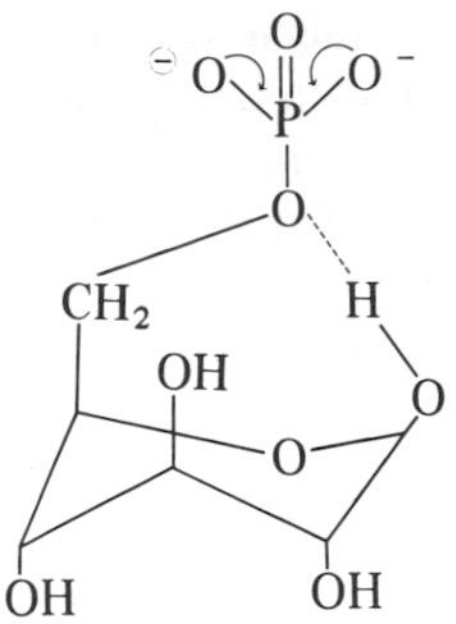

(10.18)

that of **10.13**. The adjacent hydroxyl groups of **10.11** and **10.12** probably hydrogen bond to the carbonyl group of the substrates in the transition state (which also involves nucleophilic attack by hydroxide ion), and thus stabilizes the transition state.[12]

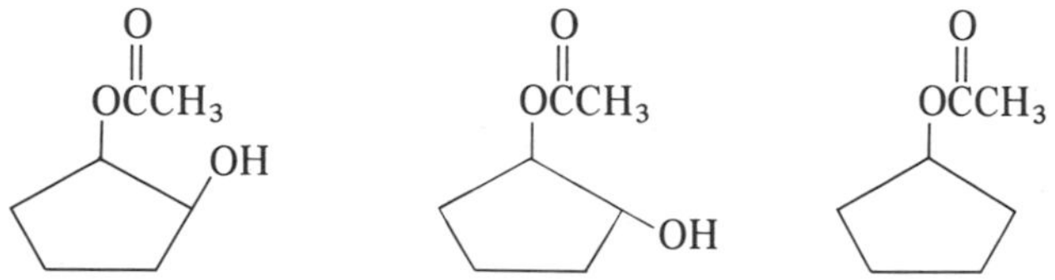

| 10.11 | : | 10.12 | : | 1 |

33   :   19   :   1

Relative rates of hydrolysis

10.1.4. Acceleration by Imidazolyl Groups

Intramolecular general base acceleration by the imidazolyl group occurs in the hydrolysis of *trans*-cinnamate ester of *endo*-5-[4′(5′)-imidazolyl]bicyclo[2.2.1]-hept-*endo*-2-yl-*trans*-cinnamate, **10.14**. This compound has a rigid structure with the imidazolyl and esterified hydroxyl groups disposed in a fashion similar to those in the serine protease, α-chymotrypsin (about 3 Å from one another). Thus **10.14** is a model of the acyl-enzyme, since catalysis by serine proteases also involves general base catalysis by the imidazolyl group of a histidine residue. General base acceleration of **10.14** was confirmed by a threefold D_2O solvent isotope effect.

On the other hand, no intramolecular participation of the imidazolyl group was not observed in the hydrolysis of the *endo-exo* isomer, *exo*-5-[4′(5′)-imidazolyl] bicyclo[2.2.1]hept-*endo*-3-yl-*trans*-cinnamate (**10.15**), in which the imidazolyl group is located far from the *trans*-cinnamoyl ester group.

The rate constant for the intramolecular general base-catalyzed hydrolysis by the imidazolyl group of **10.14** is much smaller than for the deacylation (hydrolysis) of the acyl-enzyme, *trans*-cinnamoyl-chymotrypsin.[13] However, when external benzoate ion is added, (the third component of the active site of chymotrypsin is a carboxylate ion) the rate of this model is only fourfold smaller than the enzymatic deacylation (assuming 10 *M* is the conversion factor from inter- to intramolecular catalysis[14] (see Section 10.3).

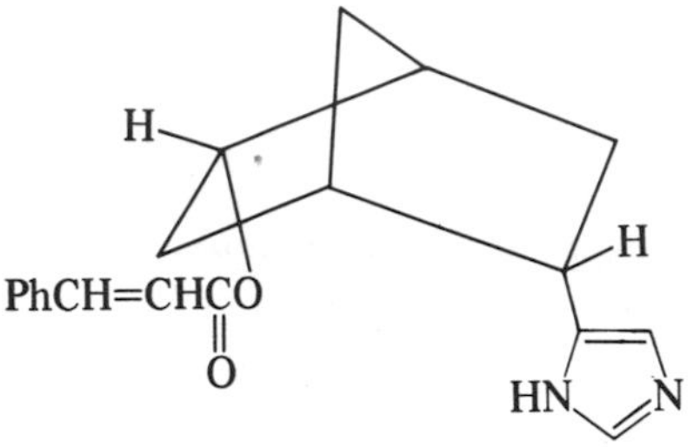

10.14

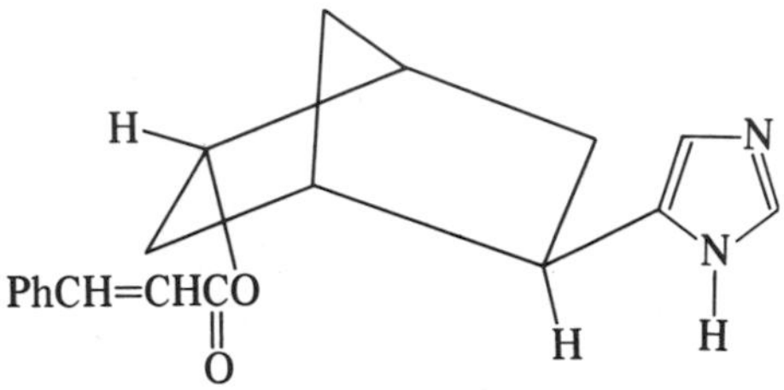

10.15

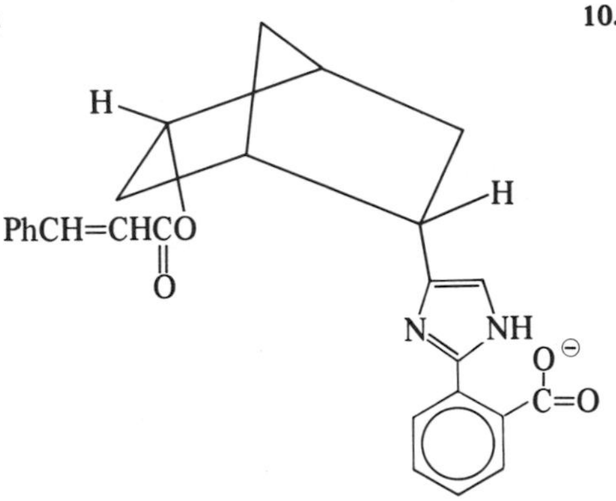

10.16

When benzoate ion is introduced intramolecularly, as in **10.16**, the model is equal to the enzymatic deacylation, clearly showing the power of intramolecular acceleration.[15]

10.2.   INTRAMOLECULAR NUCLEOPHILIC OR ELECTROPHILIC ACCELERATION

10.2.1.   Acceleration by the Carboxylate Ion and Carboxylic Acid Groups

The hydrolysis of monoaryl succinates and glutarates has provided much information concerning intramolecular nucleophilic catalysis. The pH-rate constant profiles of these reactions show dependence on a group of pK 4.5.

These reactions proceed via anhydride intermediates formed by intramolecular nucleophilic attack of a carboxylate ion at the carbonyl carbon atom of the ester function, which is followed by the hydrolysis of the anhydride intermediate [Eq. (10.19)]. The absolute rate constants of hydrolysis are quite high compared to the hydrolysis of similar acetate esters. For example, mono-p-nitrophenyl glutarate hydrolyzes about 10^5 times faster than p-nitrophenyl acetate at pH 5.

$$\begin{array}{ccccc} \overset{\displaystyle O}{\underset{\displaystyle O}{\overset{\parallel}{\underset{\parallel}{C-O\phi}}}}_{CO^{\ominus}} & \xrightarrow[-HO\phi]{k_1} & \overset{\diagup O}{\underset{\diagdown O}{\overset{C}{\underset{C}{}}}} & \xrightarrow[+H_2O]{k_2} & \overset{CO_2H}{\underset{CO_2H}{}} \end{array} \qquad (10.19)$$

The mechanism of Eq. (10.19) indicates that the carboxylate ion and the ester function must be located appropriately in order for the reaction to proceed smoothly. When the bridging between them decreases rotational freedom, the rate of the intramolecular reaction increases. The validity of this statement (and thus the importance of the conformation of the substrates in these intramolecular reactions) was clearly shown in the hydrolyses of a series of mono-p-bromo-phenyl esters of dicarboxylic acids (Table 10.1).[16] Although the rate constants of hydrolysis of the anhydride intermediates are insensitive to structure, the rate constants for anhydride formation are very sensitive to structural effects. For example, in changing from a glutarate monoester to a succinate monoester (a loss of one carbon atom and thus a loss of one freely rotatable C–C bond), the rate constant increases by 230-fold; changing from the succinate monoester to the bicyclic system (a loss of another freely rotatable C–C bond) results in a further increase of 230-fold. Introduction of two methyl groups or a double bond, which also decreases freedom of rotation, showed considerable acceleration. The two rate increases of 230-fold correspond to free-energy differences of approximately 3 kcal/mole, approximately twice the difference expected from the entropic considerations of removing one rotational degree of freedom in each change. Certainly, rotational entropy must account for a substantial part of these extraordinary rate differences, but other factors, including possibly the relief of steric strain in the last compound of Table 10.1 cannot be overlooked as appreciable contributors to the rate acceleration.[16]

In the hydrolysis of monoesters of phthalic acids, participation by both neighboring carboxylic acid and carboxylate ion occurs. When the ester contains a poor leaving group such as a methoxy, ethoxy, or chloroethoxy group, the (neutral) carboxylic acid group participates in the reaction. However, when the ester contains a good leaving group, such as the trifluoroethoxy or phenoxy group, the anionic carboxylate ion participates in the reaction. Between these two extremes, both the carboxylate ion and the carboxylic acid may participate simultaneously, as in propargyl or N-acetylserinamide esters (pK_a's of the leaving groups are about 13.5). In the hydrolysis of phenyl hydrogen phthalate,

Table 10.1.   Intramolecular Nucleophilic Catalysis in the Hydrolysis of Some Mono-*p*-Bromophenyl Esters of Dicarboxylic Acids

	$k_{\text{anhydride formation}}$ (relative)	$k_{\text{anhydride hydrolysis}}$ (relative)
COOR / COO$^{\ominus}$	1^a	1^a
COOR / COO$^{\ominus}$	20	0.07
COOR / COO$^{\ominus}$	230	1.46
COOR / COO$^{\ominus}$	10,000	11.2
COOR / CO$_2^{\ominus}$	53,000	5.2

Source: T. C. Bruice and U. K. Pandit, *Proc. Natl. Acad. Sci. USA*, **46**, 402 (1960).
^{a}The two columns cannot be compared with one another quantitatively.

the formation and decomposition of the phthalic anhydride intermediate are seen spectrophotometrically. The rate constant of the decay of the intermediate is equal to the rate constant of the hydrolysis of phthalic anhydride, proving that this hydrolysis has a nucleophilic catalysis mechanism [Eq. (10.20)].

$$(10.20)$$

The microscopic reverse of these hydrolyses should proceed by the same mechanism. This fact has been demonstrated in the reverse of the hydrolysis of succinanilic acid, the formation of succinanilic acid from aniline. The formation of succinanilic acid from these components is much faster than the formation of an anilide from a monocarboxylic acid and aniline. The rate-determining formation of succinic anhydride as intermediate in this reaction has been confirmed.

A neighboring carboxylic acid group has a profound effect on the rates of hydrolysis of β-amic acids and β-cyano acids. These molecules contain both a hydrolyzable amide or nitrile linkage and a carboxylic acid group appropriately situated with respect to one another for internal interaction. For example, the effect of the neighboring carboxylic acid group shows up in the hydrolysis of succinamic acid, succinanilic acid, and phthalamic acid. Several of these hydrolyses have been shown to be independent of external hydronium ion concentration and dependent on the unionized form of the carboxyl group. All of the reactions are quite rapid compared to the hydrolysis of ordinary amides or nitriles. These data point to the involvement of the internal carboxylic acid group in the reaction.

In order to specify the kind of effect that the carboxyl group exerts in these hydrolyses, one should consider the hydrolysis of phthalamic acid. This reaction shows dependence on an acidic group of pK 3.5. At approximately pH 3, the hydrolysis of phthalamic acid is 10^5 faster than the hydrolysis of benzamide. More importantly, at the same pH, the hydrolysis of phthalamic acid is 10^6 times faster than the hydrolysis of o-nitrobenzamide, a compound containing an $ortho$-substituent of similar electronic and steric properties. This large acceleration indicates that the carboxylic acid group cannot be affecting the reaction in the usual electronic manner, but must be participating directly in some covalent manner. In fact, the reaction involves phthalic anhydride as intermediate.

This result was proven by a double label tracer experiment. The two isotopic labels were ^{13}C and ^{18}O. Mass spectrometric analysis of the product carbon dioxide gave masses 44 and 47 for path 1 and masses 44, 45, 46, and 47 for path 2. The carbon dioxide products corresponding to path 2 were observed, indicating an intramolecular nucleophilic catalysis.[17]

(10.21)

10.2.2. Acceleration by Alkoxyl, Alkylthio, and Halogen Groups

The largest number of neighboring group participation reactions occurring in substitutions at saturated carbon involve catalysis by alkoxyl, alkylthio, and halogen groups.[18]

The relative rates of solvolysis of a series of ω-methoxyalkyl p-bromobenzene-sulfonates (Table 10.2) require that an effect of the methoxyl groups be other than electronic withdrawal, because the maximal effect is found when the number of carbon atoms in the chain is four. The large rate constants of compounds with four (and five) carbon atoms suggest nucleophilic participation proceeding through cyclic oxonium ions [Eq. (10.22)].

$$(10.22)$$

Participation by the methoxyl group occurs through five- or six-membered cyclic oxonium ion intermediates, but not when the intermediate requires three, four, or seven members. The rate constants of the hydronium ion-catalyzed hydrolysis of $CH_3OCH_2CH(OEt)_2$, and $CH_3SCH_2CH(OEt)_2$ in 50% dioxane–water at 25°C are 2×10^{-4}, and 2×10^{-2} M^{-1} sec^{-1}, respectively. If the oxygen compound is used as a measure of the inductive effect of the sulfur atom on the rate constant, then the one hundredfold higher reaction rate of the sulfur compound must be attributed to some specific interaction such as that of a sulfonium ion [Eq. (10.23)].

$$(10.23)$$

Evidence for nucleophilic participation by halogen groups is seen in the solvolysis of *trans*-2-iodocyclohexyl-*p*-toluenesulfonate in acetic acid, since it is about 1000 times faster than the unsubstituted compound. The results suggest participation by bromo- and iodo-groups proceeding according to Eq. (10.24).

Table 10.2.   The Relative Rates of
Solvolyses of ω-Methoxyalkyl
p-Bromobenzenesulfonates
by Acetic Acid

Compound	Relative Rate
Me·[CH$_2$]$_3$·OBs	1.00
MeO·[CH$_2$]$_2$·OBs	0.28
MeO·[CH$_2$]$_3$·OBs	0.63
MeO·[CH$_2$]$_4$·OBs	657.0
MeO·[CH$_2$]$_5$·OBs	123.0
MeO·[CH$_2$]$_6$·OBs	1.16

Source: S. Winstein et al., Tetrahedron,
3, 1 (1958).

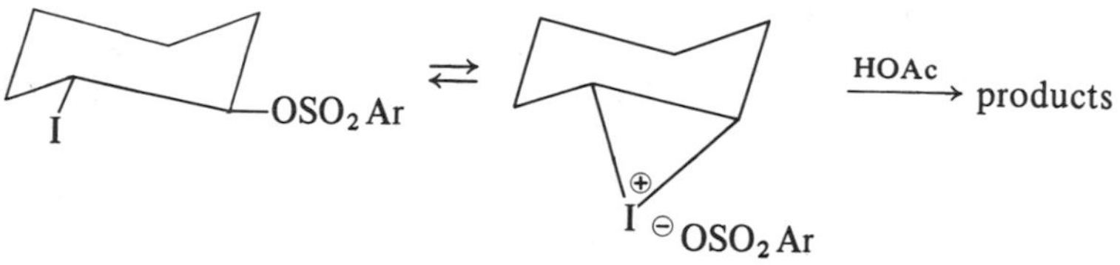

(10.24)

10.2.3.   Acceleration by Tertiary Amine Groups Including Imidazole

Since tertiary amines are better nucleophiles than carboxylate ions, one might expect better intramolecular nucleophilic catalysis with the former than the latter. Aryl esters of 4-(4'-imidazolyl) butyric acid are hydrolyzed at greatly enhanced rates in neutral solution in a reaction dependent on the concentration of neutral imidazole species. In the same manner as intermolecular nucleophilic catalysis by imidazole, the reaction was thought to proceed through (intra-molecular) nucleophilic catalysis.[19]

The loss of p-nitrophenol from the p-nitrophenyl ester, which presumably reflects lactam formation, has a half-live of 0.2 sec in 50% aqueous ethanol, and thus the reaction possesses a rate constant roughly equivalent to that for the conversion of a chymotrypsin p-nitrophenyl acetate complex to the corresponding acyl-enzyme [Eq. (10.25)].

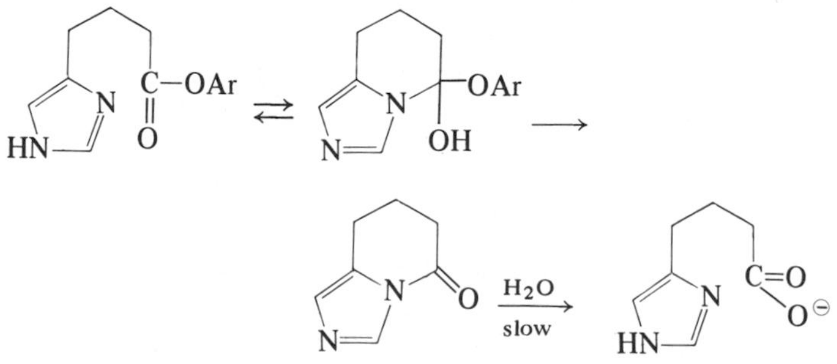

(10.25)

Although the hydrolysis of the methyl ester of 4-(4'-imidazolyl) butyric acid is not subject to intramolecular imidazole participation, the corresponding propyl-thiol ester does show such participation. In fact, the reaction of the propylthiol ester to form the lactam intermediate is approximately 10^6–10^7 times as fast as the hydrolysis of a thiol ester with hydroxide ion at neutrality. In this reaction, addition of external thiol reduces the rate of disappearance of the thiol ester and therefore requires the presence of a lactam intermediate.

The amide of 4-(4'-imidazolyl) butyric acid is hydrolyzed with participation by the protonated imidazole group. This reaction probably proceeds via the protonation of the amide group and the simultaneous attack by the imidazolyl group (the four-center nucleophilic–electrophilic mechanism) leading to a six-membered ring, as shown in Eq. (10.26).

$$ (10.26) $$

In the reaction of trimethylamine with p-nitrophenyl acetate, nitrophenol release was interpreted to occur via nucleophilic catalysis by analogy with the corresponding imidazole reaction. The lack of a deuterium oxide solvent isotope effect was in agreement with this hypothesis. The intramolecular analog of this reaction was investigated in the hydrolysis of p-substituted-phenyl 4-(N, N-dimethylamino)butyrates and valerates. Although no direct evidence for an acyltrialkylammonium ion intermediate was found in these reactions, they probably proceed by the same nucleophilic pathway as the corresponding intermolecular reaction since: (1) the Hammett rho constant had the same high value (+2.2–2.5) for both intermolecular and intramolecular processes; (2) the enthalpy of activation is identical for both intermolecular and intramolecular processes; and (3) $T\Delta S^{\ddagger}$ is 4–5 kcal/mole higher for the intramolecular reaction than for the intermolecular one. Confirmation of a nucleophilic mechanism for this process comes from the fact that the rate constant for the intramolecular reaction calculated on the basis of a general acid–hydroxide ion reaction is close to diffusion–controlled (10^7 M^{-1} sec^{-1}).

10.2.4.   Acceleration by Amide Groups

Intramolecular amide groups can exert profound nucleophilic catalysis. The amide anion is an ambident nucleophile; that is, either the oxygen atom or nitrogen atom may serve as nucleophile. When the nitrogen atom of the amide is the nucleophile, the intermediate is an imide.[20] The alkaline hydrolyses of succinamide, maleamide, phthalamide and 1,2-cis-cyclohexanedicarboxamide are considerably faster than the hydrolysis of acetamide. An imide intermediate was detected spectrophotometrically in the hydrolysis of the diamides. On this basis, Eq. (10.27) can be written. The hydrolysis of methyl N-methylphthalamate proceeds in the same fashion via an imide intermediate 10.17. The rate constant of the conversion of methyl N-methylphthalamate to the corresponding imide with hydroxide ion is 12,400 M^{-1} sec^{-1}. However, the imide hydrolyzes to the final product 200–300 times more slowly than does the parent ester with hydroxide ion.

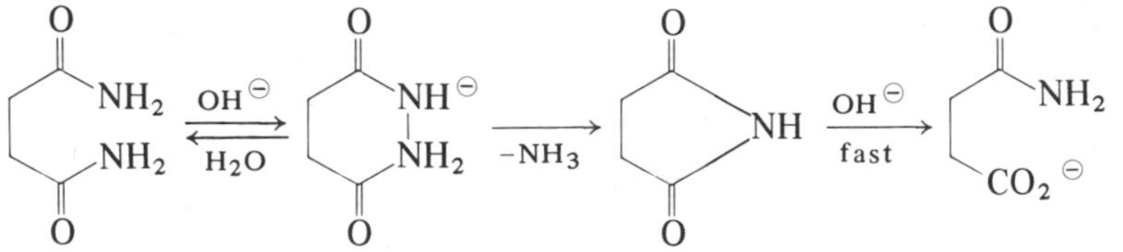

$$\text{(10.27)}$$

10.17

Nucleophilic attack by the internal amide groups is important in the racemization of optically active amino acids. The amide oxygen atom 1 of N-acylamino acids can attack the carbonyl carbon atom 5 intramolecularly, resulting in the formation of oxazolone **10.18** (with the release of HX) [Eq. (10.28)]. Furthermore, oxazolone **10.18** can easily lose the proton at the C-4 carbon atom in the presence of base and becomes its anion, since the negative charge is stabilized by resonance, as shown by Eq. (10.29). Racemization then can proceed quite easily through the anion. Formation of oxazolone has been confirmed by much experimental evidence and accepted as an important route of the racemization of amino acids. It should be noted, however, that direct proton abstraction at the α-carbon atom by base can also give rise to the racemization of amino acids.

$$R'-CONHCH-COX \rightleftharpoons \underset{\substack{| \\ 3N\ \ 2\ \ O1 \\ \diagdown C \diagup \\ | \\ R'}}{\overset{\substack{R \\ |}}{4CH-5C=O}} + HX$$

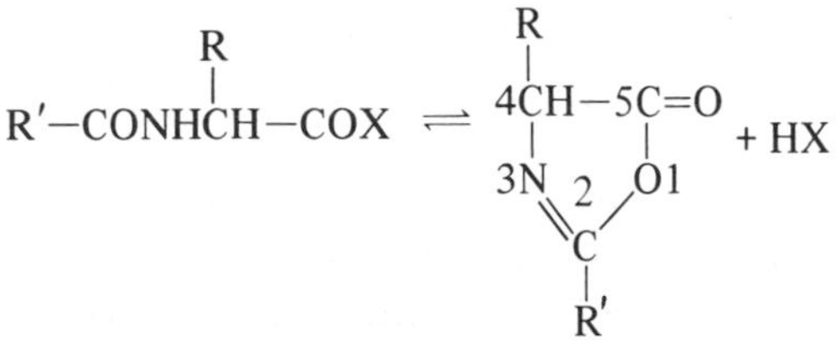

$$\text{(10.28)}$$

10.18

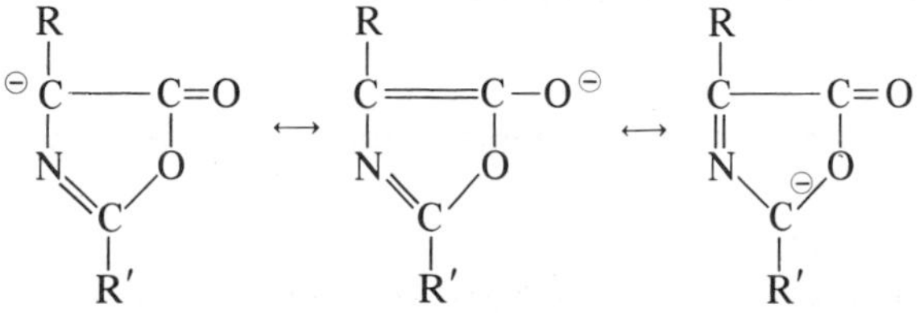

$$\text{(10.29)}$$

The hydrolysis of the diamide shown in Eq. (10.30) proceeds approximately 10^4 times faster than that of N, N'-dicyclohexylbenzamide in acetic acid solution. The *ortho*-benzamido group probably functions as an intramolecular nucleophilic catalyst on the basis of (1) the isolation of the intermediate benzoylanthranil from the reaction mixture in dry dioxane and (2) the formation of the hydrolysis product upon rapid decomposition of the benzoylanthranil intermediate in acetic acid solution (**10.19**).[21]

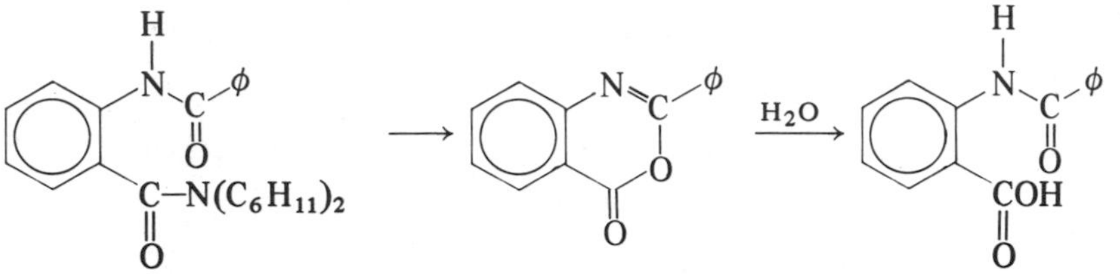

$$(10.30)$$

10.19

The alkaline hydrolysis of phosphoric acid triesters containing amido groups proceeds via a Δ^2-oxazoline intermediate **10.20** [Eq. (10.31)]. In aqueous solution around neutrality, an intermediate was detected spectrophotometrically and shown to be identical to an authentic sample of the Δ^2-oxazoline. In addition, the rate of decomposition of the intermediate was shown to be identical to that of the hydrolysis of the synthetic Δ^2-oxazoline.

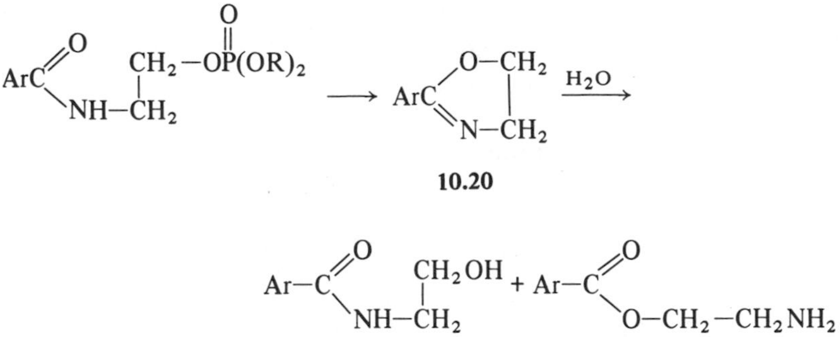

$$(10.31)$$

10.20

Intramolecular nucleophilic catalysis by amido groups is seen to be widespread and to occur through several pathways. A comparison of amide catalysis via imide or oxazoline formation indicates that the former is in general much faster than the latter, assuming the same ring size. The reason for this difference is not apparent. However, the facile interaction of intramolecular amido groups must be kept in mind in analyzing the reactions of enzymes, all of which are polyamides.

10.2.5. Acceleration by Hydroxyl and Primary Amino Groups

Diesters of phosphoric acid containing a hydroxyl group on the β-carbon atom are particularly susceptible to hydrolysis compared to unsubstituted phosphate diesters. The alkaline hydrolysis of methyl 2-hydroxycyclohexyl phosphate, for example, is approximately 10 times faster than the hydrolysis of dimethyl phosphate. Furthermore, the *cis*-ester hydrolyzes considerably faster than the *trans*-ester. These results suggest nucleophilic participation by neighboring hydroxyl groups, forming a five-membered cyclic phosphate intermediate, which is subsequently decomposed [Eq. (10.32)].

$$ (10.32) $$

The alkaline hydrolyses of *p*-nitrophenyl β-glycosides exhibit greatly enhanced rates with respect to the corresponding 2-*O*-methyl derivatives. Substitution of hydroperoxide anion for hydroxide ion reduces the rate of hydrolysis of the glycosides, indicating that the hydroxide ion acts as a base, rather than as a nucleophile. These observations suggest Eq. (10.33), which involve an epoxide intermediate (**10.21**).

$$ (10.33) $$

10.21

The alkaline hydrolysis of diphenyl β-aminoethyl phosphate is a rapid process yielding two moles of phenol. In contrast, the alkaline hydrolysis of diphenyl alkyl phosphates is a slow reaction, yielding one mole of phenol since the resulting anion of the diester is resistant to further hydrolysis. In addition, the first mole of phenol is produced more rapidly than the second in the former substrate. These observations lead to Eq. (10.34).

$$ (10.34) $$

$$ \text{HOP} \xrightarrow{\text{H}_2\text{O}} (\text{HO})_2\overset{\text{O}}{\overset{\|}{\text{P}}}\text{OCH}_2\text{CH}_2\text{NH}_2 $$

10.2.6. Acceleration by Electrophilic Groups

The hydrolysis of the diazonium salt of o-aminophenyl-2,6-dimethylbenzoate is interesting. Although the corresponding *para*-diazonium salt is stable, the *ortho*-derivative is unstable, decomposing to 2,6-dimethylbenzoic acid and *ortho*-hydroxybenzenediazonium ion. This reaction can only be explained by an electrophilic attack of the diazonium ion on the ethereal oxygen of the ester [Eq. (10.35)].[22]

(10.35)

10.2.7. Intramolecular Enzyme Models

As stated earlier, intramolecular catalysis in many regards is the essence of enzymatic catalysis. It is not surprising, therefore, that several attempts have been made to mimic enzymatic catalyses by means of intramolecular catalysis. One of these examples involves deacylation of an acyl-chymotrypsin (described in Section 10.1.4). Another is a model for a vitamin B_{12} dependent enzyme; specifically, an intramolecular model for the enzymatic insertion of coenzyme B_{12} into unactivated carbon–hydrogen bonds was developed.[23] As already noted, it is important to recognize the fact that in enzyme reactions the catalyst is regenerated. The regeneration occurs in a fashion similar to that described in our initial salicylic acid example (Section 10.1.1).

10.3. COMPARISON OF INTERMOLECULAR AND INTRAMOLECULAR CATALYSIS

The sparse data that exist on intramolecular catalysis indicate that the same laws governing the relative efficacy of intermolecular general acid–base and nucleophilic–electrophilic catalyst hold in intramolecular systems. Thus general bases of greater basicity, general acids of greater acidity, nucleophiles of greater nucleophilic–electrophilic catalysis hold in intramolecular systems. Thus general better intramolecular catalysts. However, this generalization must be tempered by the realization that intramolecular interactions place much more stringent requirements on the stereochemistry of the system than do intermolecular interactions. Therefore, exceptions to the preceding generalizations can be expected.

The proximity of the catalyst to the reaction center usually leads to a predominance of intramolecular nucleophilic over intramolecular general base catalyses. For example, there is only one well-documented case of general base catalysis by imidazole in an intramolecular reaction whereas there are several in intermolecular situations. In addition, there are many more examples of catalysis by a carboxylic acid group acting mechanistically as a hydronium ion–nucleophilic catalyst in intramolecular systems than in intermolecular systems. The list could be extended considerably to indicate the increased importance of nucleophilic catalysis in intramolecular systems.

A kinetic comparison of corresponding intramolecular and intermolecular catalyses reveals some profound differences. These catalyses correspond to first-order and second-order processes, respectively, and therefore are not amenable to straightforward comparison. Nevertheless, a comparison can be made by calculating what concentration of the intermolecular catalyst is necessary for equivalent rates of reaction of the intramolecular and inter-molecular catalyses (assuming equivalent concentrations of the two substrates). This calculation leads to a concentration of external catalyst that may be equated to the effective (local) concentration of the internal catalyst.[7]

For example, the enolization of o-isobutyrylbenzoate ion may be compared with the benzoate ion–catalyzed enolization of acetophenone. This comparison indicates that the internal carboxylate ion is equivalent to 50 M of external carboxylate ion, this factor making the second-order rate constant of the intermolecular reaction equal to the first-order rate constant of the intra-molecular reaction. This result can be alternatively expressed by saying that the effective concentration of the o-carboxylate ion in o-isobutyrylbenzoate is 50 M. In the less rigid aliphatic system involving levulinic acid, the efficiency of intramolecular catalysis is less, and the internal carboxylate ion is equivalent to only 1 M of external carboxylate ion.

One of the problems with this comparison is the possibility that the change from intermolecular to intramolecular catalysis will result in a change in mechanism. For example, whereas the intramolecular catalyses of hydrolysis of monoaryl glutarates by carboxylate ion are nucleophilic, some of the inter-molecular catalyses of the hydrolyses of aryl acetates by acetate ion are general basic, some nucleophilic, and some mixed.

The intramolecular–intermolecular comparison is further confused by the fact, discussed earlier, that the conformation of a molecule undergoing intramolecular catalysis can have a large effect on the catalytic rate constant. For example, if one were to compare an intermolecular catalysis of phenyl acetate by acetate ion with the intramolecular catalysis in monophenyl glutarate, should one use the unsubstituted glutarate or the β, β'-dimethyl derivative, which has a more favorable conformation for intramolecular interaction?

Notwithstanding these problems, we shall attempt to make several general comparisons. Comparison of the intermolecular catalysis in the hydrolysis of

Table 10.3. Ratios of Intra:
Intermolecular Catalytic Rate
Constants for the Hydrolysis of
m- and p-Substituted Phenyl Esters

Substituent	Me_2N^a (M)	Imidazole[b]
H	1260	24
p-Cl	1080	23
m-NO₂	1700	32
p-NO₂	5370	9.4

Source: T. C. Bruice and S. J. Benkovic,
J. Am. Chem. Soc., **85,** 1 (1963). © 1963 by
the American Chemical Society. Reprinted
by permission of the copyright owner.
[a]intramolecular reaction: γ-(N,N-dimethyl-
amino) butyrates.
[b]Intramolecular reaction: γ-(4-imidazolyl)-
butyrates.

p-nitrophenyl acetate with acetate ion and the intramolecular hydrolysis of
mono-p-nitrophenyl glutarate indicates that the internal catalyst is equivalent to
approximately 600 M of the external catalyst. Since the intermolecular reaction
has been shown to proceed by nucleophilic catalysis to the extent of 50%,
comparison is good to a factor of 2.

Table 10.3 shows calculations comparing intermolecular and intramolecular
reactions in nucleophilic catalyses by imidazole and by the dimethylamino
group. The table indicates that an internal dimethylamino catalyst is equivalent
to 5370 M of the corresponding intermolecular catalyst in the hydrolysis of a
p-nitrophenyl ester. On the other hand, an internal imidazole catalyst is
equivalent to only 9.4 M of the corresponding intermolecular catalyst in the same
reaction. The differences may reflect differences in conformation of the two
systems, possibly due to differences in solvent.

Generally 10 M is a reasonable value for the local concentration of
intramolecular catalyses, which is the average of many intra- and intermolecular
ratios.

The thermodynamic reason for the difference between intermolecular and
intramolecular catalysis has been determined for catalyses by the dimethylamino
group. Table 10.4 shows that the enthalpies of activation of the intermolecular
and intramolecular reactions are roughly constant, but that the entropy terms of
activation for the intramolecular catalyses are approximately 4–5 kcal/mole
more favorable than for the intermolecular catalyses. This result is in keeping
with the expectation that intramolecular processes do not lose the translational
degrees of freedom in going from the ground to the transition state that
intermolecular catalyses do. In general, intramolecular reactions may be
superior either because the ground state is raised or the transition state lowered.

Table 10.4.   Activation Parameters for Nucleophilic Displacement by the Dimethylamino Group in Intermolecular and Intramolecular Catalyses in the Hydrolysis of *m*- and *p*-Substituted Phenyl Esters

Substituent	Intermolecular (kcal/mole)		Intramolecular (kcal/mole)[a]		(kcal/mole)[b]	
	$\Delta H^{\ddagger}$	$T\Delta S^{\ddagger}$	$\Delta H^{\ddagger}$	$T\Delta S^{\ddagger}$	$\Delta H^{\ddagger}$	$T\Delta S^{\ddagger}$
p-NO$_2$	12.3	−6.3	11.9	−1.9	11.5	−2.6
m-NO$_2$	12.1	−8.0	11.5	−4.3	11.8	−4.4
p-Cl	12.5	−9.1	15.9	−2.2	13.8	−4.1
H	12.9	−9.4	12.5	−5.7	12.3	−6.4
p-CH$_3$			13.7	−5.1	14.4	−5.5

Source: T. C. Bruice and S. J. Benkovic, *J. Am. Chem. Soc.*, **85**, 1 (1963). © 1963 by the American Chemical Society. Reprinted by permission of the copyright owner.
[a] Butyrates.
[b] Valerates.

Some catalyses seen in the intramolecular systems do not exist in intermolecular systems. For example, general base catalysis of the hydrolysis of a *trans*-cinnamate ester by the imidazolyl group takes place only in the intramolecular reaction of **10.14**. In the corresponding intermolecular reaction, imidazole exhibits inhibition of alkaline hydrolysis of the *trans*-cinnamate ester, which is attributable to (nonproductive) complex formation of the substrate with the catalyst.

The fact that intramolecular catalysis is more effective than intermolecular catalysis implies that its accelerations with respect to hydronium or hydroxide ion catalysis must be large. Three representative comparisons are given for intramolecular general acid catalysis, intramolecular general base catalysis, and intramolecular nucleophilic catalysis. Intramolecular general acid catalysis in the hydrolysis of *ortho*-carboxyphenyl β-D-glucoside leads to a rate constant 10^4 times greater than the hydronium ion-catalyzed hydrolysis of a glycoside at pH 3.5 and above. Intramolecular general base catalysis in the hydrolysis of *p*-nitrophenyl 5-nitrosalicylate leads to rate constants 10^3 times greater than the hydroxide ion–catalyzed hydrolysis of *p*-nitrophenyl 2-methoxy-5-nitrobenzoate at pH 6 and below. Intramolecular nucleophilic catalysis by the dimethylamino group in the hydrolysis of *p*-nitrophenyl 4-*N*, *N*-dimethylaminobutyrate leads to rate constants 10^5 times greater than the hydroxide ion–catalyzed hydrolysis of *p*-nitrophenyl acetate at pH 8 and below. These large rate enhancements around neutrality explain some of the rate enhancements of the order of 10^{10} which enzymatic catalysts show over hydronium ion and hydroxide ion rates around neutrality.[9]

The emphasis in the previous discussion has been that proximity is of overriding importance. This is certainly so, but together with this, the importance of correct orientation for facile catalysis in intramolecular systems must also be considered. The catalyst must not only have large effective local concentrations, but it must have correct stereochemical orientation.

10.4.   INTRAMOLECULAR REACTIONS (LACTONIZATION)

10.4.1.   Orbital Steering

The proposal of orbital steering theory was made on the basis of the results shown in Table 10.5. The difference of the rate of intramolecular γ-lactonization from that of the corresponding intermolecular esterification is very large, even after allegedly all the reasonable corrections are made. Corrections involve the proximity effect applicable to an intermolecular system, the torsional strain effect due to the ring, and the number of conformational isomers in ring closure in intramolecular reactions. The difference is the largest for **10.22** (2×10^4 fold). It was proposed that simple juxtaposition of reacting atoms is not sufficient for a reaction to proceed, but precise steering of their molecular orbitals in a suitable orientation must occur. The large difference in Table 10.5 can be explained by supposing that only a small portion of all possible conformations of the reacting atoms, in which their orbitals are oriented in a precise manner, can react. Intramolecular reactions are superior to the corresponding intermolecular reactions, since the orientation of the reacting atoms in them is (in some cases) appropriate for reaction and the fraction of the productive conformations of the reacting atoms is large. Thus the large rates of enzymatic catalyses can be easily understood, since all the reactants and catalytic functional groups can be located in a stereochemically correct manner in enzymatic reactions.

Orbital steering theory has not been established, however, although it is certainly attractive. The following arguments against it have been made:

1.   The thermal vibration of a molecule should give rise to a large change of orientation of reacting atoms, which contradicts the rigid orientation of their orbitals predicted by the orbital steering theory.
2.   Molecular orbital calculations predict a shallow total energy minimum for orbital alignment, although the orbital steering theory requires a steep one.

10.4.2.   Stereopopulation Control

The rate of acid-catalyzed lactonization is greatly affected by methylation as shown by **10.23**. Introduction of two methyl groups on the methylene group showed 4.4×10^3 fold acceleration, although introduction of three methyl groups only on the benzene ring showed little effect. Furthermore, introduction

Table 10.5. Relative Rates of Acid-Catalyzed Esterification

Reactions	CH_3CH_2OH + CH_3COOH	(—COOH / —OH chain)	(benzene, —COOH / —CH₂OH)	(bicyclic, —COOH / —CH₂OH) **10.24**	(bicyclic, —COOH / HO—CH₂OH) **10.22**
Uncorrected relative rates	1	79	305	6630	1,027,000
Correction factors	Proximity: 55	Torsional strain: 64, Conformational isomers: 4.5	Conformational isomers: 3	Torsional strain: 4.4, Conformational isomers: 3	
Corrected relative rates	1	413	17	1660	18,700

Source: D. R. Storm and D. E. Koshland, Jr., *Proc. Natl. Acad. Sci. USA*, **66**, 445 (1970).

of methyl groups both on the methylene group and the *ortho* position of the benzene ring exhibited an enormous (10^{11} fold) acceleration. This large acceleration is attributable to the restriction of rotation of three C–C bonds of the side chain, since steric repulsion between the two methyl groups on the methylene group and the methyl groups on the benzene ring in **10.23** inhibits rotation of these three C–C bonds. This kind of acceleration due to an increase of the population of conformers favorable for reaction is called "stereopopulation control."[24] In the present reactions, however, some of the large acceleration may come from the relief of steric repulsion that occurs on ring closure.

10.23

$$1 \quad : \quad 1.05 \quad : \quad 4.4 \times 10^3 \quad : \quad 10^{11}$$

(relative rates)

REFERENCES

1. C. K. Ingold, *Structure and Mechanism in Organic Chemistry*, 2nd ed., Cornell University Press, Ithaca, NY, 1969.
2. A. R. Fersht and A. J. Kirby, *Chem. Br.*, **16**, 136 (1980).
3. E. T. Harper and M. L. Bender, *J. Am. Chem. Soc.*, **87**, 5625 (1965).
4. L. J. Edwards, *Trans. Faraday Soc.*, **46**, 723 (1950).
5. B. Capon and M. C. Smith, *Chem. Commun.*, 523 (1965).
6. H. Morawetz and I. Oreskes, *J. Am. Chem. Soc.*, **80**, 2591 (1958).
7. M. L. Bender, *Mechanisms of Homogeneous Catalysis from Protons to Proteins*, Wiley-Interscience, New York, 1971, Chapter 9.
8. M. L. Bender and K. A. Connors, *J. Am. Chem. Soc.*, **84**, 1980 (1962).
9. T. C. Bruice and S. J. Benkovic, *Bioorganic Mechanisms*, Vol. 2, W. A. Benjamin, New York, 1966.
10. M. L. Bender, F. J. Kezdy, and B. Zerner, *J. Am. Chem. Soc.*, **85**, 3017 (1963).
11. C. A. Bunton and H. Chaimovich, *J. Am. Chem. Soc.*, **88**, 4082 (1966).
12. T. C. Bruice and T. H. Fife, *J. Am. Chem. Soc.*, **84**, 1973 (1962).
13. M. Komiyama, T. R. Roesel, M. L. Bender, M. Utaka, and A. Takeda, *Proc. Natl. Acad. Sci. USA*, **74**, 23 (1977).
14. M. Komiyama, Myron L. Bender, M. Utaka, and A. Takeda, *Proc. Natl. Acad. Sci. USA*, **74**, 2634 (1977).
15. I. M. Mallick, V. T. D'Souza, J. Lee, R. C. Gadwood, and M. L. Bender, *J. Am. Chem. Soc.*, **106**, 0000 (1984).
16. T. C. Bruice and U. K. Pandit, *Proc. Natl. Acad. Sci. USA*, **46**, 402 (1960).
17. M. L. Bender, Y.-L. Chow, and F. Chloupek, *J. Am. Chem. Soc.*, **80**, 5380 (1958).

18. The original paper was S. Winstein and E. Grunwald, *J. Am. Chem. Soc.*, **70,** 828 (1948).

19. T. C. Bruice and S. J. Benkovic, *Bioorganic Mechanisms*, Vol. 1, W. A. Benjamin, New York, 1966.

20. E. Sondheimer and R. W. Holley, *J. Am. Chem. Soc.*, **76,** 2467 (1954).

21. T. Cohen and J. Lipowitz, *J. Am. Chem. Soc.*, **83,** 4866 (1961).

22. D. J. Triggle and S. Vickers, *Chem. Comm.*, 544 (1966).

23. R. Breslow and P. L. Khanna, *J. Am. Chem. Soc.*, **98,** 1297 (1976).

24. S. Milstein, L. A. Cohen, *Proc. Natl. Acad. Sci. USA*, **67,** 1143 (1970).

11 | Multiple Catalysis

11.1. INTRODUCTION

Many terms have been applied to catalysis involving more than one catalytic entity: concerted catalysis, bifunctional catalysis, and multiple catalysis. Since the former terms imply specific mechanistic or structural descriptions while multiple catalysis does not, we prefer to use multiple catalysis as the overall term.

Multiple catalysis can combine all the individual kinds of catalysis discussed previously, and in fact, many synthetic bifunctional systems are known. For higher systems, we must look to the enzymes, which employ either general acid–base catalysis, general base–general base catalysis, nucleophilic–electrophilic catalysis, and nucleophilic–general base catalysis, but apparently not electrophilic–general acid catalysis.

One important catalytic contributor is excluded from discussion here: *binding*, either covalent or noncovalent, which introduces a catalyst into the substrate molecule or into a complex with it. Although it might be considered a logical contributor to a multiple catalysis, its special character makes it worthy of special treatment (Chapter 12).

The identification of multiple catalysis and the simultaneity of multiple catalytic actions are vexing mechanistic questions that reappear with considerable frequency. The identification of multiple catalysis depends on kinetic observations requiring the presence of two catalytic species. Unfortunately, the appearance of two catalytic species in the rate law does not always mean the action of two independent catalysts. Often kinetically indistinguishable possibilities occur in which two species associate to form one catalytically important species. Even when two catalytic species do operate, they need not act in the same step of the mechanism. If a general base catalyzes one step of a reaction and its (conjugate) general acid catalyzes a second step of the reaction, this process is usually spoken of as a single, rather than a multiple catalysis, because at any one time only one catalytic species is operative. However, if two catalysts act in two steps of a reaction in such a way that their cumulative action is synergistic in some sense, then the process must be considered a multiple, rather than a single, catalysis. If the catalytic interactions are stepwise, there nevertheless must be some *transition state* in the reaction containing *both* species **simultaneously.** This requirement does not, of course, specify the timing of attainment of the transition state.

11.2. INTERMOLECULAR MULTIPLE CATALYSIS

11.2.1. General Acid–Base Catalysis

There are few reactions in which multiple general acid–base catalysis has been clearly identified.

The rate of the mutarotation of tetramethylglucose in chloroform, ethyl acetate, pyridine, or cresol solution is negligible. However, in cresol–pyridine mixtures, the rate is appreciable.[1] In fact, in 55–92% cresol, the rate constant is so large that it is not measurable (Table 11.1). The suggestion was made on the basis of these data that both a general acid and base are operating simultaneously in the mutarotation reaction. Looking only at the glycosidic linkage, this reaction can be represented as:

$$\text{(11.1)}$$

This interpretation has been questioned, however, since tetra-*n*-butylammonium phenoxide, containing no ionizable protons whatsoever, is a powerful catalyst for the mutarotation of tetramethylglucose in benzene solution, and since pyridine–cresol mixtures may contain appreciable quantities of pyridinium cresolate. If either the pyridinium ion or the cresolate ion is a better catalyst than pyridine or cresol, the data of Table 11.1 can be explained as a single rather than a multiple catalysis.[2]

The dehydration of formaldehyde in aqueous solution may occur as shown in Eq. (11.2). These processes can be considered to involve general acid and base catalyses, since one species (water) serves as proton donor while the other (water or hydroxide ion) serves as proton acceptor. However, an alternative explanation involving single (stepwise) rather than multiple catalysis also adequately

Table 11.1.   The Mutarotation of Tetramethyl-D(+)-Glucose in Cresol–Pyridine Mixtures

Percent Cresol	$k \times 10^4 (min^{-1})^a$
100	3
95	820
92 to 55	Too fast to measure
52.5	1800
21	168
0	3

Source: T. M. Lowry and I. J. Faulkner, *J. Chem. Soc.*, **127**, 1883 (1925).
^{a}25°C.

accounts for this reaction. Thus the will-ó-the-wisp of multiple catalysis is difficult to pin down.

$$(11.2)$$

11.2.2.   General Base–General Base Catalysis

The imidazole catalysis of the hydration of *sym*-dichloroacetone [in 95% dioxane–5% water (v/v)] may also be a multiple catalysis. Kinetically, the reaction can be either first or second order in imidazole in the range of 0 to 0.6 M imidazole. Catalysis by other secondary and tertiary amines including *N*-methylimidazole shows only a first-order dependence on amine.

If a reaction is catalyzed by general bases, a second-order dependence may mean a general base catalysis of general base catalysis. Although association of imidazole in a medium of low dielectric constants clouds the issue, the transition state of the dehydration reaction probably is

11.2.3.   General Acid–General Base Catalysis

The optimal method for detecting simultaneous general acid–base catalysis is to observe kinetic dependence on each species. Some reactions, including the halogenation of ketones, the hydration of dichloroacetone, and the ketonization of oxaloacetate ion, show reaction rates dependent both on general acid and general base concentrations. Those reactions involving apparent kinetic dependence on both carboxylic acid and carboxylate ion, however, cannot be rigorously attributed to a multiple catalysis since these reactions may be due to catalysis by the kinetically indistinguishable dimeric (basic) species, HA_2, whose existence is known.

This problem, however, does not apply to the ketonization of oxaloacetate ion since the dependence of the reaction rate on general acid and general base is seen with imidazole and triethanolamine buffers. Since these catalysts are not subject to the ambiguity seen in the carboxylate catalysis, this process may be a true multiple catalysis whose mechanism involves the simultaneous action of a general acid and general base, donating and removing protons, respectively.

11.2.4. Nucleophilic–General Base Catalysis

While nucleophilic catalysis of p-nitrophenyl acetate hydrolysis by imidazole is very effective, nucleophilic catalysis of the hydrolysis of esters by imidazole with poorer leaving groups such as p-methylphenol or p-methoxyphenol is subject to a further (general base) catalysis by imidazole. Esters with even poorer leaving groups such as phenyl acetate, trifluoroethyl acetate, and acetoxime acetate show nucleophilic catalysis by imidazole subject to further general (base) catalysis by hydroxide ion.[3] The second-order reaction in imidazole is defined as a general base catalysis superimposed on a nucleophilic catalysis, since it shows a deuterium isotope effect, although the first-order reaction of imidazole and p-nitrophenyl acetate, on the other hand, shows none. Catalysis by N-methylimidazole does not exhibit second-order catalysis by N-methylimidazole, indicating that in the former reaction one imidazole may assist the other by proton abstraction [**11.1** of Eq. (11.3)]. Alternative mechanisms in which the departure of the alcohol is facilitated by removal by a proton from a tetrahedral intermediate, **11.2**, or by addition of a proton to the leaving alcohol, **11.3**, respectively, are also possible. The imidazole reaction assisted by hydroxide ion can either occur as shown or can consist of preequilibrium formation of imidazole anion that then serves as the nucleophilic entity.

(11.3)

11.1   **11.2**   **11.3**

Hydroxide ion can either act as shown, or it can act through a preequilibrium formation of the imidazole anion that then serves as the nucleophilic entity.

Nucleophilic catalysis by imidazole occurs in the formation of a tetrahedral intermediate; but general base catalysis assistance by imidazole can occur either in the formation or the decomposition of the tetrahedral intermediate. Multiple catalysis requires only that the nucleophile still be present when the general base reacts, so that a transition state containing both species occurs.

11.2.5. Nucleophilic–Electrophilic Catalysis

The amino acid serine is cleaved by the enzyme hydroxymethylase, in the presence of the coenzymes tetrahydrofolic acid and pyridoxal phosphate, forming glycine and "activated" formaldehyde. An interesting model system for this enzyme catalysis is found in the reaction of serine N,N'-diarylethylene diamine pyridoxal phosphate, and metal ion at pH 5.5[4]. However, it must be pointed out that this model does not really catalyze as the "catalytic components" are not regenerated as they are in the enzyme system. In this reaction, pyridoxal phosphate acts as an electrophilic receptor while the N,N'-diarylethylene diamine acts as a nucleophile. Omission of either of these reagents from the reaction mixture negates the reaction. A plausible mechanism for the reaction is

$$(11.4)$$

The key step in Eq. (11.4) is the one in which the diamine reacts with the imine of the amino acid. In this reaction, the diamine acts as nucleophile and the pyridoxal moiety acts as an electrophile simultaneously facilitating the cleavage of the carbon–carbon bond under very mild conditions. In enzymatic reactions, tetrahydrofolic acid (with its ethylene diamine structure) functions as the nucleophile in place of diarylethylene diamine. (See Chapter 8).

The search for multiple catalysis, while not completely barren, has not been amply repaid. In a few instances a combination of general acid and general base appears to be superior to either alone. Likewise, a combination of nucleophilic and general basic catalysis, or nucleophilic and electrophilic catalysis, appears to be superior to either alone. However, determination of the rate enhancement brought about by the introduction of the second catalyst is not easy, because quantitative data are not always available and because the reference state for such calculations is not completely apparent. Let us then look to a more fruitful field, multiple catalysis involving bifunctional molecules.

11.3. MULTIPLE CATALYSIS INVOLVING BIFUNCTIONAL MOLECULES

If a reaction is catalyzed by two components, such as nucleophilic and electrophilic catalysts, enhanced catalysis should result if two of the components, either the nucleophilic catalyst and the substrate or the electrophilic catalyst and the substrate are joined together (Chapter 10). The nucleophilic and electrophilic catalysts may be combined, creating a bifunctional catalyst. These multiple catalysis possibilities are illustrated in Fig. 11.1. The ultimate in multiple catalysis (outside an enzyme) is accomplished by joining together the substrate and both catalysts. It has been suggested that enzymes may owe a large share of their catalytic powers to the fact that in enzymes nucleophilic and electrophilic catalysts are combined in the same molecule in the proper stereochemical arrangement for optimum interaction with the substrate.[5]

11.3.1. Bifunctional Nucleophiles

Before approaching the question of bifunctional catalysis, let us briefly consider the reactions of two species, one of which is bifunctional in nature. The simplest of these reaction involves bifunctional nucleophiles and monofunctional substrates.

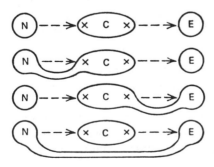

Fig. 11.1. Possible kinds of nucleophilic– electrophilic catalysis. From C. G. Swain, *J. Am. Chem. Soc.,* **72,** 4583 (1950). Copyright © 1950 by the American Chemical Society. Reprinted by permission of the copyright owner.

Bifunctional nucleophiles have led to stereoselective bifunctional catalysis as in dedeuteration. The dedeuteration of 3-pentanone-2,2,4,4-d_4 was studied in the presence of perchloric acid, sodium hydroxide, and four amines of the type RCH_2NMe_2. The Brönsted β for the amines is 0.56. The monoprotonated forms of both N,N-dimethyl-1,3-propanediamine and N,N-2,2-tetramethyl-1,3-propanediamine are bifunctional catalysts for the dedeuteration. Their primary amino group transforms the ketone to an iminium ion from which the tertiary amino group removes a deuteron internally. The monoprotonated form of the chiral catalyst is a very effective bifunctional catalyst, and it acts stereoselectively. The pro-S deuterons of the ketone are removed more rapidly than the pro-R deuterons by as much as seventyfold. Partially dedeuterated ketone containing more than 80% of the dideuterio species was isolated and found to be optically active, with a positive Cotton effect.[6,7]

A bifunctional nucleophile whose reactions have been intensively studied is catechol monoanion.[8] This species shows exceptional nucleophilicity in aqueous solution toward phosphoryl, sulfonyl and acyl halides, and phenyl chloroacetate. The catechol monoanion is the reactive species in each reaction. Not only is catechol monoanion more reactive than the dianion or diprotonated species, but it is also more reactive than monoanions of phenol, resorcinol, or hydroquinone. On the other hand, phenols with three vicinal hydroxyl groups are five times more reactive than catechol toward diisopropyl phosphorofluoridate. The nucleophilicity of catechol monoanion is equal to that of hydroxide ion in this reaction although its pK_a is much lower. Catechol forms a complex with isopropyl methylphosphonofluoridate. In the reaction of isopropyl methylphosphonofluoridate with a series of substituted catechols, kinetic evidence of complexation prior to reaction was observed. On this basis, the following mechanism has been proposed:

$$(11.5)$$

For carbonyl and sulfonyl compounds, a similar mechanism will presumably apply. In the latter reactions, however, no evidence of prior complexation exists.

The reaction of phenyl acetate with aliphatic primary amines and with diamines of structures $NH_2(CH_2)_nNH_2$ or $NH_2(CH_2)_nNH_3^{\oplus}$ (where $n = 2–6$) show kinetic dependence on the first power of the ester and the amine. The values of the rate constants of the diamines are larger than predicted from the Brönsted plot of the monoamines by a factor of up to ten, suggesting a possible contribution from the terminal amino group or ammonium ion acting as general base or acid as shown in **11.4** and **11.5** of Eq. (11.6). Alternatively, the diamines may form a separate Brönsted series.[9]

$$\tag{11.6}$$

Although n-butylamine reacts with p-nitrophenyl acetate in chlorobenzene solution by means of a process second order in amine, reaction with benzamidine is first order in amine.[10] The second-order rate constant of the aminolysis by benzamidine is at least 15,000 times larger than the second-order rate constant of the aminolysis by n-butylamine. These extraordinary results are due to the bifunctional nature of benzamidine. As shown in Eq. (11.7) reaction is facile because there is no charge formation in the transition state that resembles the tetrahedral intermediate. Nucleophilic attack by n-butylamine monomer, however, involves the creation of charge, which is inhibited in apolar solvents. In addition, both catalytic components are contained in one benzamidine molecule.

$$\tag{11.7}$$

11.3.2. Bifunctional Catalysis

The classical experiment in bifunctional catalysis is the mutarotation of α-D-tetramethylglucose in benzene solution by 2-pyridone (2-hydroxypyridine).[11] This reaction shows second-order kinetics (at low concentrations of catalyst), whereas comparable catalysis by a mixture of phenol and pyridine shows third-order kinetics. For example, at 0.05 M, the rate of the 2-hydroxypyridine reaction in benzene is approximately 50 times that of the combination of phenol and pyridine. This result ignores the difference in the kinetic order of the reactions and was first interpreted to mean that both the nitrogen and hydroxyl functions of 2-hydroxypyridine act in place of the separated pyridine and phenol catalysts. Later, however, it was found that tetramethylglucose and 2-hydroxypyridine form a complex by hydrogen bonding in benzene. The abnormally high initial specific rotations of solutions of tetramethyl-D(+)-glucose containing 2-hydroxypyridine point to complex formation. The pyranoselike hemiacetal, 2-tetrahydropyranol, partially inhibits the mutarotation catalyzed by 2-hydroxypyridine, although phenol and pyridine do not. This may be caused by competitive complexation with the catalyst. On the basis of these data, the mechanism of catalysis by 2-hydroxypyridine and related bifunctional catalysts is given by

$$(11.8)$$

This mechanism is quite similar to that proposed for the reaction of benzamidine with p-nitrophenyl acetate in chlorobenzene solution [Eq. (11.7)] and implies that any system having two electronegative atoms in a 1,3-juxtaposition, with an internal angle between the two of less than 180°, (rather than greater than 180° as in imidazole), and possessing one double bond, will satisfy the requirements for bifunctional catalysts for reactions with carbonyl substrates. Thus benzoic acid, picric acid, and 2-aminopyridine have unusually high catalytic activity.

The mutarotation of α-D-tetramethylglucose in nitromethane solution shows a pattern similar to that in benzene solution. Carboxylic acid catalysts are more effective than phenols of the same acidity by a factor of approximately 400.[12] A bifunctional catalysis by the carboxylic acid is indicated [Eq. (11.9)].

$$(11.9)$$

Although bifunctional catalysis of the acid–base variety occurs in aprotic solution (see Section 10.1.1 for examples occurring in protic solutions), its occurrence in hydroxylic solution is not known with certainty. One possible occurrence is in the hydrolysis of the iminolactone, N-phenyliminotetrahydrofuran.[13] This reaction yields aniline and butyrolactone or α-hydroxybutyranilide, depending on the breakdown of the tetrahedral intermediate [Eq. (11.10)]. The phosphate buffer markedly increased the conversion to aniline.

$$(11.10)$$

Thus it is 240 times more effective than imidazole buffer in promoting the conversion of iminolactone to aniline. This result rules out both nucleophilic and classical general base catalysis by phosphate ion since in the former imidazole far exceeds phosphate ion (although their pK's are about the same) and in the latter imidazole is equal to phosphate ion (Chapters 7 and 5, respectively). However, bifunctional catalysis by phosphate ion should be superior to that by imidazole on steric grounds. The phosphate ion enhances the transformation of the neutral tetrahedral intermediate to the zwitterionic intermediate in Eq. (11.11). The aniline yield should increase since the former leads to anilide while the latter leads to aniline, a reaction that is related to catalysis.

$$(11.11)$$

Although this mechanism is very attractive, it is extremely difficult to visualize a complex in aqueous solution having the required dissociation constant of 2×10^{-3} M between dihydrogen phosphate and the electronegative atoms of the substrate molecule.

Peptide syntheses from both activated (cyanomethyl) esters and low-energy (methyl) esters and amines are accelerated by 1,2,4-triazole as well as 2-hydroxypyridine in an aprotic medium.[14] Imidazole is somewhat effective as a nucleophilic catalyst with activated esters, but it fails completely with nonactivated esters. The results probably indicate some special property of 1,2,4-triazole and 2-hydroxypyridine that can be represented by the following nucleophilic–general acid catalysis mechanism [Eq. (11.12)].

The reaction scheme (structures with chemical mechanisms showing R–C with OCH₃ and pyridine groups):

$$R-C \underset{OCH_3}{\overset{O}{\diagdown}} \quad \xrightarrow{H-O,\ N} \quad R-\underset{OCH_3}{\overset{\overset{\displaystyle H\ O}{|\ \|}}{C}}-N \quad \rightleftharpoons \quad R-\underset{OCH_3}{\overset{O^{\ominus}}{C}}---\overset{\oplus}{N} \quad \xrightarrow{-CH_3OH}$$

$$\underset{O^{\ominus}}{\overset{\overset{\displaystyle O}{\|}}{R}C}-\overset{\oplus}{N} \quad \xrightarrow{RNH_2} \quad R\overset{O}{\overset{\|}{C}}NHR + \quad \text{(pyridine-N-OH)}$$

$$(11.12)$$

11.3.3. Bifunctional Catalysis Consisting of One Intermolecular and One Intramolecular Catalyst

In theory, bifunctional catalysis consisting of one intermolecular and one intramolecular catalyst should also show enhanced catalysis over two inter-molecular catalysts.

A multiple catalysis combining one intermolecular and one intramolecular catalyst occurs in the solvolysis (methanolysis) of diaxial 1,3-dihydroxy acetates.[15] The three acetate esters, coprostanol acetate (**11.6**), coprostane 3β, 5β-diol 3-monoacetate (**11.7**) and strophanthidin 3-acetate (**11.8**) are solvolyzed in methanol–water–chloroform solution with quite different rates. For example, in 3:1 triethylamine: triethylammonium acetate buffer (0.21 M) at 40°C, the relative rate constants of **11.6**, **11.7**, and **11.8** are 1:300:1200.

11.6   **11.7**   **11.8**

The large rate constants for **11.7** and **11.8** are attributable to the intramolecular general acid catalysis by the 5-hydroxy groups, which enhances the general base catalysis by external triethylamine (and other bases such as pyridine and N-methylimidazole) (see structure **11.9**). The external catalysis is due to general base rather than nucleophilic catalysis since esters of aliphatic alcohols are not subject to nucleophilic catalysis and since catalysis by N-methylimidazole is much poorer than that by trimethylamine (the former is a better nucleophilic catalyst, but the latter is a better general base catalyst). The fourfold rate enhancement of **11.8** over **11.7** may be due to the presence of the 19-aldehyde group

11.9

which in the form of a hemiacetal could provide an additional hydroxyl interaction, leading to a multiple internal general acid catalysis (**11.9**).[16] However, this effect is not large. A significant observation in this series is that when internal general acid catalysis is not present, the external general base catalysis is likewise not present.

11.3.4. Bifunctional Metal Ion Catalysis

Some of the metal catalyses discussed in Chapter 9 involve multiple catalysis consisting of catalysis both by metal ion and by another catalyst such as a general base or nucleophile.

An interesting example involving a metal ion plus another catalyst is the oxidative deamination of amino acids to keto acids, ammonia, and hydrogen peroxide, catalyzed by pyridoxal and manganic ion at room temperature. This reaction serves as a model for the action of some amine oxidases. α-Methylalanine, N-methylalanine, and lactic acid are not oxidized under conditions where alanine reacts readily. Other amino acids and amino acid esters and amides can replace alanine, but simple amines react slowly if at all. The rate of O_2 uptake is decreased by ethylenediaminetetraacetic acid, but is unaffected by light or by free radical inhibitors such as phenols (thus the oxidation presumably does not occur by a free radical chain mechanism). Glycine is oxidized five to six times more readily than α, α-dideuteroglycine. These results are consistent with Eq. (11.13). Intermediates **11.10** and **11.11** are similar to the intermediates proposed for other pyridoxal-catalyzed reactions of amino acids.

In this mechanism, **11.11** complexes with O_2 to give **11.12**. The transfer of a proton through the solvent and electrons through the complex (Chapter 9) can lead to **11.13**, in which oxygen has been reduced to hydrogen peroxide and the rest of the complex has lost two electrons. Compound **11.13** is in equilibrium with pyruvic acid, H_2O_2, and **11.14**. Compound **11.14** is a tautomer of the imine of ammonia with pyridoxal, and gives pyridoxal and ammonia readily.[17]

(11.13)

11.4. EVALUATION OF MULTIPLE CATALYSIS

As mentioned earlier, the search for enhanced catalysis in two-catalyst versus one-catalyst systems has not been particularly fruitful. The reason for this failure is clearly evident in Tables 11.2 and 11.3. The rate constants of several sets of singly and doubly catalyzed reactions (Table 11.2) as well as the rate constants of some sets of uncatalyzed and singly catalyzed reactions (Table 11.3) are given. In the reactions of Table 11.2, the third-order rate constant has approximately the same numerical value as the second-order rate constant. In the hydrazinolysis reactions, the value of the third-order constant is an order of magnitude larger than the second-order constant. These values mean that the third-order reaction is observable only when the catalyst concentration is of the order of 1 M for reactions of Table 11.2 and 0.1 M for reactions of Table 11.3.

What are the underlying causes of the lack of appreciable acceleration by an additional catalyst? The answer can be seen in an analysis of the thermodynamic factors in the hydrazinolysis reaction. Figure 11.2 shows the data for a series of hydrazinolysis reactions with and without catalysis (second- and third-order reactions). The catalytic reactions have considerably lower enthalpy of activation than the uncatalyzed reactions, but the entropy of activation of the catalyzed

Table 11.2. Comparison of Second- and Third-Order Rate Constants of Some Imadazole-Catalyzed Reactions

Reaction	Second-Order Rate Constant k_{Im} (M^{-1} min^{-1})	Third-Order Rate Constant k_{Im}^2 (M^{-2} min^{-1})
Hydrolysis of p-methylphenyl acetate	0.21	0.17
Hydrolysis of p-methoxyphenyl acetate	0.20	0.19
Hydration of sym-dichloroacetone	2.36	3.55

Table 11.3. Rate Constants of the Hydrazinolysis of Substituted Phenyl Acetates: Uncatalyzed, General Acid-Catalyzed and General Base-Catalyzed Reactions

Substituent	k_u (M^{-1} min^{-1})	k_{ga} (M^{-2} min^{-1})	k_{gb} (M^{-2} min^{-1})
p-NO$_2$	327		
m-NO$_2$	39.7		
H	0.245	2.62	10.75
p-CH$_3$	0.130	1.98	7.86
p-OCH$_3$	0.097	1.82	8.20

Source: T. C. Bruice and S. J. Benkovic, J. Am. Chem. Soc., **86**, 418 (1964). © 1964 by the American Chemical Society. Reprinted by permission of the copyright owner.

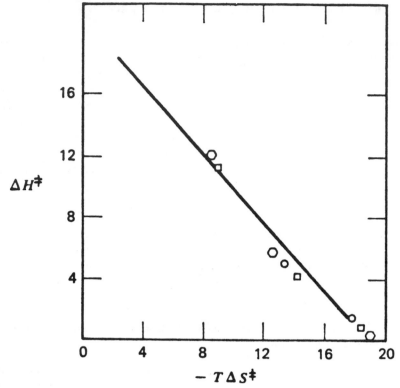

Fig. 11.2. Plot of $\Delta H^{\ddagger}$ versus $-T\Delta S^{\ddagger}$ for the reaction of hydrazine with (☐) p-methyl-, (◯) p-methoxy-, and (⦿) unsubstituted phenyl acetates depicting compensation. From T. C. Bruice and S. J. Benkovic, *J. Am. Chem. Soc.*, **86**, 418 (1964). Copyright © 1964 by the American Chemical Society. Reprinted by permission of the copyright owner.

reactions is considerably smaller, the two nearly compensating one another so that the free energies of activation of the two are not appreciably different.

These and similar data can be analyzed in terms of the ratio, $T\Delta S^{\ddagger}/$kinetic order. Table 11.4 (Ref. 18) shows a surprisingly constant value of 4–6 kcal mole^{-1} kinetic order for this ratio. This value compares favorably with a value of 5.4 kcal/mole suggested as a reasonable value for the loss of entropy from incorporation of a water molecule into a transition state. A somewhat smaller value (2–3 kcal/mole) is calculated by the Sackur–Tetrode equation[19] for the loss in translational entropy of two molecules in forming a complex. In conclusion, the introduction of any additional external catalyst must necessarily lead to unfavorable entropy of activation, even though the enthalpy of activation is made more favorable.

How then is it possible to accelerate a reaction by going to a higher-order catalytic process? The answer is to go to a catalytic reaction that increases the catalytic functionalities, but not the kinetic order of the reaction, such as catalysis by a bifunctional molecule. If the same decrease in enthalpy of

Table 11.4. A Comparison of the Value of $T\Delta S^{\ddagger}$ (kcal/mole) to the Kinetic Order of Displacement Reactions on the Phenyl Ester Bond

Reaction	Kinetic Order	$T\Delta S^{\ddagger}/$ Kinetic Order
$(CH_3)_3N$ + Ph esters	2	4
γ-$(N,N$-Dimethylamino)butyrate Ph esters	1	4
δ-$(N,N$-Dimethylamino)valerate Ph esters	1	4–5
$OH^{\ominus}$ + Ph esters	2	3
$AcO^{\ominus}$ + Ph esters	2	4
Monophenyl glutarates	1	4
H_2NNH_2 + Ph esters	2	4
$2(H_2NNH_2)$ + Ph esters	3	4–5
$H_2NNH_2 + H_2NNH_3^{\ominus}$ + Ph esters	3	6
Imidazole + Ph esters	2	6–7
2(Imidazole) + Ph esters	3	5

Source: T. C. Bruice and S. J. Benkovic, *J. Am. Chem. Soc.*, **86**, 418 (1964). © 1964 by the American Chemical Society. Reprinted by permission of the copyright owner.

activation brought about by changing from one to two catalysts is not accompanied by a compensating activation entropy loss, the catalysis should be considerably more efficient.

Multiple catalysis can be advantageous over single catalysis if the multiple catalyst is a bifunctional or polyfunctional molecule. In such a case, the entropy of activation should remain constant for the two systems, while the activation enthalpy should be more favorable for the bifunctional catalyst than for the monofunctional catalyst. Presumably an enzyme has this advantage.

11.5.   ENZYME MODELS UTILIZING MULTIPLE CATALYSIS

Many enzyme models are based on multiple catalysis. They are referred to in Section 8.33 (pyridoxal and folic acid); Section 9.22 (zinc ion + N-propyl-1,4-dihydronicotinamide); Section 10.1.4 (imidazole + benzoate ion); Section 11.3.4 (manganic ion + pyridoxal); and Section 10.1.1 (the hydration of fumaric acid monoanion).[20] Another model is that for the enzyme oritidine-5-phosphate decarboxylase, which catalyzes the last step in the biosynthesis of pyrimidine nucleotides. In this transformation, orotidylate is converted into uridylate. The model involves, in part, an acid–base catalyzed decarboxylative elimination.[21]

REFERENCES

1.   T. M. Lowry and I. J. Faulkner, *J. Chem. Soc.*, **127**, 1883 (1925).
2.   Y. Pocker, *Chem. Ind.*, 968 (1960).

3. J. F. Kirsch and W. P. Jencks, *J. Am. Chem. Soc.*, **86,** 833 (1964); T. C. Bruice and S. K. Benkovic are quoted in this paper.

4. E. Brode and L. Jaenicke, *Biochem. Z.*, **332,** 259 (1960).

5. C. G. Swain, *J. Am. Chem. Soc.*, **72,** 4578 (1950).

6. J. Hine and J. P. Zeigler, *J. Am. Chem. Soc.*, **102,** 7524 (1980).

7. J. Hine, W.-S. Li and J. P. Zeigler, *J. Am. Chem. Soc.*, **102,** 4403 (1980).

8. T. C. Bruice and S. J. Benkovic, *Bioorganic Mechanisms*, Vol. I, W. A. Benjamin, New York, 1966.

9. T. C. Bruice and R. G. Willis, *J. Am. Chem. Soc.*, **87,** 531 (1965).

10. F. M. Menger, *J. Am. Chem. Soc.*, **88,** 3081 (1966).

11. C. G. Swain and J. F. Brown, Jr., *J. Am. Chem. Soc.*, **74,** 2534 (1952).

12. E. L. Blackall and A. M. Eastham, *J. Am. Chem. Soc.*, **77,** 2184 (1955).

13. B. A. Cunningham and G. L. Schmir, *J. Am. Chem. Soc.*, **88,** 551 (1966); see also Y.-N. Lee and G. L. Schmir, *J. Am. Chem. Soc.*, **101,** 3026 (1979) for a kinetic analysis of this system which indicates that bifunctional catalysts are better than monofunctional catalysts.

14. H. C. Beyerman and W. VandenBrink, *Proc. Chem. Soc.*, 266 (1963).

15. S. M. Kupchan, S. P. Eriksen, and M. Friedman, *J. Am. Chem. Soc.*, **88,** 843 (1966).

16. S. M. Kupchan, S. P. Eriksen, and Y.-T. S. Liang, *J. Am. Chem. Soc.*, **88,** 347 (1966).

17. G. A. Hamilton and A. Revesz, *J. Am. Chem. Soc.*, **88,** 2069 (1966).

18. T. C. Bruice and S. J. Benkovic, *J. Am. Chem. Soc.*, **86,** 418 (1964).

19. W. J. Moore, *Physical Chemistry*, Prentice-Hall, New York, 1950, p. 348.

20. M. L. Bender and K. A. Connors, *J. Am. Chem. Soc.*, **84,** 1980 (1962).

21. R. B. Silverman and M. P. Groziak, *J. Am. Chem. Soc.*, **104,** 6434 (1982).

12 | Catalysis by Complexation

Many complexes are formed between organic molecules. They include: (1) covalent complexes—those considered here are rapidly and reversibly formed systems; (2) electrostatic or ionic complexes; (3) hydrogen-bonded complexes; (4) metal ion complexes; (5) apolar complexes; (6) micellar complexes; and (7) polymeric complexes. Most of these complexes can have an effect on the rates of reaction. Here we consider explicitly the effect of complexation on catalysis.

The role of complexation on catalytic action has been noted many times. In all heterogeneous catalyses, adsorption of the substrate(s) on the catalytic surface appears to be an initial event. Likewise in enzymatic catalysis, formation of an enzyme–substrate complex is usually the primary step. In many of these reactions, the catalyst–substrate complex is at a lower energy level than the uncomplexed materials. This fact is difficult to reconcile with a facilitation of reaction in which the free energy of activation must be lowered. The explanation, of course, is that complexation must lower the free energy of the transition state by an amount even greater than the ground state for catalysis to occur. This lowering can occur either through a change in reaction pathway by the complexation, or through a lowering of the transition state energy without a change in pathway, as with simple catalyses.

Complexation can juxtapose substrate and catalyst. When this occurs, the subsequent catalytic event should have the characteristics of an intramolecular reaction (Chapter 10) resulting in an even more favorable activation entropy. The initial complexation of substrate and catalyst will have an unfavorable translational entropy, however, unless compensated by other favorable energy factors, such as solvation entropy. These favorable factors occur with sufficient frequency to make complexation an important catalytic pathway.

In addition to bringing catalyst and substrate into proximity with one another, complexation must be stereospecific, so that the catalytic group(s) are oriented correctly with respect to the reactive center of the substrate. This requirement is important, because random complexation can lead to many more nonproductive than productive (catalytically active) complexes. The formation of stereospecific complexes depends on complementarity of structure between the catalyst and substrate in some sense. As we learn more about the general laws of complementarity, we will undoubtedly be able to design more specific catalysts.

12.1. THEORY OF ACCELERATION DUE TO COMPLEXATION

Complex formation between catalyst and substrate can lead to remarkable catalytic effects. The acceleration due to complex formation will be described theoretically. These theories apply to enzymatic reactions, since they also involve complex formation prior to reaction.

12.1.1. Substrate Anchoring

The acceleration due to complex formation can be attributed to contact of the catalyst and the substrate for a period longer than in intermolecular reactions. The profile of the potential energy of the reaction of A with B along the reaction coordinate is depicted in Fig. 12.1. Region I corresponds to the solvated reactants. Region II characterizes the course of the potential-energy change preceding the formation of the "activated complex" C^*, and Region III the course after the chemical transformation has occurred. Region IV corresponds to the solvated products. In aqueous solutions, the time required for a small molecule to rotate 2π radians about its molecular axis is very short (of the order of 10^{-11} sec). For water molecules, it is only about 10^{-12} sec. Therefore, if the "activated complex" dissociates instead of proceeding to products, the reactants will be rapidly solvated and reach Region I and only after a new collision will a new "activated complex" be formed.

The probability of forming an "activated complex" is a function of the lifetime of an AB pair and thus may be assumed to obey a simple exponential law. Thus the ratio of the probabilities in the two cases is given by $P(t_1)/P(t_2) = (1 - e^{\omega_1 t_1})/(1 - e^{\omega_2 t_2})$. The term ω is the half-life corresponding to the rate

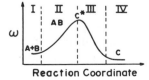

Reaction Coordinate

Fig. 12.1. The potential energy along the reaction path. From J. Reuben, *Proc. Natl. Acad. Sci. USA*, **68**, 563 (1971).

constant k. If we consider the reactions to proceed via a similar pathway, i.e., $\omega_1 = \omega_2$, and if the rate constants are such that $\omega_1 t_1$ and $\omega_2 t_2 \ll 1$, then $P(t_1)/P(t_2) = t_1/t_2$.

Complex formation of A and B should increase the lifetime of the AB pair, resulting in acceleration. For example, an acceleration of about 10^4-fold is expected if the lifetime of the AB complex is 10^{-7} sec. The effect of substrate anchoring is entropic in origin since it does not assume a change in the energy levels available to the system, but rather a shift in the fraction of time spent at each of the energy states.[1]

Substrate anchoring may thus appear to be similar to the proximity effect; however, it considers the duration of the proximity. As shown above, a long residence time should lead to rate accelerations by factors of 10^6–10^9. The rate acceleration observed by going from inter- to intramolecular reactions may be explained in a similar manner.

12.1.2. Concentration Profiles

Catalytic effects of enzymes have been interpreted in terms of several models. One of the most widely used adopts the transition-state theory and constructs from it a diagram presenting the relative free energies of species in the ground and activated states. A typical example of such a diagram is illustrated in Fig. 12.2 for a reaction involving a single species S.

The central concept of transition-state theory is that reactants are in equilibrium with the transition-state species and that the rate of reaction is proportional to the *concentration* of the transition state. On that basis, one can introduce the thermodynamic concept of an equilibrium constant $K^{\ddagger}$ between the

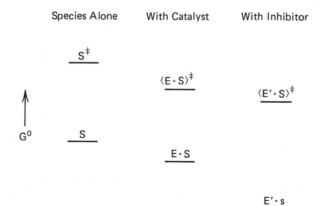

Fig. 12.2. Relative (standard) free energies of ground state and transition state species in the absence and in the presence of a catalytic complexing molecule such as an enzyme. Substrate denoted by S, enzyme by E, inhibitor by E'. From I. M. Klotz, *J. Chem. Educ.*, **53**, 159 (1976). Copyright © 1976 American Chemical Society. Reprinted by permission of the copyright owner.

transition-state species and the reactants. Consequently, one can calculate a standard free-energy change $\Delta G^{\circ\ddagger}$ for the activation step.

To a novice this implies that the catalytic effect of an enzyme E on a substrate S is due to the fact that the energy level $G^{\circ\ddagger}$ for the enzyme–substrate complex $(E \cdot S)^{\ddagger}$ is below that for the corresponding nonenzymatic species $S^{\ddagger}$. The use of the letter E predisposes one to presume a catalytic effect on a reaction rate. If S′ were a molecule that combined strongly with E but slowed down the chemical transformation (i.e., if S′ were an inhibitor), the free-energy level of $E \cdot S'$ would be below that of S and hence that of $(E \cdot S)^{\ddagger}$ could also lie below that of S′. Indeed, no acceleration in rate occurs (in the simplest, single-species reaction) unless $\Delta G^{\circ\ddagger}$ for $E \cdot S \to (E \cdot S')$ is smaller than $\Delta G_S^{\circ\ddagger}$ for $S \to S^{\ddagger}$. Thus a free-energy diagram can actually be misleading. In any event, it does not reveal the nature of enzyme catalytic effects. That is, the transition-state theory obscures the facts.

On the other hand, if one considers again the central concept of transition-state theory, it is clear that the relative *concentrations* of participating species are needed. What we need is a *concentration profile*.[2]

Let us consider a simple single-substrate reaction (Fig. 12.3). In the non-enzymatic reaction (left side of Fig. 12.3), the *concentration* of substrate in the ground state is represented by the height of the bar S_0; that of the transition state species $S^{\ddagger}$, which is in equilibrium with S, is shown by the top of the bar $S_0^{\ddagger}$.

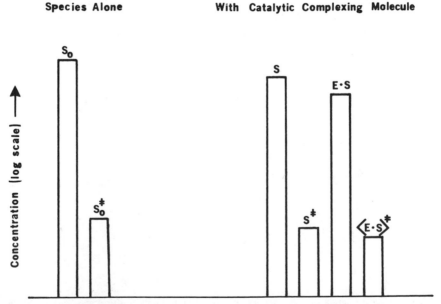

Fig. 12.3.   Concentration profile presenting concentrations of substrate S in ground states and in transition states under various circumstances. This profile is for a reaction involving only a single solute species. From I. M. Klotz, *J. Chem. Educ.,* **53,** 159 (1976). Copyright © 1976 American Chemical Society. Reprinted by permission of the copyright owner.

The rate of reaction is proportional to the *concentration* of S_0. The substrate may be bound (in the ground state) to produce the species $E \cdot S$ in the presence of an enzyme or catalytic complexing molecule E. The *concentration* of the $E \cdot S$ will depend on the magnitude of the binding constant K_{ES}. If this constant is appreciable, a significant *concentration* of $E \cdot S$ will be produced at equilibrium and that of S will drop perceptibly. Insofar as the crucial issue, the rate of change of S, is concerned, the crucial question is, what is the *concentration* of $(E \cdot S)^{\ddagger}$?

If the equilibrium constant for the formation of $(E \cdot S)^{\ddagger}$ from $E \cdot S'$ is the same as that for the formation of $S_0^{\ddagger}$ from S_0 in the absence of a complexing molecule, then no acceleration in the rate will be effected by the complexing molecule. For under these circumstances

$$K_{E}^{\ddagger} = \frac{[(E \cdot S)^{\ddagger}}{(E \cdot S)} = \frac{(S^{\ddagger})}{(S)} = K^{\ddagger} = \frac{(S_0^{\ddagger})}{(S_0)} \qquad (12.1)$$

and hence the ratio of concentration of the transition-state species to that of the ground state species is given by

$$\frac{(E \cdot S)^{\ddagger} + (S^{\ddagger})}{(E \cdot S) + (S)} = \frac{K_E^{\ddagger}(E \cdot S) + K_E^{\ddagger}(S)}{(E \cdot S) + (S)} = K^{\ddagger} = \frac{(S_0^{\ddagger})}{(S_0)} \qquad (12.2)$$

$$[(E \cdot S)^{\ddagger}] + (S^{\ddagger}) = (S_0^{\ddagger}) \qquad (12.3)$$

Thus if the total concentration of S is the same in the presence of a complexing molecule as in its absence, then the fraction of activated species is the same in the presence as in the absence of a complexing molecule and hence

$$[(E \cdot S)^{\ddagger}] + (S^{\ddagger}) > (S_0^{\ddagger}) \qquad (12.4)$$

In a single-substrate reaction, a complexing molecule must provide an environment conducive to the formation of activated species if the complexing molecule is to effect an acceleration in rate of reaction.

If the complexing molecule does favor formation of the activated species, then the *concentration* of the latter depends not only on $K_E^{\ddagger}$ but also on the association constant K_{ES} for the formation of $E \cdot S$

$$E + S \rightleftharpoons E \cdot S, \qquad K_{ES} = \frac{(E \cdot S)}{(E)(S)} \qquad (12.5)$$

Algebraic analysis leads to the following expression for the fraction $\alpha^{\ddagger}$ of activated molecules

$$\alpha^{\ddagger} \equiv \frac{[(E \cdot S)^{\ddagger}] + (S^{\ddagger})}{(E \cdot S) + (S)} = \frac{K_E^{\ddagger} K_{ES}(E) + K^{\ddagger}}{K_{ES}(E) + 1} \qquad (12.6)$$

Thus the fraction of molecules activated and hence the rate of reaction depend on the *concentration* of the complexing molecule (E) and on its affinity for substrate, as expressed in K_{ES}. If $K_E^{\ddagger}$ is greater than $K^{\ddagger}$ and is independent of K_{ES}, then the stronger the binding of S, the more accelerated its rate of reaction will be. It is also evident from Eq. (12.6), or Fig. 12.2, that at very high (E) or as K_{ES} becomes increasingly large, $\alpha^{\ddagger}$ tends toward a limit and the velocity of the reaction will attain a plateau value.

Thus the velocity of the transformation is increased in the presence of a complexing molecule even if $\Delta G^{\circ\ddagger}$ is unchanged. If, in addition, the complexing molecule provides an environment conducive to the existence of activated species, then the velocity will be increased even further.

Therefore, we find that to increase the *concentration* of transition-state species, a catalytic complexing molecule must bind substrates as a minimum. Furthermore, if a complexing molecule provides an environment facilitating activation, this will result in further rate accelerations. A *concentration* profile (Fig. 12.3) illustrates the interplay between interrelated species and reveals explicitly the *concentrations* of activated molecules in the transition state, which determine the reaction rate.[2]

12.1.3. The Effect of Complex Formation on the Activation Parameters of Bimolecular Reactions

In the intermolecular reaction of A + B, large amounts of translational and rotational entropy are lost in going from the reactants to the transition state. However, when the reaction involves complex formation and a subsequent intramolecular reaction, this entropic loss is (partly) completed in the reactant states, since the reactants (one of them can be catalyst) lose translational and rotational entropy when complexed with each other.

The loss of translational and (overall) rotational entropy for a 2 → 1 reaction in aqueous solution is ordinarily on the order of 45 entropy units (e.u.) (standard state 1 M, 25°C); the translational entropy is usually greater than 8 e.u. (corresponding to 55 M). These values are large enough to explain the large acceleration observed in the reactions involving complex formation as well as that observed in enzymatic reactions.[3]

The preceding argument indicates that the effect of complex formation should show up in favorable activation entropy. Further study, however, suggests that this effect can appear in $\Delta H^{\ddagger}$ rather than $\Delta S^{\ddagger}$ when the structural change of the solvent, water, in the reaction is taken into consideration.

In a reaction in which two molecules form one, there is a large gain in entropy due to the destruction of one of the water "cages." This entropic gain is partly canceled by an unfavorable enthalpy change arising from a decrease of hydrogen bonding. Thus, when the complex is formed in aqueous solution, entropic loss due to the loss of translational and rotational entropy is almost canceled by the entropic gain due to the structural change in the water. Consequently, the effect of complex formation can show up in $\Delta H^{\ddagger}$.[4]

12.2. COVALENT COMPLEXES

Many of the nucleophilic and electrophilic catalyses described in Chapter 6 proceed through the formation of intermediates, for example, imines, which facilitate reaction. The catalyses described here operate in the same general fashion, but differ in one respect: a functional group is formed in the intermediate that serves as an intramolecular catalyst to complete the reaction.

12.2.1. Carbonyl Complexes

Rapid and reversible additions to carbonyl groups make substrates containing such groups susceptible to catalyses occurring through adducts of carbonyl compounds. Nitrogen and oxygen nucleophiles, in particular, add to carbonyl groups readily to form tetrahedral adducts and/or imines. If such an adduct or an imine contains a functional group that can interact with the reaction center, a specific and powerful catalysis can occur.

The hydrolysis of esters, containing carbonyl substituents is susceptible to this catalysis. For example, the alkaline hydrolysis of methyl *ortho*-formylbenzoate is 10^5 times faster than the hydrolysis of methyl *ortho*-nitrobenzoate, a compound of similar steric and electronic properties.[5] These reactions, which are first order in hydroxide ion concentration, may be explained by the addition of hydroxide ion to the carbonyl group of the substrate, producing an adduct whose oxyanion can function as an intramolecular nucleophilic catalyst [Eq. (12.7)].

$$(12.7)$$

Nucleophiles other than hydroxide ion which are capable of addition to the carbonyl group should also catalyze the hydrolysis of the esters described above. This is indeed the case. Morpholine catalyzes the hydrolysis of methyl *ortho*-formylbenzoate. Figure 12.4 illustrates a typical spectrophotometric curve of the facile morpholine-catalyzed hydrolysis of methyl *ortho*-formylbenzoate, using a stopped-flow mixing device. 3-Morpholinophthalide can be isolated from the

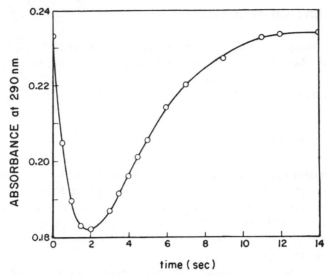

Fig. 12.4. The morpholine-catalyzed (0.085*M*) hydrolysis of methyl *ortho*-formylbenzoate. The reaction was followed using a spectrophotometer equipped with a stopped-flow mixing device. From M. L. Bender, J. A. Reinstein, M. S. Silver, and R. Mikulak, *J. Am. Chem. Soc.,* **87,** 4549 (1965). Copyright © 1965 by the American Chemical Society. Reprinted by permission of the copyright owner.

reaction mixture after a few seconds of contact time and is hydrolyzed to *ortho*-formylbenzoate ion at exactly the same rate as the overall reaction. This combination of isolation and kinetic evidence clearly indicates the following pathway for this catalysis.[5]

$$(12.8)$$

A similar mechanism can explain the ready saponification of compound **12.1**. **12.1** can be completely hydrolyzed to the corresponding keto-acid by heating under reflux for 15 min with 1% methanolic KOH, whereas the corresponding compound containing no γ-keto group is recovered unchanged under the same conditions. The hydrolysis of **12.1** proceeds as shown by Eq. (12.9).

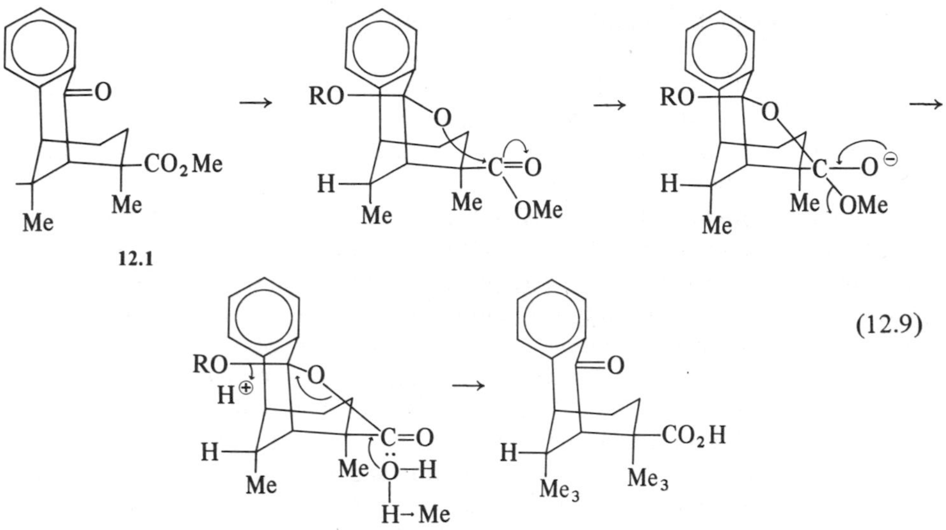

12.1

(12.9)

12.2.2. Amine Complexes

Since amines can catalyze the hydrolysis of esters containing carbonyl groups (12.2.1), carbonyl compounds should be able to catalyze the hydrolysis of esters containing amino groups. This prediction was confirmed in the benzaldehyde-catalyzed hydrolysis of p-nitrophenyl leucinate.[6] The rate constant for this catalysis is six times larger than the rate constant for the corresponding imidazole catalysis, even though imidazole is 10^{14} more basic than benzaldehyde. As predicted, benzaldehyde has no catalytic effect on an ester without an amino group such as p-nitrophenyl acetate. By analogy with Eq. (12.8), the mechanism of this catalysis can be postulated to proceed through the addition of the amino ester to benzaldehyde, forming an adduct capable of intramolecular catalysis of ester hydrolysis, as shown in Eq. (12.10).

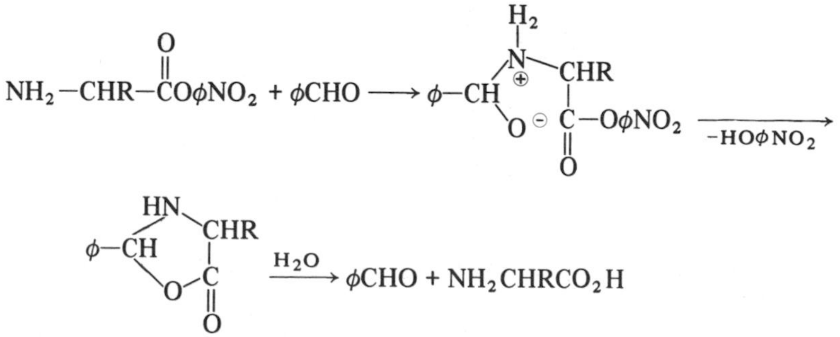

(12.10)

Although the details of this process are still speculative, the analogy between Eqs. (12.8) and (12.10) is a reasonable one, and certainly the rate enhancement is profound.

12.2.3. Nucleophiles Containing a Catalytic Group

A facile nucleophilic coupling of a group containing a second nucleophile, followed by a second intramolecular (catalytic) reaction with the free nucleophile, and final decomposition to give the products, regenerating the nucleophile, can lead to effective catalysis.

The transformation of phenylglyoxal to mandelic acid can be catalyzed in a highly specific manner by a catalyst that adds to the aldehyde group and then facilitates the hydride transfer necessary for this internal oxidation–reduction reaction. One of the best catalysts is 2-dimethylaminoethanethiol. A mixture of ethyl mercaptan and diethylamine or triethylamine catalyzes the transformation much less effectively than 2-diethylaminoethanethiol.[7] When the reaction is carried out in deuterium oxide, no carbon-bound deuterium is found in the product, indicating intramolecular hydride transfer. When the reaction is carried out at 0° C in methanol, an adduct of the catalyst and substrate can be isolated, which contains no free thiol group, indicating the thiol group of this bifunctional (ambident) catalyst adds to the substrate. On the basis of these experiments, the mechanism shown in Section 8.3.9 was postulated. In enzymatic transformations, glutathione serves as coenzyme in reactions which may be similar mechanistically (Chapter 8).

The catechol monoanion–catalyzed hydrolysis of phenyl chloroacetate is another example of covalent complex formation.[8] In this reaction the catechol monoanion functions as an effective bifunctional catalyst (Chapter 11). In the hydrolysis of phenyl monochloroacetate, the intermediate is easily decomposed, the rate being faster than the overall hydrolytic rate, since the intermediate has a new general acid or base catalyst. The failure of phenol, guaiacol, resorcinol, or hydroquinone to exhibit positive catalysis is consistent with Eq. (12.11).

(12.11)

When the original ester contains a very good leaving group, the intermediate in the reaction can be isolated.

A similar reaction in which each step can be observed directly is the reaction of *para*-nitrophenyl acetate with *ortho*-mercaptobenzoic acid. The dianion of *ortho*-mercaptobenzoic acid reacts with the ester producing thioaspirin. Thioaspirin then hydrolyzes to give acetate ion regenerating *ortho*-mercaptobenzoic acid. These two steps constitute an overall catalysis of ester hydrolysis brought about by *ortho*-mercaptobenzoate dianion, the rate-determining step being the second step. Whereas catalysis by catechol monoanion involves nucleophilic attack followed by intramolecular general basic catalysis by phenolic anion, catalysis by *ortho*-mercaptobenzoate dianion involves nucleophilic attack followed by intramolecular catalysis by the *ortho*-carboxylate ion, similar to the hydrolysis of aspirin (Chapter 10).

$$(12.12)$$

The sluggish hydrolysis of thioaspirin makes Eq. (12.12) a relatively inefficient process.

12.2.4. Electrophiles Containing a Catalytic Group

By analogy with nucleophilic reactions, a facile electrophilic reaction, followed by a second (catalytic) reaction effected by a group on the original electrophile, and then decomposition to give products, regenerating the electrophile, should also lead to substantial catalysis.

The electrophile 8-quinolineboronic acid is a catalyst for the hydrolysis of chloroethanol and 3-chloro-1-propanol in dimethylformamide solution containing water and collidine.[9] In the absence of 8-quinolineboronic acid, the chloroalcohols undergo slow solvolysis to products that are not glycols. In the presence of the catalyst, alcoholic portions of these substrates are not available for cyclization, a fact that can be rationalized with the formation of borate esters from the alcohols. Inhibition of the catalytic reaction by both water and ethylene glycol support this suggestion. The reaction occurs with inversion of configuration, ruling out direct participation by the tertiary amine, that should result in net retention of configuration as the consequence of two inversion processes. These results are consistent with a mechanism for the catalysis by 8-quinolineboronic

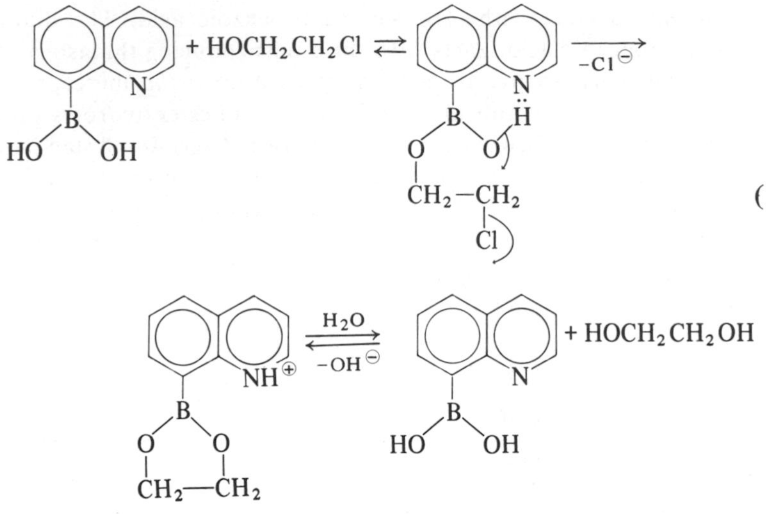

(12.13)

acid, in which reversible esterification of the boronic acid group by the alcoholic groups of the substrate is followed by displacement of the halogen atom of the substrate by a water molecule that may or may not be bonded covalently to the boron atom. This catalyst is in essence bifunctional, but instead of consisting of two functions that interact with the reaction center, one function binds the substrate stereospecifically and the other carries out the actual chemical transformation. As implied by this description, neither a boronic acid nor a nitrogen base alone can carry out this catalysis.

Phenyl salicylate is hydrolyzed abnormally rapidly in borate buffers. The catalytic constants for the hydrolysis of phenyl salicylate in borate buffers are more than a hundredfold greater than those for phenyl *ortho*-methoxybenzoate or phenyl benzoate, while in imidazole buffers they are only two- to threefold greater than in borate buffers. Borate ion normally has only a very weak catalytic effect on ester hydrolysis, and with *p*-nitrophenyl acetate, no effect at all is observed. Catalysis does not occur with phenyl salicylate in phosphate buffers, nor with catechol monobenzoate in borate buffers. The most likely explanation of these results is that complex formation of the kind shown in Eq. (12.14) facilitates the hydrolytic reaction.

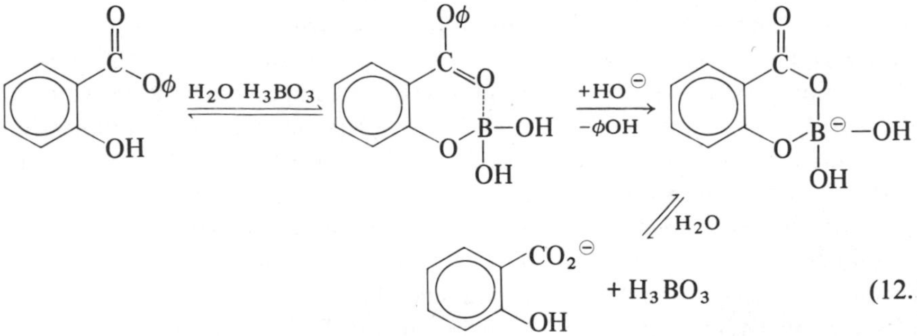

(12.14)

Since the complex from a salicylate involves a six-membered ring while the complex from catechol monobenzoate would involve a less favored seven-membered ring, the catalytic specificity can be explained.

12.3.   NONCOVALENT COMPLEXES

One of the special features of enzymatic catalysts is their ability to form adsorptive, usually noncovalent, complexes. Ideally, model reactions for enzymatic processes should include such an association. The forces responsible for complexes are varied including ionic, hydrogen-bonded, and apolar forces. Of these, the last is probably the most important. Apolar complexes take many forms such as pi molecular, micellar, and inclusion complexes.

When complexing occurs between substrate and catalyst, two pathways can be written:

$$C + S \underset{K}{\rightleftharpoons} C \cdot S \xrightarrow{k_{cat}} C + P \tag{12.15}$$

$$C \cdot S \underset{K}{\rightleftharpoons} C + S \xrightarrow{k_{cat}'} C + P \tag{12.16}$$

These two equations correspond to complexing leading to catalysis and complexing preventing catalysis, respectively. The complex in Eq. (12.15) is referred to as a productive complex, while that in Eq. (12.16) is referred to as a nonproductive complex.

The minimal experimental evidence for complex formation in a catalytic (or other) process is adherence to a kinetic scheme showing a "saturation" phenomenon (see Figs. 1.5 and 1.6). Kinetic analysis of Eq. (12.15) may be carried out in the simplest fashion by assuming a fast preequilibrium formation of the complex and $[S]_0 \gg [C]_0$. Under these conditions

$$K = \frac{[C][S]}{[C \cdot S]} \tag{12.17}$$

and

$$[C]_0 = [C] + [C \cdot S] \quad \text{and} \quad [S]_0 = [S] \tag{12.18}$$

On substitution,

$$[C]_0 = \frac{K[C \cdot S]}{[S]_0} + [C \cdot S] = [C \cdot S]\left(\frac{K}{[S]_0} + 1\right) \tag{12.19}$$

Therefore,

$$[C \cdot S] = \frac{[C]_0}{\dfrac{K}{[S]_0} + 1} \tag{12.20}$$

$$\text{rate} = k_{cat}[C \cdot S] = \frac{k_{cat}[C]_0}{\dfrac{K}{[S]_0} + 1} = \frac{k_{cat}[C]_0[S]_0}{K + [S]_0} \tag{12.21}$$

Equation (12.21) shows a direct dependence of the velocity $[S]_0$ at values of $[S]_0 \ll K$, but a zero-order dependence on $[S]_0$ at values of $[S]_0 \gg K$, thus showing a "saturation" of the catalyst by the substrate. The equilibrium constant of the dissociation of the $C \cdot S$ complex, K, is experimentally equal to the concentration of substrate at which the rate is half its maximal value. The rate constant k_{cat} is the maximal rate constant. This derivation is, of course, a complete description of Michaelis–Menten kinetics first developed for enzymatic systems.

Equation (12.21) has another kinetic equation of equivalent form. Using the same assumptions as above, except that the substrate is saturated by the catalyst instead of vice versa, the rate corresponding to Eq. (12.21) is

$$\text{rate} = \frac{K_{cat} K [C]_0 [S]_0}{K + [C]_0} \tag{12.22}$$

12.3.1. Ionic and Hydrogen Bonding Complexes

On the basis of electrostatic theory, ionic interactions are expected to be of importance in media of low dielectric constant. Since we are particularly concerned with aqueous solutions rather than media of low dielectric constant, the possibility of forming simple ionic complexes of catalytic importance is remote.

Hydrogen bond interactions between solutes are expected to be of importance in apolar media containing no hydrogen bond donors or acceptors. The importance of the solvent is vividly seen in Fig. 12.5, which depicts the hydrogen-bonded association of N-methylacetamide.

Catalysis occurring in aprotic media containing no proton donors, for example, benzene, cyclohexane, or carbon tetrachloride, may then occur through prior hydrogen-bonded complexes. This catalysis will be of special importance for those reactions that are susceptible to general acid–base catalysis. Of particular importance will be those reactions in aprotic media that are susceptible to multiple general acid–base catalysis (Chapter 11).

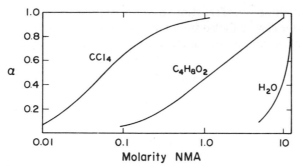

Fig. 12.5. Variation of degree of association α with concentration of N-methyl acetamide in carbon tetrachloride, dioxane, and water, respectively; α is defined by the ratio of the concentration of H-bonded N-methyl acetamide (of N-H groups hydrogen-bonded to O=C) to the total concentration. From I. M. Klotz and J. S. Franzen, *J. Am. Chem. Soc.*, **84**, 3461 (1962). Copyright © 1962 by the American Chemical Society. Reprinted by permission of the copyright owner.

12.3.2.   Apolar Complexes

In aqueous solution, complexes are formed between hydrophobic, usually hydrocarbon-containing, compounds. We will consider the effects of some of these complexes on the kinetics of chemical reactions, with particular emphasis on catalysis. Several different forms of apolar complexes can be discussed; pi molecular complexes, complexes of aliphatic hydrocarbons, and inclusion complexes.

A.   Pi Molecular Complexes

The formation of complexes between aromatic, hydrophilic compounds, and other compounds such as heterocyclics, other aromatic compounds or tetra-cyanoethylene is well established. These complexes are called pi-complexes, donor–acceptor complexes, or charge transfer complexes.

   If one is to make any generalization on the effect of pi molecular complexes on the rates of organic reactions, it would be that such complexes decelerate rather than accelerate reaction. This inhibition can be viewed in terms of Eq. (12.16). That is, the complexed substrate does not react to give product while the uncomplexed one does. These complexes are of course nonproductive ones. The simplest explanation of the inhibition of reaction through complex formation is that complex formation prevents access of some reagent to the substrate and thus prevents reaction.

   Inhibition by pi molecular complex formation includes the inhibition of the alkaline hydrolysis of ethyl p-aminobenzoate by caffeine. Likewise, the alkaline hydrolysis of methyl *trans*-cinnamate is inhibited by several imidazole, purine, and xanthine compounds such as imidazole, benzimidazole, purine, uracil, theophylline, caffeine, and guanine. Finally, the hydrolysis and aminolysis of *trans*-indoleacryloylimidazole and p-nitrophenyl *trans*-indoleacrylate are inhibited by formation of a pi molecular complex with 3,5-dinitrobenzoate ion.[10]

The kinetics of all three sets of reactions are consistent with a scheme in which the uncomplexed ester can be hydrolyzed but the complexed ester cannot. In each case, the binding constant of the complex determined kinetically agrees with that determined by independent physical measurement, such as solubility or spectrophotometric measurements, although some systems are more complicated than simple 1:1 complexes.

A charge transfer description of the indoleacrylate ester-3,5-dinitrobenzoate ion complex predicts that the complexed ester would be more reactive toward nucleophiles because of the loss of electron density to the complexing agent. The fact that the opposite is observed can be explained in terms of steric interference to reaction on complexation.

A slight rate enhancement has been found in the alkaline hydrolysis of 4-nitrophthalimide when acenaphthene is used as complexing agent. The charge transfer description of this complex would predict a higher electron density in the 4-nitrophthalimide complex and thus a lower rate constant of reaction. These two observations suggest that the charge transfer explanation might not be completely correct for these systems.

However, a charge transfer description of complex formation may afford a ready explanation for the enhancement of acetolysis of 2,4,7-trinitro-9-fluorenyl p-toluene-sulfonate (**12.2**) [Eq. (12.23)]

$$(12.23)$$

by the addition of aromatic donor molecules such as anthracene or phenanthrene. The 1:1 complex of **12.2** with phenanthrene is 21–27 times more reactive than uncomplexed **12.2** at various temperatures.[11] On the other hand, hexaethylbenzene shows no rate enhancement, presumably because it cannot form a complex. If the carbonium ion-like transition state of this reaction is complexed to a greater extent than is the ground state, the reaction rate should be higher. This attractive possibility, however, is difficult to reconcile with the fact that the rate enhancement is due solely to the *activation entropy* (Table 12.1).

Pi molecular complex formation also apparently leads to rate enhancement in nucleophilic displacement (Table 12.2).

Phenoxide ion is a threefold better nucleophile than hydroxide ion toward benzyldimethylsulfonium ion, although it is a slightly poorer nucleophile toward

Table 12.1.   Activation Parameters of the Acetolysis of 12.2

Substrate	$\Delta H^{\ddagger}$ (kcal/mole)	$\Delta S^{\ddagger}$ (e.u.)
Free **12.2**	25.7	-11.7
The **12.2**-phenanthrene complex	26	-5

Source: A. K. Colter, S. S. Wang, G. H. Megerle, and P. S. Ossip, *J. Am. Chem. Soc.*, **86**, 3106 (1964). © 1964 by the American Chemical Society. Reprinted with permission by the copyright owner.

methyl bromide and trimethylsulfonium ion. This inversion of relative nucleophilicity is attributed to complex formation between phenoxide ion and benzyldimethylsulfonium ion, which stabilizes the transition state of the reaction. Methyl bromide and trimethylsulfonium ion do not form π-complexes with phenoxide ion, whereas benzyldimethylsulfonium ion does.[12]

Imidazole catalyzes the reversible transamination of aminophenylacetic acid by pyridoxal in aqueous solution near neutrality. The catalysis is associated with the tautomerism between the two imines [aldimine (**12.3**) and ketimine (**12.4**) in Eq. (12.24)].

(12.24)

The most interesting aspect of the catalysis is the change in the dependence of catalytic rate (the rate of the formation of the intermediate **12.4** from an amino acid and pyridoxal) from the square of the total imidazole concentration at low imidazole concentrations to independence of the total imidazole concentration at high imidazole concentrations (Fig. 12.6). However, morpholine and carbonate buffers showed measurable catalyses only at high concentrations of reactants. The efficiency of imidazole as catalyst was ascribed to its ability to

Table 12.2.   Relative Rates of Nucleophilic Displacement Reactions

Reactions	Relative rates
$\phi O^{\ominus} + \phi CH_2 S^{\oplus} Me_2$	3.4
$HO^{\ominus} + \phi CH_2 S^{\oplus} Me_2$	1
$\phi O^{\ominus} + (CH_3)_3 S^{\oplus}$	0.5
$HO^{\ominus} + (CH_3)_3 S^{\oplus}$	1
$\phi O^{\ominus} + MeBr$	0.7
$HO^{\ominus} + MeBr$	1

Source: C. G. Swain and L. J. Taylor, *J. Am. Chem. Soc.*, **84**, 2456 (1962). © 1962 by the American Chemical Society. Reprinted with permission of the copyright owner.

form complexes, as shown in the kinetics, with reactants and intermediates, as shown in Eq. (12.25) and Fig. 12.6.[13]

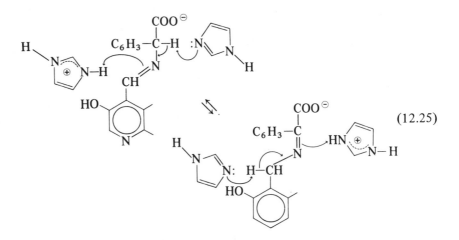

$$(12.25)$$

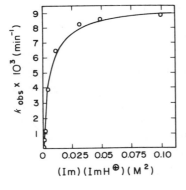

Fig. 12.6.   Dependence of the rate constant of the appearance of intermediate **12.4** on the product of imidazole and imidazolium ion concentrations in the transamination by pyridoxal. From J. C. Bruice and R. M. Topping, *J. Am. Chem. Soc.*, **85**, 1488 (1963). Copyright © 1963 by the American Chemical Society. Reprinted by permission of the copyright owner.

B.   COMPLEXES OF ALIPHATIC HYDROCARBONS

The reason that alkanes aggregate in aqueous solution is largely because such an assembly represents a minimal disruption of the interaction of water molecules. The internal cohesion of water is likely to be the main reason for the low solubility of apolar solutes in water. The implication then is that water forces the hydrocarbons together and not that they attract each other. However, this is not to say that the dipoles of polar hydrocarbons don't play a role in their aggregation in aqueous solution—in fact, they do.

In an attempt to provide a more quantitative explanation of why hydrocarbons interact in solution, let us provide a thermodynamic picture of such aggregation. In the process of solubilizing a nonpolar solute in water, we must first separate a number of water molecules from each other to provide a "hole" for the solute. This is associated with a positive ΔH. Once inserted in this hole, the water molecules surround the solute in such a way as to optimize their hydrogen bonding. Under normal circumstances, that is, in the absence of such solutes, water molecules can maintain their full complement of four hydrogen bonds by assuming a variety of different orientations. However, since hydrogen bonding between the water molecules and the alkanes cannot occur, the number of possible orientations in which the water molecules can maintain full hydrogen bonding is limited; that is, a decrease in entropy, ΔS, occurs. The water around the hydrocarbon thus becomes more structured.

So far, then, the insertion of the hydrocarbon into an aqueous environment has been associated with a positive ΔH and a negative ΔS. However, even though creating the solvent "hole" cost us some enthalpy, the hydrogen bonds reform around the solute although in a more restricted way. And we regain most of our enthalpy losses. The net free-energy change for this process is still positive because of the negative entropy component in the process.

Imagine a situation with many such hydrocarbon molecules in aqueous solution. If they were to associate, "shedding" the structured, high entropy water between the adjoining faces (water associated with a negative ΔS), this process would thus be more favorable than the original insertion of the hydrocarbon into water. Furthermore, if the molecules have appropriate dipoles, the ΔH of association will be negative, making the process even more favorable. Thus one can potentially explain a favorable interaction between organic molecules in aqueous solution.

C.   INCLUSION COMPLEXES

Inclusion complexes possess high stability constants and potentially high stereospecificity as well, since the binding involves multiple interactions. These complexes resemble enzymatic complexes in these important respects; therefore, it is worthwhile to consider in detail the catalytic properties of inclusion complexes.

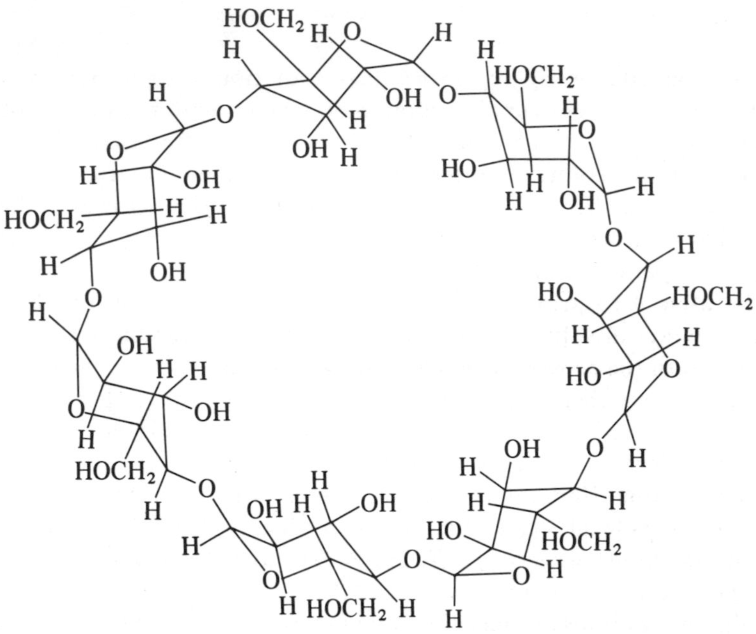

Fig. 12.7. Cycloheptaamylose using skeletal models.

Cycloamyloses. In 1891, a group of unusual nonreducing oligosaccharides was isolated from *Bacillus macerans* grown on a medium rich in amylose. However, it was not until sometime later that the definitive structural elucidation of these compounds was accomplished, showing them to be *cyclic* oligosaccharides containing from six to twelve α-1,4 linked glucose units[14] (Fig. 12.7 and Table 12.3).

The important structural features in these compounds are their toroidal shape, hydrophobic cavity, outer surface, and hydrophilic faces (Fig. 12.8). Because of apparent lack of free rotation about the glycosidic bond connecting the glucose units, the cycloamyloses are not perfectly cylindrical molecules but are cone-shaped. The 6-hydroxyl face is the narrow side, while the 2,3-hydroxyl

Table 12.3. Molecular Dimensions of Cycloamyloses

Cycloamylose	Number of Glucose Residues	Cavity Dimensions (Å)	
		Diameter	Depth
Cyclohexaamylose (α)	6	4.5	6.7
Cycloheptoamylose (β)	7	~7.0	~7.0
Cyclooctaamylose (γ)	8	~8.5	~7.0

Source: M. L. Bender and M. Komiyama, *Cyclodextrin Chemistry*, Springer Verlag, Berlin, 1978.

Fig. 12.8. Front and side view of cyclohexaamylose using space-filling models.

face is somewhat wider. X-ray studies have provided information about the diameters of several of these oligosaccharides (Table 12.3).

The most interesting characteristic of these cyclic oligosaccharides is their ability to complex a variety of guest molecules in their cavities.[4] These guest molecules range from small noble gases to large acyl coenzyme A compounds; the stability of the resulting complexes vary with the size of both the guest and the host. If a substrate is too large, it simply will not fit into the cavity; therefore, it cannot bind to the cycloamylose. Conversely, if the substrate is too small, it will pass in and out of the cavity with little apparent binding.

The fact that the guest molecule is actually contained within the cavity of the host was first shown by X ray.[8] However, further proof was required before it could be established that this was the case in solution. This was provided by NMR data. The 3 and 5 methine protons of the glucose units of the cycloamyloses point inside the cavity, while the 2 and 4 methine protons point outside the cavity. Only the 3 and 5 protons and not the 2 and 4 protons, as observed in their NMR spectra, are strongly shielded on complexing an aromatic substrate, indicating that they are in the magnetic field of the aromatic pi electron cloud (Fig. 12.9).

For such shielding to occur, of course, the aromatic substrate must be contained *within* the cavity and in a geometry in which the H-6 methylenes of cyclohexaamylose are not in the aromatic pi cloud. This signifies only partial penetration of the cavity by the substrate.

The essential steps in catalyses by cycloamylose (C) takes place through inclusion complex formation of substrates (S) with cycloamylose. Thus the reaction scheme is given by

$$S + C \underset{K}{\rightleftharpoons} C \cdot S \xrightarrow{k_{cat}} \text{Products}$$

$$\downarrow k_{un}$$

$$\text{Products}$$

(12.26)

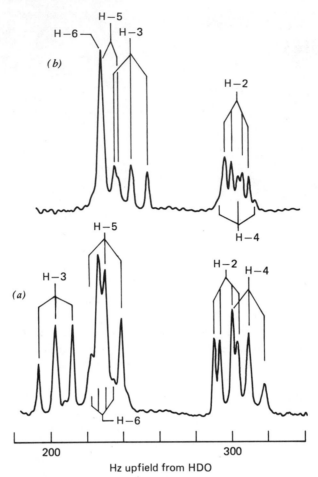

Fig. 12.9. 1_H correlation NMR spectra, 250 MHz, of (a) 0.005 M cyclohexaamylose and (b) 0.005 M sodium p-nitrophenolate complex. Both samples were prepared using pD 11 phosphate buffer in D_2O. The resolution in both spectra was digitally enhanced; spectrum (b) is presented with a smaller vertical scaling factor. From R. J. Bergeron and R. Rowan, *Bioorganic Chem*, **5**, 425 (1975).

which is identical with Eq. (12.17) except that this reaction scheme contains an uncatalyzed reaction (k_{un}). This was shown by the following facts:

1.   The reaction rate is not a linear function of cycloamylose concentration, but approaches a maximum value asymptotically with an increase of cycloamylose concentration.

2.   Competitive inhibition was found by addition of organic compounds, which competitively bind in the cycloamylose cavity.

3. α-Methyl glucoside, a monomolecular analog of cycloamyloses, exhibited a much smaller effect on the reaction rates.

Equation (12.26) is identical with that in many enzymatic reactions. The rate constant k_{cat} and the dissociation constant of the C · S complex, K, which is equal to k_{-1}/k_1 can be determined by the use of Lineweaver–Burk-type plots [Eq. (12.27)] under conditions that $[C]_0 \gg [S]_0$, where $[C]_0$ and $[S]_0$ are the initial concentrations of cycloamylose and substrate, respectively. Plots of

$$\frac{1}{k_{obs} - k_{un}} = \frac{K}{(k_{cat} - k_{un})[C]_0} + \frac{1}{k_{cat} - k_{un}} \tag{12.27}$$

$1/(k_{obs} - k_{un})$ versus $1/[C]_0$ give values of k_{cat} and K from the intercept and slope, respectively. When $[S]_0 \gg [C]_0$, $[C]_0$ in Eq. (12.27) is simply replaced by $[S]_0$. Fig. 12.10 shows typical plots using Eq. (12.27).

Cycloamylose-catalyzed reactions show many of the kinetic features shown by enzymatic reactions, including saturation, and stereospecific catalyses, as well as substrate–catalyst complex formation and competitive inhibition already described. Thus cycloamyloses serve as models of many enzymes.

Cycloamylose-catalyzed reactions can be classified in the following two categories:

1. Covalent catalyses, in which cycloamyloses catalyze reactions via formation of covalent intermediates.

2. Noncovalent catalyses, in which cycloamylose provide their cavities as apolar or sterically restricted reaction fields without the subsequent formation of any covalent intermediates.

Table 12.4 lists the reactions accelerated by cycloamyloses together with the kind of catalysis involved. The classic example of a cycloamylose catalyzed reaction is seen in the hydrolysis of nitrophenyl acetates.

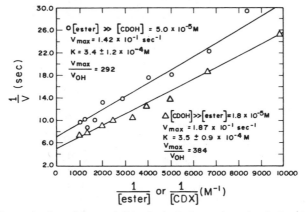

Fig. 12.10. Determination of k_{cat} and K in the hydrolyses of m-t-butyl-phenyl acetate catalyzed by cycloheptaamylose: $\bigcirc$, (ester) $\gg$ (catalyst); $\triangle$ (catalyst) $\gg$ (ester); 10% acetonitrile–water, pH 10.6. From M. L. Bender, *Trans. NY Acad. Sci.*, **29**, 301 (1967).

Table 12.4.   Reactions Accelerated by Cycloamyloses

Reactions	Substrates	Acceleration factor[a]	Kind of catalysis[b]
Cleavage of esters	Phenyl esters	300	C
	Mandelic acid esters	138	U
Cleavage of amides	Penicillins	89	C
	N-Acylimidazoles	50	C
	Acetanilides	16	C
Cleavage of organophosphates	Pyrophosphates	>200.	C
	Methyl phosphonates	66.1	C
Cleavage of carbonates	Aryl carbonates	7.45	C
Cleavage of sulfates	Aryl sulfates	18.7	N
Intramolecular acyl migration	2-Hydroxylmethyl-4-nitrophenyl trimethylacetate	6	N
Decarboxylation	Cyanoacetate anions	44.2	N
	α-Ketoacetate anions	3.95	N
Oxidation	α-Hydroxyketones	3.3	N

Source: M. L. Bender and M. Komiyama, *Cyclodextrin Chemistry*, Springer Verlag, Berlin, 1978.
[a] Ratio of the rate catalyzed by cyclodextrin to the uncatalyzed reaction rate.
[b] C, N, and U, respectively, refer to covalent catalysis, noncovalent catalysis, and unknown.

In these reactions there is a rate differentiation between *meta-* and *para-* substituted esters independent of electronic effects, a manifestation of steric effects. Furthermore, since simply alkaline hydrolysis does not show this effect, the steric effect must be associated with the influence of the cycloamylose.

Table 12.5 shows the values of k_{cat} and K, obtained from plots such as those depicted in Fig. 12.10. The binding constants vary from 10^{-2} to $10^{-3} M$, values not far different from some enzymatic binding constants. The rate constants are normalized in the middle column of Table 12.5 by the uncatalyzed rate constants,

Table 12.5.   Catalytic Rate Constants and Accelerations in the Cyclohexaamylose-Catalyzed Hydrolysis of Phenyl Acetates[a]

Acetate	$k_{cat} \times 10^2 (sec^{-1})$	k_{cat}/k_{un}	$K \times 10^3 (M)$
p-t-Butylphenyl	0.075	1.2	7.7
p-Tolyl	0.26	4	14
ortho-Tolyl	1.4	37	48
Phenyl	5.5	69	70
m-Chlorophenyl	26	140	6
m-Tolyl	11	160	34
m-t-Butylphenyl	13	260	2.0
3,5-Dimethylphenyl	14	250	20

Source: M. L. Bender, *Trans. NY Acad. Sci.*, **29**, 301 (1967).
[a] 25°C 1.5%, (v/v) acetonitrile–water, pH 10.6.

thus showing the maximal accelerations imposed by the cycloamylose. These accelerations vary from 20% for *p-t*-butylphenyl acetate to 260-fold for *m-t*-butylphenyl acetate, again showing the clear specificity for *meta*-substituted compounds. The explanation for this specificity must reside in complex formation.

On the basis of binding of the guest molecule within the cavity of the host, and more particularly, the binding of the apolar (hydrocarbon) portion of the guest within the cavity of the host, one may attempt to explain the specificity between *meta*- and *para*-substituted phenyl acetates. Several facts should be kept in mind: (1) *meta*/*para* specificity is greatest for substituents of largest bulk; (2) the reactivity of an unsubstituted ester is midway between the *meta*- and *para*-substituted compounds; thus the *para*-substituted compounds have a negative specificity while the *meta*-substituted compounds have a positive specificity. From these considerations, the models of Fig. 12.11 were assembled. The difference seen is that the carbonyl linkage of the *para*-ester (small acceleration) is a considerable distance from the cycloamylose moiety, whereas the carbonyl linkage of the *meta*-ester (large acceleration) is contiguous to the ring of secondary hydroxyl groups surrounding the cavity of the cycloamylose. Hence the acceleration must stem from a stereospecific interaction of one of the hydroxyls of the cycloamylose with the carbonyl group of the ester (stereospecific binding).

A hydroxyl group can accelerate reaction by general base, general acid, or nucleophilic catalysis. Nucleophilic catalysis is indicated here on the basis of experiments on the hydrolysis of *m*-nitrophenyl benzoate with cyclohexaamylose. Observations of this reaction at 390 nm show a rapid liberation of *m*-nitrophenol (approximately 20 sec for completion), whereas observations at

Fig. 12.11. Pauling–Corey–Koltun models of the α-cyclohexaamylose-*p-t*-butylphenyl acetate complex (left) and the α-cyclodextrin-*m-t*-butylphenyl acetate complex (right). From M. L. Bender, *Trans. NY Acad. Sci., Series II,* **29,** No. 3, 301 (1967). Copyright © The New York Academy of Sciences 1967. Reprinted by permission.

245 nm show the slow formation of benzoate ion (thousands of seconds for complete reaction). Thus the reaction must proceed in two steps with the formation of m-nitrophenol preceding the formation of benzoate ion. This conclusion is reinforced by kinetic experiments, which show that the rate constants for the liberation of benzoic acid from the three benzoate esters, m-t-butylphenyl, m-chlorophenyl, and m-nitrophenyl benzoate are identical, indicating the formation of a common intermediate whose decomposition is rate determining. This intermediate may thus be identified as benzoyl-cyclohexa-amylose. It has, in fact, been isolated.

The mechanism of the cycloamylose (C) reaction can then be given as

$$
\underset{}{C} + R\overset{\displaystyle O}{\overset{\|}{-C}}-OAr \rightleftharpoons C \cdot R\overset{\displaystyle O}{\overset{\|}{-C}}-OAr \xrightarrow{-ArO^{\ominus}}
$$

$$
\underset{}{C}\overset{\displaystyle O}{\overset{\|}{-C}}-R \xrightarrow{OH^{\ominus}} C + R\overset{\displaystyle O}{\overset{\|}{-C}}-O^{\ominus} \qquad (12.28)
$$

This pathway is formally analogous to that of the chymotrypsin and trypsin reactions. All covalent catalyses by cycloamyloses proceed by this or similar schemes.

In Chapter 10, the proposition was made that intramolecular catalysis is superior to the corresponding intermolecular catalysis. An implication was that this superiority could be carried over to an intracomplex catalysis of an enzyme–substrate reaction. This prediction can be tested by comparing the intracomplex reaction of the cyclohexaamylose-m-t-butylphenyl acetate complex with the corresponding intermolecular reaction using the lower limit of the interconversion between corresponding intermolecular and intramolecular reactions developed in Chapter 10 (10 M). As shown in Table 12.6, the calculated intracomplex rate constant (48 sec^{-1}) is in reasonable agreement with the experimentally observed k_{cat} (lim) of 13 sec^{-1}.

Thus, in at least one instance, the intramolecular–intermolecular comparison

Table 12.6.   Kinetic Factors Responsible for the Difference in Rate of Liberation of m-t-Butylphenol from m-t-Butylphenyl Acetate by Hydroxide Ion and Cyclohexaamylose

Rate constant of hydroxide ion catalysis	1.2 M^{-1} sec^{-1}
Conversion to rate constant of alkoxide ion reaction of pK 12.1 (fourfold)	4.8 M^{-1} sec^{-1}
Conversion to an intramolecular reaction from an intermolecular reaction (10 M)	48 sec^{-1}
Experimental cyclohexaamylose k_{cat} (lim)[a]	13 sec^{-1}

Source: M. L. Bender, *Trans. NY Acad. Sci.*, **29**, 301 (1967). © New York Academy of Sciences. Reprinted by permission.
[a] The limiting k_{cat} value corresponding to the complete ionization of the hydroxyl groups of cyclohexaamylose.

in Chapter 10 is mirrored by a quite similar intracomplex–intermolecular comparison.

Since the cyclohexaamylose and chymotrypsin pathways are formally identical, it is of interest to determine the relationship of each of their catalytic rate constants to the rate constants for the alkaline hydrolysis of the same substrates. Table 12.7 shows such calculations. In each set, the second-order hydroxide ion rate constants are compared to the second-order catalytic complexation process, which may be defined as k_{cat}/K (the rate constants when $[S]_0 \gg K$ or $[E]_0 \gg K$). The cycloamylose reactions are seen to be 5×10^3–10^5 times faster than hydroxide ion reactions while the chymotrypsin reactions are 3×10^4–10^6 times faster than hydroxide ion reactions. The conclusion that may be drawn from these comparisons is that the cycloamylose reactions show roughly the same rate enhancement with respect to hydroxide ion reactions as do the chymotrypsin reactions. There is one proviso, however. The rate constant for the chymotrypsin reaction was determined at pH 8, its maximum, whereas the rate constant for the cycloamylose reaction was determined at pH 13, its maximum. Therefore, the two sets are not strictly comparable.

Recently it was discovered that cycloamyloses form complexes in dimethyl sulfoxide-water solution as well as water. This allowed considerable latitude in finding an organic substrate for complexation and catalysis by cycloamyloses. It

Table 12.7.   A Comparison of Second-Order Rate Constants

Cycloamylose + m-t-butylphenyl acetate		
α-C + m-t-BPA[d]	$6.5 \times 10^3\ M^{-1}\sec^{-1}$	$M^{-1}\sec^{-1}$
$OH^{\ominus}$ + m-t-BPA[d]	$1.2\ M^{-1}\sec^{-1}$	
α-C/$OH^{\ominus}$		1.7×10^3
β-C + m-t-BPA[d]	$3.1 \times 10^4\ M^{-1}\sec^{-1}$	
$OH^{\ominus}$ + m-t-BPA[d]	$1.2\ M^{-1}\sec^{-1}$	
β-C/$OH^{\ominus}$		2.5×10^4
Chymotrypsin + Substrates		
Chymotrypsin + ATrA[a]	$12.6\ M^{-1}\sec^{-1}$	
OH^- + ATrA	$3 \times 10^{-4}\ M^{-1}\sec^{-1}$	
Chymol/OH^-		4×10^4
Chymotrypsin + ATrEE[b]	$4 \times 10^5\ M^{-1}\sec^{-1}$	
OH^- + ATrEE	$0.61\ M^{-1}\sec^{-1}$	
Chymo/OH^-		$\sim 10^6$
Chymotrypsin + ATyEE[c]	$1.2 \times 10^4\ M^{-1}\sec^{-1}$	
$OH^{\ominus}$ + ATyEE	$0.45\ M^{-1}\sec^{-1}$	
Chymo/$OH^{\ominus}$		3×10^4

Source: M. I. Bender, *Trans. NY Acad. Sci.*, **29**, 301 (1967). © The New York Academy of Sciences.
[a] Acetyl-L-tryptophanamide.
[b] Acetyl-L-tryptophan ethyl ester.
[c] Acetyl-L-tyrosine ethyl ester.
[d] m-t-butylphenyl acetate

was then found that very fast acylation of cycloheptaamylose by bound *p*-nitrophenyl ferrocinnamate (ferrocene acrylate) (**12.5**) occurred in dimethyl-sulfoxide–water mixture (60/40 v/v).

12.5

In the cycloamylose-**12.5** complex, the acyl group of **12.5** rather than the leaving group is bound to the cycloamylose. The acylation of cycloheptaamylose by **12.5** is accelerated by 750,000 fold and the rate achieved is comparable with that for acylation of the enzyme chymotrypsin by *p*-nitrophenyl acetate. The much larger acceleration of the cleavage of **12.5** by cycloamylose than that of *m*-*t*-butylphenyl acetate by cycloheptaamylose (260-fold) is explained in terms of the structural change of the complex from the initial state to the transition state. Molecular models suggest that **12.5** can go to the tetrahedral intermediate with full retention of the optimum binding geometry in the cycloamylose cavity, whereas this is not the case for the reaction of *m*-*tert*-butylphenyl acetate (the aromatic ring is pulled somewhat out of the cavity).[15]

Catalyses by cycloamyloses do not always involve formation of covalent intermediates. Instead, cycloamyloses can simply provide an apolar and sterically restricted cavity for the included substrate, whose cavity can then serve as reaction medium, resulting in acceleration or deceleration of reaction. These catalyses are defined as noncovalent catalyses. The reaction scheme for noncovalent catalyses is identical with that for covalent catalyses [Eq. (12.28)] except that they do not involve covalent intermediates.

Noncovalent catalyses by cycloamyloses are attributable to (1) microsolvent effects due to the apolar character of the cycloamylose cavity, and (2) conformation effects due to the geometric requirements of inclusion. Hydrogen bonding between the substrate and the hydroxyl groups of the cycloamylose can be considered to be a part of the microsolvent effect.

Since the cavities of cycloamyloses are spatially restricted, cycloamyloses can include one conformational isomer of a substrate as a guest more favorably than other conformational isomers. When the conformer, which is more favorably included, is more reactive than the other conformers, cycloamyloses exhibit acceleration; the opposite also holds true.

Conversion from **12.6** to **12.7** through intramolecular acyl migration is accelerated (sixfold) by cyclohexaamylose through this conformational effect [Eq. (12.29)]. The reaction proceeds through intramolecular nucleophilic attack by the hydroxyl group on the carbonyl group, which is more probable in (**12.6b**) than in (**12.6a**). The acceleration by cycloamylose has been shown to follow an increase in activation entropy ($\Delta\Delta S^{\ddagger} = 4.3$ e.u.); there is no appreciable change in activation enthalpy ($\Delta\Delta H^{\ddagger} = 0.2$ kcal/mole). These values are consistent with the freezing of a rotational degree of freedom. Thus a portion of the free energy gained from the formation of an inclusion complex is used to increase the population of the reactive isomer (12.6b).

$$(12.29)$$

Cycloamyloses modified by other functional groups can be good catalysts, since cycloamyloses and the functional groups attached thereto provide both a binding site and a catalytic site. For example, introduction of two imidazolyl groups on the primary side made cycloheptaamylose a good model of the enzyme, ribonuclease,[16] since the imidazolyl groups worked cooperatively both as a general acid catalyst (in the acidic form) and as a general base catalyst (in the neutral form).[16] A bell-shaped pH-rate profile was seen in the hydrolysis of a synthetic cyclic phosphate by cyclohexaamylose bis(imidazole) [Eq. (12.30)] (like the enzyme ribonuclease), while a smaller sigmoid curve was seen with a monoimidazole derivative. This result can be interpreted in terms of the mechanism shown in Eq. (12.30), in which the substrate is bound in the cavity of the cycloamylose derivative; subsequently one imidazole acts as a general base while the other acts as a general acid.

$$(12.30)$$

Another enzyme that has been successfully mimicked with a cycloamylose derivative is carbonic anhydrase. To do this, a compound that would bind CO_2, complex a zinc ion, and in addition function as general base was needed. To accomplish all of this, bis(N-histamino)cycloheptaamylose (**12.8**) and bis(N-imidazolyl)cycloheptaamylose (**12.9**) were devised (12.31).

(12.31)

12.8   **12.9**

The complexes of **12.8** and **12.9** with zinc ion showed larger rates of hydration of carbon dioxide than the compounds without the cycloamylose derivative, although the absolute catalytic rate constants are lower than their enzymatic counterparts by many orders of magnitude, as shown in Table 12.8.

Still another example of a cycloamylose enzyme mimic involves a cyclo-aheptaamylose–pyridoxamine artificial enzyme.[29] In general, any coenzyme can be attached to cycloamylose leading to a facile reaction, the latter providing the binding and the former providing the chemistry. When pyridoxamine was covalently attached to cycloheptaamylose, indolepyruvic acid was converted to tryptophan 200 times faster than its reaction with pyridoxamine alone.

From all these examples, it seems clear that cycloamylose-based catalysts have the potential to act as artificial enzymes with accelerations of enzymatic magnitudes.

Table 12.8.   Rates of Hydration of CO_2 Catalyzed by Imidazole Derivatives (pH 7.50 at 25°C)[a]

Catalyst	$10^{-3} M$	Second order k_{cat} $(M^{-1} sec^{-1})$
Imidazole	4.0	
Histamide	4.0	14.9
(Imidazole)$_2$Zn[11]	2.0	2.0
10-Zn[11]	2.0	16.2
(Histamine)$_2$Zn[11]	2.0	57.9
9-Zn[11]	2.0	166
Human carbonic anhydrase B (k_{cat}/K_m)		1×10^7

Source: I. Tabushi, Y. Kuroda, and A. Mochizuki, *J. Am. Chem. Soc.*, **102**, 1152 (1980). Copyright © 1980 by the American Chemical Society. Reprinted by permission of the copyright owner.
[a]The initial concentration of CO_2 was $1.25 \times 1^{-2} M$;
[b]Negligibly small.

12.3.3. Polymeric Complexes

Since enzymatic catalysts are polymers and show multiple interactions toward their substrates, the use of polymeric catalysts that can also show multiple interactions toward substrates has been of considerable interest.

The reasons for the generally observed superiority of polymeric over monomeric catalysts are:

1. The local concentration of catalytic groups are larger on the polymer than in the bulk of the solution. Therefore, if the substrate is bound to the polymer, its reaction should be catalyzed at a rate larger than by an equivalent, homogeneously dispersed catalyst.

2. The high local concentration of the catalytic groups can cause multiple catalysis, which is also a factor in effective catalysis (Chapter 11).

3. The electric field of the polymeric catalyst can stabilize (destabilize) the transition state, or increase (decrease) the local concentrations of charged species near the polymer through coulombic interactions.

4. The polymeric catalyst fixes the substrate at a position suitable for catalysis through multiple interactions between the substrate and the catalyst.

Apolar binding of the substrate to the polymer is a factor of considerable importance in this kind of catalysis. For example, the hydrolysis of butyl acetate by a homogeneous polymeric sulfonic acid is tenfold faster than by a comparable concentration of hydrochloric acid, but the hydrolysis of ethyl acetate shows no rate enhancement.

Electrostatic interactions between substrate and polymer also lead to binding. A water soluble polystyrenesulfonic acid hydrolyzes 2-amino-2-deoxy-β-D-glucopyranoside hydrochloride thirtyfold and the diethylaminoethyl ether hydrochloride of starch twentyfold faster than an equivalent concentration of hydrochloric acid. These specific catalyses are likely due to one factor: the selective binding of the cationic substrates to the polymeric catalyst.

In special circumstances, pi molecular complexes can lead to binding of substrate to an ion exchange resin. Although the rate of hydrolysis of propyl acetate decreases monotonically by increasing the concentration of silver ion on a sulfonic acid resin, the rate of hydrolysis of allyl acetate goes through a maximum at about 50% as the concentration of silver ion on the resin increases. A twofold specificity for the hydrolysis of the olefinic ester is produced because of its increased concentration near the polymer surface.

A partially protonated poly(4-vinylpyridine) in ethanol–water solution was shown to serve as a particularly effective catalyst, relative to 4-picoline, nonprotonated polymer, or highly protonated polymer, for the solvolysis of a nitrophenyl ester substrate bearing a negative charge. Toward the neutral substrate, 2,4-dinitrophenyl acetate, the monomer is a better catalyst than the

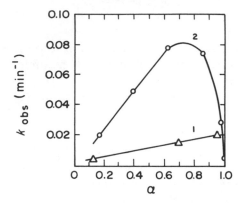

Fig. 12.12. Solvolysis of potassium 3-nitro-4-acetoxybenzenesulfonate catalyzed by 0.0157 M 4-picoline, $\triangle$ (curve 1) and by 0.010 base M poly(4-vinylpyridine), $\bigcirc$ (curve 2). From R. L. Letsinger and T. J. Savereide, *J. Am. Chem. Soc.*, **84**, 114 (1962). Copyright © 1962 by the American Chemical Society. Reprinted by permission of the copyright owner.

polymer, perhaps because of a steric effect in the latter. However, the polymer is a better catalyst than the monomer toward the anionic ester, 3-nitro-4-acetoxy-benzenesulfonate ion, at essentially all degrees of ionization (α), (Fig. 12.12).[17] The catalytic activity approaches a maximum when the polymer is partially protonated (very close to a bell-shaped curve, implying that both protonated pyridinium ions and neutral pyridine groups are necessary for this catalysis). The pyridine groups serve as nucleophilic catalysts, while the pyridinium cations serve as electrostatic binding agents increasing the local concentration of the anionic substrate in the region of the polymeric coil. At its maximum, the polymer is a ninefold better catalyst than the monomer toward 3-nitro-4-acetoxybenzenesulfonate ion. The asymmetry of the rate constant-α (degree of protonation)-profile is attributed to an unfavorable conformational change (for catalysis) when the acidity of the medium increases.

Multiple catalysis by two functional groups in different ionic states is seen in the hydrolysis of p-nitrophenyl acetate by poly-4(5)-vinylimidazole at high pH. The reaction is first order with respect to the polymeric catalyst. Imidazolyl groups can exist in three states, that is, cationic, neutral, and anionic [Eq. (12.32)]. As shown in Fig. 12.13, the rate constant of the hydrolysis catalyzed by

$$ \text{HN}\overset{\oplus}{\text{NH}} \underset{}{\overset{K_1}{\rightleftarrows}} \text{HN}\quad\text{N} \underset{}{\overset{K_2}{\rightleftarrows}} \text{N}\overset{\ominus}{\quad}\text{N} \qquad (12.32) $$

$$ \alpha_{\oplus} \qquad\qquad\qquad \alpha_0 \qquad\qquad\qquad \alpha_{\ominus} $$

the polymeric catalyst exponentially increases with the increase of the fraction of neutral and anionic species (α), although that catalyzed by the monomeric imidazole increases only linearly. This effect was attributed to cooperation (multiple catalysis, Chapter 11) of a neutral imidazolyl group and an anionic imidazolyl group, which are in proximity to each other on the polymer. While the neutral imidazoles could assist the attack of the anions behaving as a weak acid (**12.10**), another possibility, that is, a nucleophilic attack of neutral imidazole catalyzed by the anions as general bases (**12.11**) looks more reasonable.[18]

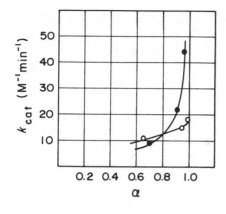

Fig. 12.13. Solvolysis of *p*-nitrophenyl acetate catalyzed by poly-4(5)-vinylimidazole (●) and imidazole (O). From C. G. Overberger, T. St. Pierre, N. Vorchheimer, J. Lee, and S. Yaroslavsky, *J. Am. Chem. Soc.*, **87**, 296 (1965). Copyright © 1965 by the American Chemical Society. Reprinted by permission of the copyright owner.

(12.33)

12.10

12.11

Cooperation is also found between two functional groups in copolymeric catalysts. At high pH values, the copolymer of 4(d5)-vinylimidazole and *p*-vinylphenol is a more superior catalyst toward *p*-nitrophenyl acetate than any of its monomeric or polymeric analogs (Table 12.9). This is attributed to

Table 12.9.   The First-Order Rate Constant of Solvolysis of *p*-Nitrophenyl Acetate

	$k_{obs} \times 10^4$ (min^{-1})		
pH	Copolymer of 4(5)-Vinylimidazole and *p*-Vinylphenol	Poly-4(5)-vinylimidazole	Imidazole
7.4	3.0	2.1	2.6
8.2	5.1	3.0	2.4
9.1	28.6	3.2	2.7

Source: C. G. Overberger, J. C. Salamone, and S. Yaroslavsky, *J. Am. Chem. Soc.*, **89**, 6231 (1967). In 80% Ethanol-H$_2$O, I = 0.02 M.

bifunctional catalysis by imidazole and phenoxide ion. One probable mechanism is the nucleophilic attack of imidazole on the substrate, followed by reaction of phenoxide ion acting as a general base on the tetrahedral intermediate **12.11**. Considerable rate enhancements were also observed for reactions catalyzed by copolymers of 4(5)-vinylimidazole and vinyl alcohol.

The effect of an electric field in the vicinity of a polymeric catalyst is also important. An anionic polymer such as polystyrenesulfonic acid shows a larger (about tenfold) acceleration of the benzidine rearrangement of hydrazobenzene over that monomeric hydrochloric and benzenesulfonic acids show. This reaction proceeds as shown by Eq. (12.34).

$$\text{⟨O⟩}-NH=NH-\text{⟨O⟩} \xrightarrow{2H^{\oplus}} \text{⟨O⟩}-{}^{\oplus}NH_2-{}^{\oplus}NH_2-\text{⟨O⟩}$$

(12.34)

$$\longrightarrow H_2N-\text{⟨O⟩}-\text{⟨O⟩}-NH_2$$

Acceleration by the anionic polymeric catalyst is due to an increase in the local concentration of monoprotonated hydrazobenzene and protons near the polymer caused by electrostatic attraction.

The hydrolysis of p-nitrophenyl acetate by polymers of incompletely alkylated poly(4-vinylpyridine) are of particular importance mechanistically compared to hydrolysis by the enzyme α-chymotrypsin. The alkylation was carried out by treatment of the polymer with benzyl chloride in a mixture of nitromethane and methanol. The molecular weight of the polymer was approximately 21,000. Kinetic measurements showed that the rate of hydrolysis of p-nitrophenyl acetate by the polymer (calculated per mole of free pyridine nuclei) was 2–2.5 orders of magnitude greater than the rate by the corresponding monomer, 4-ethylpyridine. Separate presteady-state and steady-state portions of the hydrolysis were seen, indicating discrete acylation and deacylation steps such as those in enzymatic reactions. This behavior implies the formation and decomposition of an acyl-polymer intermediate by analogy with an acyl-enzyme intermediate in the enzyme-catalyzed reaction. The rate constant of the formation of the intermediate k_2 and that of its decomposition k_3 are shown in Table 12.10. A comparison of the polymeric catalyst with α-chymotrypsin indicates that in the rate-determining k_3 step, the polymer is only about twentyfold less efficient than the enzyme chymotrypsin.[19]

A remarkable catalysis was found in the synthetic polymer based on poly(ethyleneimine) (PEI), which is water soluble, containing pendant dodecyl groups and methyleneimidazolyl groups for use with an apolar substrate, p-nitrophenyl caproate. Water solubility of a polymeric catalyst is important when an apolar substrate is used, because then a driving force for binding can be found

Table 12.10.   The Hydrolysis of *p*-Nitrophenyl Acetate Catalyzed by Benzylated Poly(4-vinylpyridine) and Chymotrypsin[a]

Catalyst[b]	$k_m(M)$	$k_2(\text{sec}^{-1})$	$k_2(\text{sec}^{-1})$
PC-4	2.2×10^{-4}	1.6×10^{-3}	4.2×10^{-4}
PC-7	1.1×10^{-4}	4.4×10^{-3}	4.0×10^{-4}
PC-10	1.1×10^{-4}	4.4×10^{-3}	4.4×10^{-4}
PC-26	1.1×10^{-4}	4.4×10^{-3}	4.2×10^{-4}
PC-50	1.1×10^{-4}	4.4×10^{-3}	4.1×10^{-4}
α-Chymotrypsin	1.2×10^{-3}	4	8.0×10^{-3}

Source: Y. E. Kirsh, V. A. Kabanov and V. A. Kargin, *Dokl. Akad. Nauk SSSR* **177**, 112 (1967).
[a] Catalysis conditions: for PC, 25° C, pH 8, [Tris—HCl] = 10^{-3} M, l = 10^{-2} M; for α-chymotrypsin, 25° C, pH 8.6, [Tris—HCl] = 2×10^{-2} M, l = 0.2 M.
[b] PC-*n* = polymeric catalyst where *n* is percent benzylation.

by introducing pendant apolar groups on the polymer. Plots of ester concentration, versus time are consistent with a two-step pathway analogous to that of a hydrolytic enzyme such as chymotrypsin, in which an initial acylation burst is followed by a rate-determining deacylation reaction. Preceding all of this is undoubtedly a preequilibrium binding step for which much evidence has been given before. The rate of hydrolysis of *p*-nitrophenyl caproate by these polymers approaches that by the enzyme α-chymotrypsin as is seen in Table 12.11, and thus these polymers have been called synzymes (synthetic enzymes).[20]

The ultimate polymeric catalyst, where the polymer is not water soluble, is one which has a stereospecific cavity for substrate recognition in addition to the catalytic functionalities. The question as to how to generate a stereospecific cavity in these systems has already been successfully addressed. A stereospecific 4-nitrophenyl α-D-pyranoside-2,3-4,6-di-O-(4-vinylphenylboronate) [see Eq.

Table 12.11.   Relative Effectiveness of Various Catalysts in the Cleavage of *p*-Nitrophenyl Caproate[a]

Catalyst	Catalytic Constant k $(M^{-1}\text{ min}^{-1})$
Imidazole	10
α-Chymotrypsin	10,000
PEI-D(10%)-Im(15%)[b]	2,700

Source: I. M. Klotz, G. P. Royer, and I. S. Scarpa, *Proc. Natl. Acad. Sci., USA*, **68**, 263 (1971).
[a] The substrate used was *p*-nitrophenyl caproate, except for the first two reactions, in which *p*-nitrophenyl acetate was used at a pH near neutrality.
[b] D, dodecyl group; Im, methyleneimidazole group.

(12.35)] was copolymerized to generate a macroporous polymer. The 4-nitrophenyl α-D-mannopyranoside was then split out leaving a chiral hole, which was shown to selectively remove one enantiomer from a D,L-pair.[22]

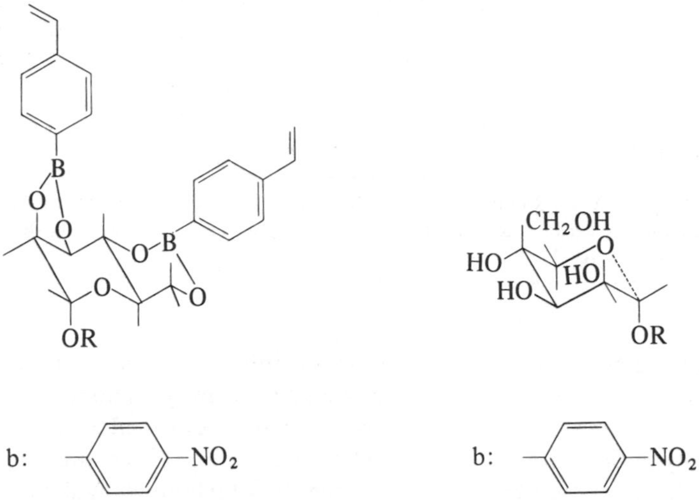

$$(12.35)$$

b: —NO$_2$

b: —NO$_2$

The fundamental assumption of this section is that the proper matrix for bringing catalyst and substrate together is one of the critical factors in devising an efficient catalyst. This assumption implies that the matrix does not take an active part in the catalysis other than holding the substrate and catalyst rigidly in proximity to one another. However a matrix, especially a macromolecular one, such as one of the polymers we have been discussing or an enzyme, can in the binding reaction raise the ground state energy of the substrate by some process other than rigidification, for example, by distortion of the substrate.

Certainly proximity and orientation effects are very important in catalysis. Random collision between molecules is too haphazard a process to bring about specific and efficient catalysis. Direction must be given. This direction can be brought about chemically by stereospecific complexing. At the same time that this direction is being imposed on the substrate, the unfavorable translational entropy of bringing two molecules together must be overcome by any of the forces described earlier. If the complex possesses, in addition to strength, correct stereochemistry between catalyst and substrate, then the favorable entropy of activation associated with intramolecular reactions and the favorable enthalpy associated with catalyses can combine, leading to overall efficiency. In addition, the stereochemistry will result in specificity. These are the two goals of catalysis.

12.3.4. Micellar Complexes

Micellar catalysis can have pronounced effects on reaction rates. Micelles are aggregates of a large number of soap or detergent molecules loosely bound mainly through hydrophobic (apolar) interactions. When the concentration of a detergent is gradually increased in aqueous solution, many physical properties such as surface tension, density, pH, and conductivity also change gradually. However, there is a point where this change is discontinuous; that is, a sudden increase or decrease occurs in the property with a small increase in detergent concentration. The concentration at which this occurs is called the critical micelle concentration or CMC. Micelles are usually formed in aqueous solution; the polar and apolar groups are found at the surface and in the interior of micelles, respectively. However, inverse micelles, surfactant aggregates in apolar solvents, are also known, in which the polar and apolar groups are located in the interior and surface of the micelles, respectively. Micelles bind a variety of organic substrates through apolar interaction, resulting in catalyses of reaction (or sometimes just the deceleration of reaction). Reactions catalyzed by micelles usually take place on the surface of the micelles. Furthermore, micellar catalyses exhibit enzymelike features such as Michaelis–Menten type of kinetics and stereospecificity. Thus reactions in micelles can be used as models of enzymatic reactions.[23]

At least three factors can account for alteration of the rate when the reactant or reactants are incorporated into or onto a micelle: (1) proximity, (2) electrostatic effects; and (3) medium effects. Proximity effects arise if reactants are concentrated or diluted by incorporation into the micellar phase. Alterations in stability or reactants or transition states by electrostatic effects due to the micellar charges or by short-range interactions involving the molecules that constitute the micelle can also affect reaction rates.

The following generalizations may be made about electrostatic effects in micelles. Cationic micelle-forming detergents accelerate the reactions of neutral organic molecules with anionic reagents, but decelerate the reactions of neutral organic molecules with cationic reagents. On the other hand, anionic micelle-forming detergents accelerate the reactions of neutral organic molecules with cationic reagents but decelerate the reactions of neutral organic molecules with anionic reagents. These generalizations hold in a surprisingly large number of instances. For example, the cationic detergent cetyltrimethylammonium bromide (CTAB) accelerates the reactions of dyes with hydroxide ion four- to fiftyfold, and the alkaline hydrolysis of p-nitrophenyl hexanoate up to fivefold. However, the (acid) hydrolysis of methyl orthobenzoate is inhibited by CTAB (Fig. 12.14). On the other hand, anionic detergents such as sodium lauryl sulfate (NaLS) or sodium oleyl sulfate (NaOS) accelerate the acid hydrolysis of methyl orthobenzoate (Fig. 12.14) up to eightyfold.[24]

In the region below the critical micelle concentration of NaLS ($0.0016\,M$) (not shown in Fig. 12.14), the hydrolytic rate constant of methyl orthobenzoate is proportional to the fourth power of the detergent and increases only slowly with increasing detergent concentration. In this region, NaLS is presumably a monomer. The great sensitivity of the rate constant to the detergent concentration suggests that the substrate induces the formation of micelles, containing a 4:1 ratio of NaLS: substrate.

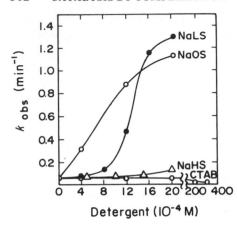

Fig. 12.14. First-order rate constants for the hydrolysis of methyl orthobenzoate in aqueous solution at 25°C and pH 4.76 plotted as a function of the concentration of sodium lauryl sulfate (NaLS) (●), sodium oleyl sulfate (NaOS) (O), sodium heptadecyl sulfate (NaHS) (Δ), and cetyltrimethyl ammonium bromide (CTAB) (☐). From M. T. A. Behme, J. G. Fullington, R. Noel, and E. H. Cordes, *J. Am. Chem. Soc.,* **87**, 266 (1965). Copyright © 1965 by the American Chemical Society. Reprinted by permission of the copyright owner.

The alkaline hydrolysis of methyl 1-naphthoate in 50% dioxane–water was shown to be sensitive to both medium and electrostatic effects. When the reaction is carried out in lauryltrimethylammonium chloride, a slight acceleration is observed, whereas sodium lauryl sulfate produces a substantial deceleration. Hydrocarbons such as naphthalene or *cis*-decalin show a substantial decelerating effect. Let us consider the hydrocarbon effects first. In a system involving methyl 1-naphthoate and a hydrocarbon, the probability of finding a hydrocarbon adjacent to the ester is significantly greater than the corresponding mole fraction. Therefore, the microscopic environment of the ester is more apolar in the presence of the hydrocarbon than in its absence. That is, the dielectric constant of the microscopic environment is lower. Under these conditions, the attack of hydroxide ion on the ester would be expected to be impeded. When the added hydrocarbon bears an electric charge, the same sorting of solvent molecules will persist, and therefore a lower dielectric constant will again surround the ester substrate. But in addition, the electric charge will have an effect on the reaction, a negative charge repelling hydroxide ion and a positive charge attracting it. Thus, with salts of organic anions, both the environmental effect and the electrostatic effect work in the same direction to produce a substantial deceleration, while with organic cations, the environmental and electrostatic effects work in opposite directions, leading from slight to large increase if the latter effect is more prominent than the former.

Effects of substrate structure are understandable on the basis of variable micelle interactions. For example, the hydrolysis of *p*-nitrophenyl hexanoate is accelerated by cetyltrimethylammonium bromide to a greater extent than the hydrolysis of *p*-nitrophenyl acetate, presumably because of the greater incorporation of the former ester into the micelle. Likewise, the hydrolysis of ethyl orthovalerate and ethyl orthopropionate are accelerated by sodium lauryl sulfate, but the hydrolysis of ethyl orthoformate is not.[24]

The effect of structure is beautifully illustrated by the rate constants of acid hydrolysis of the straight chain alkyl sulfates given in Table 12.12. The acid-catalyzed hydrolysis proceeds through preequilibrium protonation of the sulfate

Table 12.12.   The Rate Constants of the Acid Hydrolysis of Some Sodium Alkyl Sulfates[a]

Sodium Alkyl Sulfate	$k_H \times 10^5 (M^{-1} sec^{-1})$
Methyl	5
Ethyl	5
Amyl	5.25
Decyl	41
Dodecyl	198
Tetradecyl	280
Hexadecyl	347
Octadecyl	505

Source: J. L. Kurz, *J. Phys. Chem.*, **66**, 2239 (1962).
© 1962 by the American Chemical Society. Reprinted by permission of the copyright owner.
[a] 90°, [sodium alkyl sulfate] = [HClO₄] = 0.04 M.

moiety followed by attack of water at the sulfur atom. [Eqs. (12.36) and (12.37)]

$$ROSO_3^{\ominus} + H^{\oplus} \rightleftharpoons ROSO_3H \tag{12.36}$$

$$ROSO_3H + H_2O \longrightarrow ROH + HSO_3^{\ominus} + H^{\oplus} \tag{12.37}$$

Although the lower esters (from methyl to amyl) have essentially identical hydrolytic rate constants, the higher esters (from decyl to octadecyl) increase markedly with chain length. The higher esters form micelles themselves and do not require added detergent. In essence, the sulfate moieties become stronger bases due to the presence of a negative potential on micelle, which enhances prior protonation [Eq. (12.36)]. The rate constant of the hydrolysis of sodium decyl sulfate is independent of substrate concentration and equivalent to the rate constant of hydrolysis of sodium amyl sulfate below the CMC of sodium[25] decyl sulfate. But above the CMC, the hydrolytic rate constant increases markedly (Fig. 12.15).

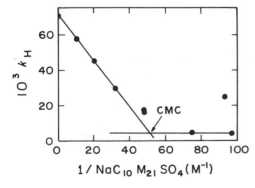

Fig. 12.15. k_H at 90° C for sodium decyl sulfate as a function of its reciprocal concentration; the horizontal line is defined by the value for k_H for sodium amyl sulfate under the same conditions [I = 0.51, (HClO₄) = 0.02 M]. From J. L. Kurz, *J. Phys. Chem.*, **66**, 2239 (1962). Copyright © 1962 by the American Chemical Society. Reprinted by permission of the copyright owner.

The catalytic action in reversed micelles has been reviewed and discussed by Fendler and Fendler.[26] Dramatic rate enhancements were observed in the polar cavity, for example, in the aquation of the tris (oxalato) chromium (III) anion in the alkylammonium carboxylate–benzene reversed micelle. As the role of metallic ions is very important in enzymatic transformations, much attention has been given to the behavior of these ions in reversed micelles.[23]

12.4. ENZYME MODELS INVOLVING COMPLEXATION

Covalent complexes are interesting from a chemical point of view, but from an enzymatic point of view, they are not very interesting. Noncovalent complexes, such as those involving cycloamylose, have in general shown the most enzymelike characteristics. Cycloamyloses and derivatized cycloamyloses have been shown to be excellent models for the enzymes chymotrypsin, ribonuclease, transaminase[28] and carbonic anhydrase.[29] Polymeric complexes, especially those involving PEI polymers, have exhibited large rate enhancements.[20] Reversed micellar catalyses have been shown to approximate enzymatic rates.[26] Catalysis via complexation will be an important area of future study.

One of the most exciting aspects of chemistry in the near future will be the design and synthesis of organic model enzymes that can catalyze reactions stereospecifically at rates in excess of those exhibited by natural enzymes.[27] This book, and especially its last chapters, will assist in that enterprise.

REFERENCES

1. J. Reuben, *Proc. Natl. Acad. Sci. USA*, **68**, 563 (1971).
2. I. M. Klotz, *J. Chem. Educ.*, **53**, 159 (1976).
3. M. I. Page and W. P. Jencks, *Proc. Natl. Acad. Sci. USA*, **68**, 1678 (1971).
4. I. M. Klotz, *Science*, **128**, 815 (1958).
5. M. L. Bender and M. S. Silver, *J. Am. Chem. Soc.*, **84**, 4589 (1962).
6. B. Capon and R. Capon, *J. Chem. Soc. Chem. Commun.*, **1965**, 502.
7. V. Franzen, *Chem. Ber.*, **88**, 1361 (1955).
8. E. J. Fuller, *J. Am. Chem. Soc.*, **85**, 1777 (1963).
9. R. L. Letsinger, S. Dandegaonker, W. J. Vulo, and J. D. Morrison, *J. Am. Chem. Soc.*, **85**, 2223 (1963).
10. F. M. Menger and M. L. Bender, *J. Am. Chem. Soc.*, **88**, 131 (1966).
11. A. K. Colter, S. S. Wang, G. H. Megerle, and P. S. Ossip, *J. Am. Chem. Soc.*, **86**, 3106 (1964).
12. C. G. Swain and L. J. Taylor, *J. Am. Chem. Soc.*, **84**, 2456 (1962).
13. T. C. Bruice and R. M. Topping, *J. Am. Chem. Soc.*, **85**, 1488 (1963).
14. M. L. Bender and M. Komiyama, *Cyclodextrin Chemistry*, Springer Verlag, Berlin, 1978.
15. M. F. Czarniecki and R. Breslow, *J. Am. Chem. Soc.*, **100**, 7771 (1978).
16. R. Breslow, M. C. Czarniecki, J. Emert, and H. Hamaguchi, *J. Am. Chem. Soc.*, **102**, 762 (1980).

17. R. L. Letsinger and T. J. Saveride, *J. Am. Chem. Soc.*, **84,** 114 (1962).

18. C. G. Overberger, T. St. Pierre, N. Vorchheimer, J. Lee, and S. Yaroslavsky, *J. Am. Chem. Soc.*, **87,** 296 (1965).

19. Y. E. Kirsh, V. A. Kabanov, and V. A. Kargin, *Dokl. Akad. Nauk. SSSR*, **177,** 112 (1967).

20. I. M. Klotz, G. P. Royer, and I. S. Scarpa, *Proc. Natl. Acad. Sci. USA*, **68,** 263 (1971).

21. R. Breslow, in *Biomimetic Chemistry*, D. Dolphin, C. McKenna, Y. Murakami, and I. Tabushi, Eds., Advances in Chemistry Series, Washington, No. 191, 1980, p. 1.

22. G. Wulff et al., *Makromol. Chem.*, **178,** 2799 (1977).

23. J. B. Nagy, in *Solution Behavior of Surfactants, Theoretical and Practical Aspects*, K. L. Mittal and E. J. Fendler, Eds., Vol. 2, Plenum Press, New York, 1982, p. 760.

24. M. T. A. Behme, J. G. Fullington, R. Noel, and E. H. Cordes, *J. Am. Chem. Soc.*, **87,** 266 (1965).

25. J. L. Kurz, *J. Phys. Chem.*, **66,** 2239 (1962).

26. J. H. Fendler and E. Fendler, *Catalysis in Micellar and Macromolecular Reactions*, Academic Press, New York, 1975.

27. F. H. Westheimer, Speech at Inauguration of Center on Bioinorganic and Bioorganic Chemistry, University of Chicago, 1982.

28. R. Breslow, H. W. Ezanich, *J. Am. Chem. Soc.*, **105,** 1390 (1983).

29. R. Breslow, *Chem. Brit.*, **19,** 126 (1983).

Index